SCANNING ELECTRON MICROSCOPY AND X-RAY MICROANALYSIS

Robert Edward Lee

Electron Microscopy Center
Department of Anatomy and Neurobiology
Colorado State University

P T R Prentice Hall, Englewood Cliffs, New Jersey 07632

Library of Congress Cataloging-in-Publication Data

Lee, Robert Edward.
   Scanning electron microscopy and X-Ray microanalysis / Robert
Edward Lee.
     p. cm.
   Includes bibliographical references and index.
   ISBN 0-13-813759-5
   1. Scanning electron microscopy. 2. X-ray microanalysis.
I. Title.
   [DNLM: 1. Electron Probe Microanalysis. 2. Microscopy, Electron,
Scanning. QH 212.S3 L479s]
QH212.S3L44 1993
502'.8'25—dc20
DNLM/DLC
for Library of Congress
                                                     91-45512
                                                       CIP

Editorial/production supervision
  and interior design: *Harriet Tellem*
Cover design: *Wanda Lubelska Design*
Prepress buyer: *Mary McCartney*
Manufacturing buyer: *Susan Brunke*
Acquisitions editor: *Betty Sun*
Editorial assistant: *Maureen Diana*
Page layout: *Debbie Toymil*

 © 1993 by P T R Prentice-Hall, Inc.
A Simon & Schuster Company
Englewood Cliffs, New Jersey 07632

The publisher offers discounts on this book when ordered
in bulk quantities. For more information, write:

Special Sales/Professional Marketing
Prentice-Hall, Inc.
Professional Technical Reference Division
Englewood Cliffs, New Jersey 07632

Printed in the United States of America
10 9 8 7 6 5 4 3 2 1

ISBN 0-13-813759-5

Prentice-Hall International (UK) Limited, *London*
Prentice-Hall of Australia Pty. Limited, *Sydney*
Prentice-Hall Canada Inc., *Toronto*
Prentice-Hall Hispanoamericana, S.A., *Mexico*
Prentice-Hall of India Private Limited, *New Delhi*
Prentice-Hall of Japan, Inc., *Tokyo*
Simon & Schuster Asia Pte. Ltd., *Singapore*
Editora Prentice-Hall do Brasil, Ltda., *Rio de Janeiro*

To my son, Christian Robert Lee

Contents

Acknowledgments

I would like to thank my colleagues John Rash and John Walrond for helpful discussions on certain aspects of electron microscopy that were not clear to me. I also want to thank Ted Bowden and Bob Anderholt of Philips Electronics, who took time to help me with some points on electron optics.

I would particularly like to thank John Foerster for the considerable amount of time he took to show me how to produce computer-generated drawings.

Finally, I wish to thank my family for the support they have offered through the years, which has allowed me to pursue my interest in science, often at the expense of family obligations.

1

The Scanning Electron Microscope

The human eye is able to resolve objects that are about 0.1 mm apart, if the objects have sufficient contrast differences. Two white beads, 0.1 mm apart on a black background would be resolved because of the large contrast difference. However, two black beads, 0.1 mm apart on a black background, would probably not be seen as two separate objects by the unaided eye.

Microscopes are devices that magnify details that are not visible to the unaided eye. The resolution of a light microscope is dictated by Abbe's law:

$$d = \frac{0.612\lambda}{n \ \sin\alpha}$$

where d = resolution
λ = wavelength of the radiation used to view the specimen
n = index of refraction of the medium through which the radiation travels
α = aperture angle

Abbe's equation states that the resolution of a microscope is approximately equal to one-half the wavelength of the radiation used to view the specimen. The best possible resolution of a light microscope using blue-green light (400-nm wave-

TABLE 1-1 UNITS OF LENGTH COMMONLY USED
IN SCANNING ELECTRON MICROSCOPY

1 millimeter (mm) = 1000 micrometers (μm)
1 μm = 1000 nanometers (nm)
1 nm = 10 angstroms (Å)

length) is, therefore, 200 nm (0.2 μm) (Table 1-1). Attempting to resolve objects
closer together than 200 nm results in the fusing of image details (Fig. 1-1).

The ultimate resolution of the microscope was increased in the 1930s by
using electrons, instead of visible light, as a source of radiation. Electrons are of
a much shorter wavelength (about 0.1 nm) than light photons and are able to
produce better resolution images. The first transmission electron microscope was
developed in Germany in 1932 by Max Knoll and Ernst Ruska. Today, transmission electron microscopes are able to produce images of thin specimens with a
resolution of 0.2 nm. Transmission electron microscopes, however, have some
disadvantages. First, the specimen has to be a thin section or foil so that the
electron beam can be transmitted through it. Second, the transmission electron
microscope has a relatively short depth of field (in micrometers). Third, it is not
possible to directly view the surface topography of a specimen. Further development of microscopes to overcome these disadvantages resulted in the fabrication
of the scanning electron microscope.

200 nm
Apart

150 nm
Apart

100 nm
Apart

Figure 1-1   Two objects 200 nm apart can be just resolved as two separate objects
in the light microscope. Two objects closer together than 200 nm will appear to be
fused when viewed in the light microscope.

HOW THE SCANNING ELECTRON MICROSCOPE PRODUCES AN IMAGE

The transmission electron microscope forms a true image of the specimen, as does
the light microscope. The true image is formed when the specimen is illuminated
with the radiation (electrons or light photons), with some of the radiation transmitted through the specimen, while some of the radiation is absorbed by the specimen
(Fig. 1-2).

The scanning electron microscope does not produce a true image of the
specimen, as does the light and transmission electron microscope. Instead, the
scanning electron microscope produces a point by point reconstruction of the
specimen, basically the same way a television set reconstructs an image from

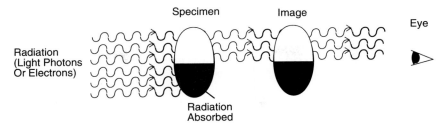

Figure 1-2   The light and electron microscope produce a true image of the specimen by selectively absorbing or transmitting light photons or electrons.

signals sent by the television station transmitter. The only real difference between image reconstruction in a television set and a scanning electron microscope is that the scanning electron microscope reconstructs the image of the specimen from a signal emitted from the specimen when it is illuminated by the high-energy electron beam, instead of a signal from the television transmitter.

The television vacuum tube is a cathode ray tube containing an internal electron beam that is focused to a pinpoint on the phosphor of the viewing face (Fig. 1-3). The pinpoint of electrons is scanned over the viewing face in a raster of 525 lines. Each of the 525 lines contains 1000 tiny dots called **pixels**. Thus, an image of 525,000 pixels (525 lines × 1000 pixels) is produced every time the electron beam is rastered over the viewing face of the television vacuum tube. The gray level (black to white) of the pixels produces the image on the viewing face of the television vacuum tube. Areas of the viewing face that are black contain dark pixels, while white areas contain light pixels. The gray level of each pixel is determined by the intensity of the signal from the television transmitter at the particular moment that the electron beam in the television vacuum tube is resting on that

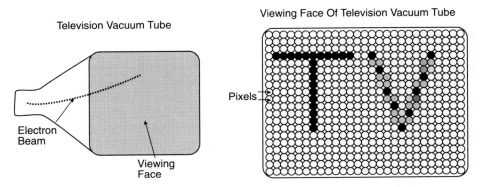

Figure 1-3   The image in a television vacuum tube is constructed on a pixel by pixel basis. The electron beam in the vacuum tube illuminates the phosphor on the viewing face. The stronger the electron beam, the brighter is the phosphor at each pixel position. The gray level of the individual pixels produces the image on the viewing face.

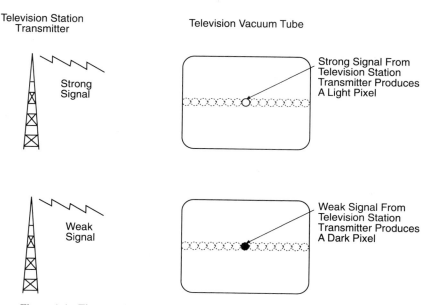

Figure 1-4 The gray level of a pixel on the viewing face of a television vacuum tube is determined by the intensity of the signal from the transmitter of the television station at that particular moment in time. The stronger the signal from the transmitter, the lighter is the pixel.

pixel. A strong signal from the television station transmitter produces light pixels; a weak signal produces dark pixels (Fig. 1-4).

The image in the scanning electron microscope is produced in basically the same way, except that, instead of the signal coming from the television station transmitter, the signal now comes from the interaction of an electron beam with the specimen in the column of the scanning electron microscope (Fig. 1-5). The electron beam in the column of the scanning electron microscope is focused to a pinpoint of electrons on the specimen. The pinpoint of electrons is scanned across the specimen in a series of lines in synchrony with the electron beam inside the cathode ray tube (that is used to view the specimen). Thus, if the electron beam in the cathode ray tube is set to scan 1000 lines, the electron beam in the column will produce a scan of 1000 lines over the specimen. If, at a particular point in time, the electron beam in the cathode ray tube is at pixel position 275 on line 450, then the electron beam in the column will be resting at pixel position 275 on line 450 on the specimen. The gray level of the pixels in the cathode ray tube is determined by the number of electrons that are knocked out of the specimen atoms by the electron beam that is illuminating the same pixel position on the specimen in the column. If a large number of electrons are knocked out of the specimen atoms at that moment in time, a large number of electrons are received by the detector,

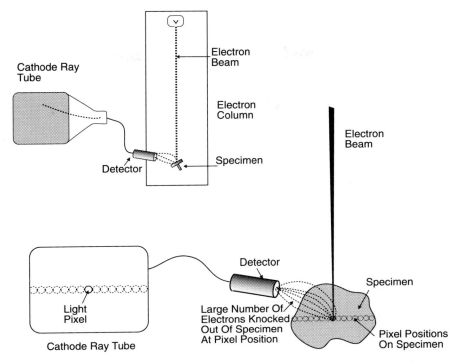

Figure 1-5   The gray level of each pixel on the viewing face of a cathode ray tube of a scanning electron microscope is determined by the number of electrons that are ejected from specimen atoms as the atoms are being illuminated by the electron beam in the column. The larger the number of electrons knocked out of the specimen atoms, the lighter is the gray level of the pixel.

a strong signal is sent through the circuitry, and a *light* pixel is produced on the cathode ray tube. If the opposite happens, few electrons are knocked out of the specimen atoms at that moment in time, few electrons are received by the detector, a weak signal goes through the circuitry, and a *dark* pixel is produced on the cathode ray tube. With a raster of 1000 lines, the image on the cathode ray tube is produced by the different gray levels of 1 million pixels (1000 lines × 1000 pixels per line).

The number of electrons that escape from a pixel position on the specimen when it is illuminated by the electron beam is determined by the **topography** of the specimen surface and the **atomic number of the elements** in the specimen at that pixel position. The more irregular the topography of the specimen surface is, the more electrons that escape from that area of the specimen (Fig. 1-6). Thus, portions of the specimen with irregular surfaces will have a large number of electrons escaping from the surface of the specimen, resulting in the production of light

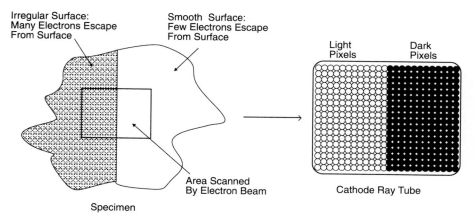

Figure 1-6   The topographic contrast of a specimen depends on the roughness of the surface of the specimen. The rougher the specimen surface, the more electrons that escape and the brighter are the pixels of this area on the cathode ray tube.

pixels on the same area of the cathode ray tube. Fewer electrons escape from a specimen surface that is smooth when it is bombarded by the high-energy electrons in the beam, resulting in dark pixels in the same area on the cathode ray tube.

Within a specimen area with the same surface topography (for example, a smooth surface), the higher the atomic number of the elements in that area of the specimen is, the more electrons that escape from the surface (Fig. 1-7). Thus, a specimen area containing high-atomic number elements appears light in the cathode ray tube, while an area with low-atomic number elements appears dark.

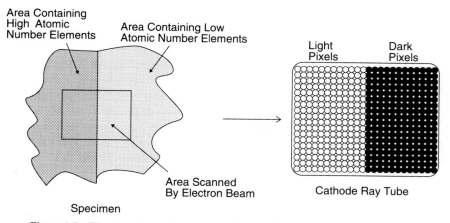

Figure 1-7   The atomic number contrast of a specimen depends on the atomic number of the elements in a particular portion of the specimen. The higher the atomic number of the elements in a portion of the specimen, the more electrons that are ejected and the brigher are the pixels of that area on the cathode ray tube.

CHARACTERISTICS OF THE IMAGE PRODUCED BY THE SCANNING ELECTRON MICROSCOPE

Abbe's law does not directly apply in determining the ultimate resolution of the scanning electron microscope, since a true image of the specimen is not formed. Instead, the ultimate resolution of the scanning electron microscope is determined by the escape volume of the radiation (usually electrons) escaping from the specimen surface at that pixel position. The resolution would be the size of an atom if it were possible to sequentially stimulate atoms in the specimen to produce radiation (as occurs in the scanning tunneling microscope). However, at the present time such a resolution is not possible with the scanning electron microscope. Instead, the resolution of the scanning electron microscope is usually determined by the minimum escape volume of electrons emitted from the specimen at a pixel position. The escape volume of electrons is, in turn, determined by the diameter of the electron beam that is used to illuminate the specimen. Three to five nanometers is the minimum diameter of the electron beam that still has enough electrons to produce an image. This limits the ultimate resolution of the scanning electron microscope to an escape volume slightly larger than this and a potential magnification of about 50,000 times normal, or about one-twentieth the potential magnification of the transmission electron microscope.

The scanning electron microscope produces an image of the surface of the specimen with a depth of focus about 500 times greater than a light microscope. This is because of the very narrow electron beam that illuminates the specimen in the scanning electron microscope. The great depth of focus produces an image in which all, or most, of the specimen is in focus. Such an image has a great deal of three-dimensional depth.

DEVELOPMENT OF THE SCANNING ELECTRON MICROSCOPE

The development of the scanning electron microscope began in 1935 with the work of Max Knoll at the Technical University in Berlin. He placed specimens into a modified cathode ray tube and scanned the specimens with an electron beam of a diameter between 0.1 and 1 mm. The relatively large diameter of the beam (0.1 mm = 100 μm = 100,000 nm) precluded the ability of Knoll to obtain high-magnification images (today's scanning electron microscopes can produce electron beam diameters of 3 to 10 nm).

In 1938, Manfred von Ardenne (Fig. 1-8), working in his private laboratory in Berlin, published an important theoretical paper that clearly explained the principles of scanning electron microscopy. Von Ardenne added scan coils to a transmission electron microscope that had demagnifying lenses in order to produce an electron beam of a relatively small diameter. The image was produced by passing the rastered pinpoint of electrons through a thin specimen, with the image recorded on film that was mechanically rastered in synchrony with the electron beam. As such, the instrument was really a scanning *transmission* electron microscope.

Figure 1-8 Professor Dr. h.c. mult. Manfred von Ardenne. Born Jan. 20, 1907 in Hamburg, Germany. After graduating from the University of Berlin, he worked in a laboratory in his parents' apartment where he developed cheap, resistance amplifiers that were produced in the millions by the Loewe Company. In 1928, he founded the Institute for Electron-Physics in Berlin-Lichterfeld. Here he worked on cathode ray tubes, television, and radar techniques. From 1937–1939 he worked out the theory and construction of a scanning electron beam for an electron microscope. After the Second World War, he accepted an invitation by the U.S.S.R. to head an atomic research laboratory near Suchumi in Soviet Georgia. For his research, he was awarded the Stalin Prize in 1953. He then moved to Dresden. At the Dresden Private Institute, he worked on electron rays, vacuum ovens, and plasma sputtering of metals. In 1960, he changed his research from physics to medicine, working on cancer therapy at the Von Ardenne Institute for Applied Medical Research. (*Photograph courtesy of Dr. Von Ardenne*)

However, the principle of scanning a small-diameter electron beam over the specimen had been demonstrated.

In 1942, V. K. Zworykin (Fig. 1-9), J. Hiller, and R. L. Snyder in the RCA Laboratories in the United States developed the first scanning electron microscope that could examine bulk specimens. They recognized that secondary electron emission from a specimen surface illuminated with a high-energy electron beam could be used to produce topographic contrast. The detector they used was an electron multiplier that was biased to +50 V to collect the secondary electrons. Three electrostatic lenses were used to produce an electron beam with a diameter of 50 nm. The images that were obtained were noisy by today's standards. The

Figure 1-9 Vladimir Kosma Zworykin. Born July 30, 1889, in Mourom, Russia. He received his degree in electrical engineering from the Petrograd Institute of Technology. He did graduate work on X-rays at the College de France in Paris. From 1914 to 1917 he served in the Russian Army as an officer in the radio corps. In 1919 he went to the United States, where he worked as a bookkeeper while he learned English. He began working at Westinghouse Electric Corporation in 1920. In 1923, while still working for Westinghouse, he enrolled at the University of Pittsburgh to continue his graduate work, receiving his Ph.D. in physics in 1926. In 1929, he became director of electronic research at the RCA Manufacturing Company. From 1942 to 1945 he was associate research director of the RCA Laboratories, moving up to become director of electronic research in 1946. In 1947, he became a vice-president of the Radio Corporation of America. Photograph from J. D. Ryder and D. G. Fink, *Engineers and Electrons*, IEEE Press, New York, 1984. (*Photograph courtesy of RCA*)

images, however, demonstrated that the concept of a scanning electron microscope was viable. World War II resulted in a break in the development of the scanning electron microscope (and resulted in the destruction of von Ardenne's microscope in Berlin).

The modern scanning electron microscope was developed by Sir Charles W. Oatley (Fig. 1-10) and his students at Cambridge University in England from 1948 to 1961. In 1948, Oatley assigned a Ph.D. project to his student Dennis McMullan (Fig. 1-11) to construct a scanning electron microscope. McMullan had several years experience in industry in the design of radar and television circuitry, so he had the knowledge to build a slow-scan version of a television camera chain.

Figure 1-10   Sir Charles W. Oatley. Born February 14, 1904, in Frome, Somerset, England. Educated at St. John's College, Cambridge, graduating with a degree in physics in 1925. He joined the teaching staff of Kings College, London, in 1927. During World War II he was a member of the British Army's Radar Research and Development Establishment. In 1945 he was elected to a fellowship at Trinity College, Cambridge, and a lectureship in the University's engineering department, where in 1960 he succeeded to the chair of electrical engineering and became head of the electrical division. He retired in 1971, but continued to carry out research. He was knighted in 1974. (*Photograph courtesy of Sir Charles W. Oatley*)

McMullan modified an existing transmission electron microscope and in 1951 produced his first images from the scanning electron microscope. Subsequently, the electron beam diameter was reduced to 20 nm and the signal to noise problem was reduced. In 1953, this instrument (Fig. 1-12) was taken over by another student, K. C. A. Smith (Fig. 1-11), who replaced the electrostatic lenses with electromagnetic lenses and improved the scanning system by introducing double-deflection scanning coils and an objective lens stigmator. Smith also used signal processing to improve the image. T. E. Everhart (Fig. 1-13) built a detector designed to produce a relatively noise free signal of secondary electrons by accelerating the low-energy secondary electrons into a scintillator biased to about 10,000 V. R. F. M. Thornley completed the design of this system, which is almost universally used today as the Everhart–Thornley detector on scanning electron microscopes. In the early 1960s, it became apparent that the scanning electron microscope was a valuable instrument that was ripe for commercial production (even though a well-known research institute stated that only three to five scanning electron microscopes would be

Figure 1-11 Dennis McMullan (left) and K. C. A. Smith (right). Dennis McMullan was born in Reading, England, in 1923 and graduated in Mechanical Sciences at Cambridge University. After some years in industry working on radar and the development of cathode ray tubes for television, he returned to Cambridge University in 1948 to work on the development of the scanning electron microscope and received his Ph.D. in 1953. After a further six years in government service (guided missile development) in Canada and with the Rank Organization in London, he joined the staff of the High Energy Physics Group at Imperial College, London. Later he engaged in research on photoelectric imaging devices in the Applied Physics Group at Imperial College. In 1969, he went to the Royal Greenwich Observatory to develop electronographic cameras for stellar field astronomy; from 1972 he was the head of the Instrumental Science Division until 1980 when he joined the Microstructural Physics Group at the Cavindish Laboratory at Cambridge University. K. C. A. Smith improved the microscope built by McMullan. He altered the position of the electron collector and kept it at a positive potential, ensuring that secondary electrons, as well as backscattered electrons, were collected. This improved the signal to noise ratio and contrast of the image. (*Photographs courtesy of Dr. McMullan and Dr. Smith*)

sold in the United States in the first year and that an overall total of ten would saturate the market). This led to the first commercial production of a scanning electron microscope (the Stereoscan, Fig. 1-14) by Cambridge Instrument Company in 1965. The instrument was an immediate success and won the Queen's Award for Technology for that year.

Figure 1-12 A replica of the scanning electron microscope that was produced at Cambridge University by Dennis McMullan and Sir Charles Oatley in 1952. (*From: Proceedings Royal Microscopical Society*, 21:203 [1986]; *Courtesy of The Royal Microscopical Society*)

Figure 1-13 Thomas E. Everhart. Born February 15, 1932, in Kansas City, Missouri. Everhart received an A.B. in physics from Harvard University in 1953, an M.Sc. in applied physics from the University of California at Los Angeles in 1955, and a Ph.D. in engineering from Clare College, Cambridge University, in 1958. From 1958 to 1978 he was in the Department of Electrical Engineering and Computer Science at the University of California, Berkeley, where he rose from the position of assistant professor to departmental chairman. From 1979 to 1984 he was the dean of the College of Engineering at Cornell University. From 1984 to 1987 he served as the chancellor of the University of Illinois at Urbana–Champaign. In 1988 he became the president of the California Institute of Technology. (*Photograph courtesy of Dr. Everhart*)

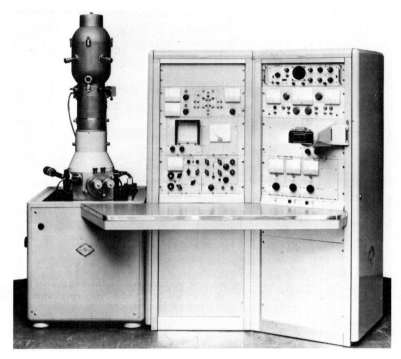

Figure 1-14   The first commercial scanning electron microscope, the Stereoscan MK1, made by Cambridge Instruments. (*Photograph courtesy of Cambridge Instruments*)

THE MODERN SCANNING ELECTRON MICROSCOPE

The remaining chapters of this book describe the construction and operation of the scanning electron microscope (Figs. 1-15 and 1-16).

Electron gun (Chapter 2): The electron gun is the source of the electron beam. Electrons are drawn from a negative cathode and accelerated toward an anode at ground potential.

Electron lenses (Chapter 3): The configuration of the electron beam is manipulated by lenses. Electron lenses in modern scanning electron microscopes are electromagnets.

Lens aberrations and stigmators (Chapter 4): The electron beam has certain aberrations that limit the resolution of the scanning electron microscope. Stigmators are devices that correct for astigmatism in the electron beam.

Assembled column of the scanning electron microscope (Chapter 5): The final configuration of the electron beam is determined by the manner in which the

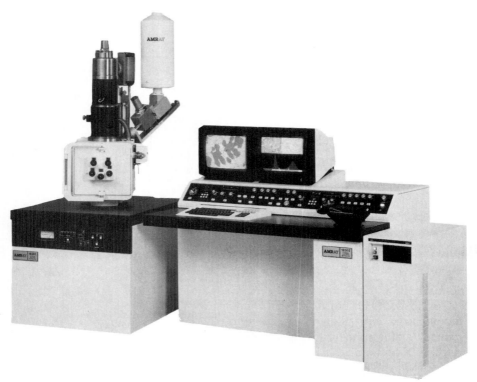

Figure 1-15   A digital scanning electron microscope, the AMRAY 1830I. (*Photograph courtesy of AMRAY*)

components of the scanning electron microscope are assembled and operated. These components include the electron gun, anode, lenses, apertures, objective lens stigmator, and scanning coils.

Electron beam–specimen interactions (Chapter 6): The image in the scanning electron microscope is constructed from radiation emitted from the specimen during rastering of the specimen by the electron beam. Secondary and backscattered electrons are usually used to image the specimen, although X-rays, Auger electrons, and cathodoluminescence can also be used.

Detectors (Chapter 7): The particular detector system in use in the scanning electron microscope determines the type of radiation that is used to produce the image of the specimen.

Image reconstruction (Chapter 8): The image of the specimen is reconstructed on a pixel by pixel basis on the cathode ray tube of the scanning electron microscope.

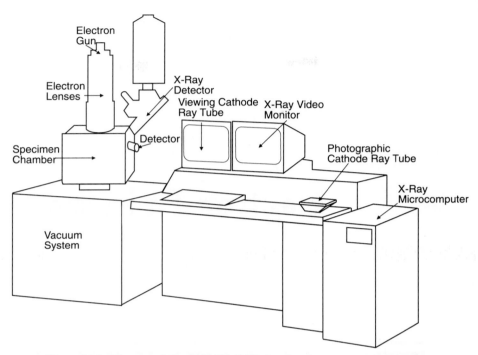

Figure 1-16 Diagram of the AMRAY 1830I showing the component parts of a scanning electron microscope fitted with X-ray microanalytical equipment.

Image processing (Chapter 9): The image produced by the scanning electron microscope can be manipulated using modern image-processing techniques to maximize the information in the image.

Vacuum system (Chapter 10): The electron beam has to be generated and accelerated in a vacuum to prevent oxidation of the cathode and to maintain a coherent electron beam over a distance of about 1 m. An understanding of vacuum generation is an essential part of operating a scanning electron microscope.

Specimen preparation (Chapter 11): The specimen has to be placed in a high vacuum before it can be examined. Removing volatile components from specimens (particularly biological specimens) requires complex preparation steps to ensure a minimum amount of distortion and shrinkage of the specimen during drying.

X-ray microanalysis (Chapter 12): X-rays emitted on interaction of the electron beam with the specimen have energies that are characteristic of the elements from which the X-rays are generated. The collection of these X-rays enables the investigator to determine the elemental composition of the specimen being examined in the scanning electron microscope.

2

Electron Emission

TYPES OF ELECTRON EMISSION

The cathode of the gun is the source of electrons for the beam in the electron microscope. The cathode in almost all electron microscopes is either a tungsten wire or a lanthanum hexaboride rod, although other substances have been used, mostly in experimental systems. Whatever the composition of the cathode, its main function is to supply electrons for the generation of the electron beam by electron emission from the cathode. The emission of electrons from a substance can be induced by application of heat (**thermionic emission**), strong electrical fields (**high-field emission** or **field emission**), electromagnetic radiation (**photoelectric emission**), or atomic particles (**secondary emission**). Only the former two, thermionic emission and field emission, are used to generate the electron beam in electron microscopes.

In the cathode, the negative electrons are held in orbits around the positive nuclei of the atoms (Table 2-1). The strength of the positive force of the atomic nucleus holding the electrons in orbit is greatest for those electrons next to the nucleus (K and L orbitals) and least for those electrons farthest from the atomic nucleus (M and N orbitals). The electrons in the outer orbit of the atom are the **valence electrons**. The valence electrons have the least holding force exerted on them by the atomic nucleus and are, therefore, the easiest electrons to remove

TABLE 2-1 CONFIGURATION OF THE ORBITAL ELECTRONS IN LANTHANUM
AND TUNGSTEN ATOMS

	Atomic number	K	L		M			N				O			P
		s	s	p	s	p	d	s	p	d	f	s	p	d	s
Lanthanum	57	2	2	6	2	6	10	2	6	10		2	6	1	2
Tungsten	74	2	2	6	2	6	10	2	6	10	14	2	6	4	2

The electrons that are circled are far from the atomic nucleus and are the main contributors to
thermionic emission.

from the atom. The electrons in the lowest orbital (K shell electrons) have the
lowest energy, while those in the outer orbitals have the most energy. The electrons in the outer orbitals, however, do not have sufficient kinetic energy to escape
from the surface of the metal. The barrier of potential energy at the surface of the
metal (Fig. 2-1) that prevents the escape of electrons from the surface is due to
two factors: (1) the polarization field at the outer surface layer of atoms and (2)
the image field beyond this.

Polarization Field.   The polarization field is due to atoms in the surface layer
having other atoms exerting forces on them from underneath, but no forces from
on top. This results in an asymmetric field with a negative electron layer at the
surface and a positive surface layer under it. When a negative electron enters into
the field between the positive and negative layers (Fig. 2-2a), it is repelled by the
negative layer and attracted to the positive layer, resulting in a higher requirement
of energy for the electron to escape through this field and leave the surface of the
metal.

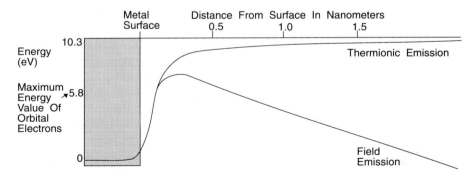

Figure 2-1 Potential energy barrier for electron emission from the surface of
tungsten. In thermionic emission, the electrons have to be raised, by heating, from
their maximum value at room temperature (5.8 eV) to a value greater than 10.3 eV
so that they can escape from the surface of the metal. In field emission, the electrons are able to tunnel through the potential energy barrier and escape from the
surface of the metal.

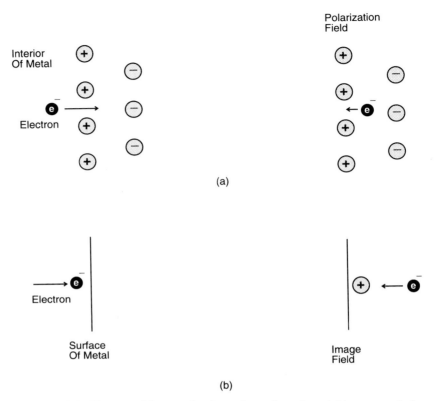

Figure 2-2   The potential energy barrier at the surface of a metal is composed of the polarization field (a) and the image field (b).

Image Field.   The image field results after an electron has passed through the polarization field. On leaving the surface of the metal, the electron induces a positive charge at the surface of the metal that is an electrostatic image of the electron (Fig. 2-2b). This positive electrostatic image pulls the electron back toward the surface of the metal. For the electron to completely escape from the surface of the metal, it must have sufficient energy to also overcome the image field.

An electron must have sufficient energy to escape through the polarization and image fields if the electron is to escape from the surface of the metal. This potential energy barrier set up by the polarization and image fields is greatest at the surface and becomes progressively less farther from the surface of the metal (Fig. 2-1). At room temperature, the escape of an electron from the surface of a metal is a rare event. Tungsten, a common cathodic material, will emit only one electron every 10^{14} years at room temperature. This is because of the energy difference between the energy of the outer orbital electrons (the most energetic

TABLE 2-2 CHARACTERISTICS OF POTENTIAL ELECTRON EMITTERS

Material	Work function (V)	Melting point (K)	Operating temperature (K)	Comments
Sodium	1.9	371		Vaporize at too low a
Calcium	2.5	1123		temperature to use as
Barium	2.0	990		thermionic emitters
Thoriated tungsten	2.6		1900	Can be used, but is easily poisoned
Oxide-coated nickel	1.0		1000	Can be used, but is easily poisoned
Lanthanum hexaboride	2.4		1900	Produces bright beam, but is easily poisoned
Tantalum	4.1	3250		Can be used
Tungsten	4.5	3653	2650	Most widely used

Adapted from Wischnitzer, 1970

electrons) and the amount of energy necessary to escape the surface of the metal. This energy difference is called the work function (Table 2-2). The **work function** *is the potential in volts that must be overcome for the electron to escape the surface of the metal.* The work function is a measure of the work needed to remove the electron from its orbital within the atom, move it to the surface of the metal, and overcome the positive charge left on the surface of the metal after the electron has escaped. *A metal with a low work function requires less energy to remove electrons and therefore is capable of emitting more electrons at a given temperature than a metal with a high work function.* A material with low work function usually has a large atomic radius, or has a large spacing between the atoms in the crystal structure. The work function is dependent on the texture of the metal surface (rough versus smooth) and on the cleanliness of the surface. The rougher and cleaner the surface is, the greater the emission of electrons.

THERMIONIC EMISSION

In thermionic emission the atoms in the metal are heated in order to raise the energy of the orbital electrons through the work function value so that the electrons can escape from the surface of the metal. In the case of tungsten (Fig. 2-3), the work function is 4.5 electron volts (eV). Once the outer orbital electrons have had their energy increased from 5.8 eV (their energy at room temperature) to 10.3 eV by heating, the electrons are able to escape from the surface of the metal, and electron emission occurs. The increase in electron emission from a tungsten cathode with increased temperature is shown in Table 2-3.

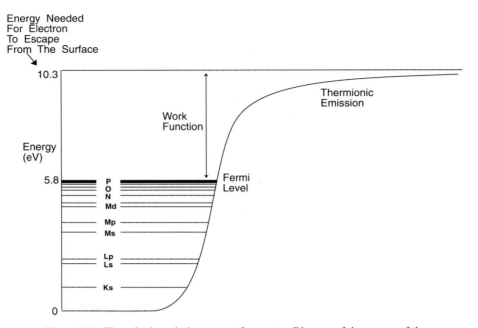

Figure 2-3   Thermionic emission curve of tungsten. Diagram of the energy of the orbital electrons of tungsten showing that the work potential is the difference between the energy necessary for the electrons to escape from the surface of the metal and the energy of the highest-orbital electrons.

TABLE 2-3.   EFFECT OF TEMPERATURE ON THERMIONIC ELECTRON EMISSION FROM A TUNGSTEN CATHODE

Temperature		Electron emission
K	°C	(A/cm²)
1000	727	1.1×10^{-15}
1200	927	9.7×10^{-12}
1400	1127	6.6×10^{-9}
1600	1327	9.3×10^{-7}
1800	1527	4.5×10^{-5}
2000	1727	1.0×10^{-3}
2200	1927	1.3×10^{-2}
2400	2127	0.1
2600	2327	0.7
2800	2527	3.6

The range of materials that can serve as thermionic emitters in cathodes is fairly narrow. The materials need to have a sufficiently low work function so that a large supply of electrons is available to form the electron beam. Also, the material must have a high melting point to maximize the number of electrons emitted in thermionic emission. The material must also be stable at high temperatures and maintain its form. Many materials with low work functions, such as sodium, have very low melting points, and hence attainable thermionic emission is too small to be useful. A long operating life means that the metal does not undergo excessive evaporation under operating conditions. Tungsten is a metal that fulfills these criteria and, because it is economical, it is commonly used as a cathode in electron microscopes.

Thoriated tungsten (tungsten containing about 1% thoria, ThO_2) and oxide-coated nickel, with work functions lower than tungsten, have been used as cathodes in electron microscopes. These two substances have lower work functions because of the structure of the elements in these cathodes. Thoriated tungsten has a layer of thorium on the tungsten. Thorium is electropositive with respect to tungsten, thereby aiding the escape of electrons from the surface of the cathode and reducing the work function. Oxide-coated nickel has an electropositive layer of barium or strontium oxide on the surface of the nickel that acts in a similar manner to the thorium on the tungsten cathode. However, because of oxide poisoning of these cathodes, they are not used extensively in electron microscopes and, instead, tungsten is usually the cathode of choice. Tungsten suffers from poisoning, also, but not to the extent of the other two substances. In **poisoning** of cathodes, thin layers of water vapor and electronegative gases, such as oxygen (but not inert gases), are adsorbed to the surface of the cathode. These adsorbed layers result in a reduction in electron emission since the electrons have to pass through the adsorbed layers to leave the surface of the cathode.

Cathodes that use thermionic emission can be heated directly or indirectly. In **direct heating**, an electrical current is passed through a cathode, such as tungsten, which is drawn out into a wire. The resistance of the wire to the electrical current causes it to heat up. In **indirect heating**, the cathode is heated by an outside source. For example, a lanthanum hexaboride rod can be heated by a hot tungsten wire wrapped around the cathode.

ELECTRON GUNS

The electron gun is the source of electrons in electron microscopes. The electron gun is normally composed of three parts: the **cathode**, the **shield** or **Wehnelt cylinder**, and the **anode** (Fig. 2-4). Almost all electron microscopes use one of three types of electron guns: (1) electron gun with thermionic emission from a tungsten cathode, (2) electron gun with thermionic emission from a lanthanum hexaboride cathode, or (3) electron gun with field emission from a tungsten cathode.

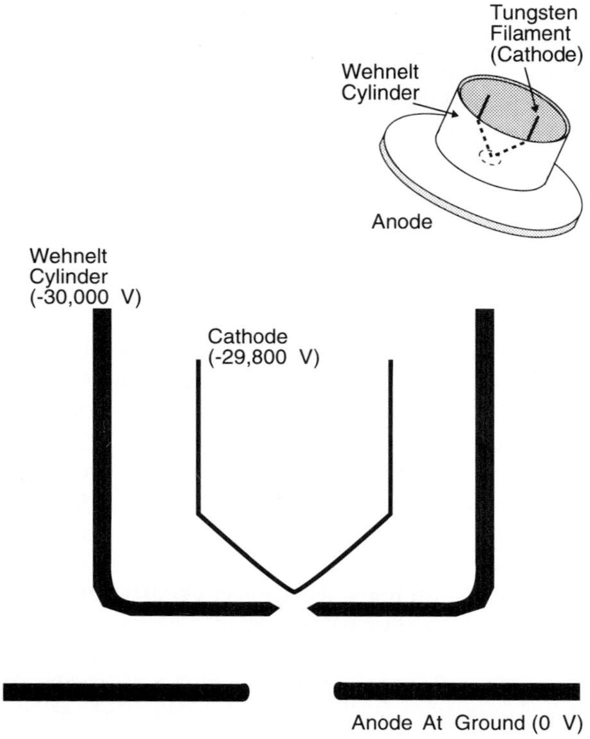

Figure 2-4 Biased electron guns, showing the position of the cathode, Wehnelt cylinder, and anode.

ELECTRON GUN WITH THERMIONIC EMISSION FROM A TUNGSTEN CATHODE

An electron gun with thermionic emission from a tungsten cathode (filament) is the most common type. The tungsten cathode is a reliable source of electrons and it is relatively economical to manufacture. Although tungsten has a relatively high work function, it has a very high melting point (3370°C, 3643 K) and a long operating life. The tungsten cathode is a wire approximately 0.01 cm in diameter that is bent into a V-shape with a tip about 100 μm in radius. The tungsten cathode is heated by directly passing an electrical current of about 2.5 A at 1 V through the wire. The natural resistance of the thin wire to electrical current causes the wire to heat to incandescence (the electrical resistance of the wire depends on the diameter and length of the wire; the thicker and shorter the wire, the less the resistance). The apex of the V becomes the hottest part and therefore the effective electron source. In some older electron microscopes, it was possible to check the integrity of the filament by looking through a small window in the column to see the visible light given off from the glowing tungsten cathode.

The more current that passes through the wire, the hotter the tungsten cathode is. The tungsten cathode is normally operated at a temperature of about 2700 K (2427°C), which results in an electron emission of about 1.75 A/cm². The voltage of the tungsten cathode can be varied from − 1000 to − 30,000 V in most scanning electron microscopes. The electrons emitted from the tungsten filament are accelerated to an anode (approximately 1 cm from the cathode) at ground potential placed under the cathode. Initially, the electrons are almost stationary as they leave the filament; their energies range from just above zero to 1.5 eV (Fig. 2-5). The kinetic energy of the electrons is initially determined by the temperature of the cathode. On leaving the tungsten cathode, the electrons diffuse until they reach the high electrostatic field within the opening (aperture) of the Wehnelt cylinder, whereupon the electrons are accelerated to a very high velocity by the potential difference between the tungsten filament/Wehnelt cylinder and the anode. If the tungsten cathode is at − 30,000 V (with respect to the anode at ground potential), the emitted electrons are accelerated to − 30,000 V as they are drawn to the anode. After leaving the electron gun, the electrons pass down the column of the electron microscope at a constant velocity that is not affected by the fields of the magnetic lenses. The **accelerating voltage**, the difference in potential between the cathode and anode, is 30,000 V.

In designing the electron gun circuits, the engineer has the option of making the cathode very negative and keeping the anode at ground potential or of keeping the cathode at ground potential and making the anode very positive. Choosing the latter would mean that the whole microscope would have to be maintained at the accelerating voltage and the microscope would have to be insulated from the operator, a difficult task. Choosing the former means that only the cathode, Wehnelt cylinder, and anode have to be insulated to maintain the high negative voltage relative to the rest of the microscope. Air is not a sufficiently good insulator to

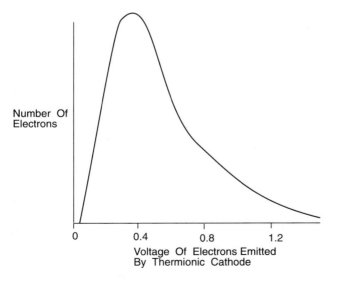

Figure 2-5 The voltage of electrons emitted by thermionic emission from a tungsten cathode.

Number Of Electrons

0   0.4   0.8   1.2

Voltage Of Electrons Emitted By Thermionic Cathode

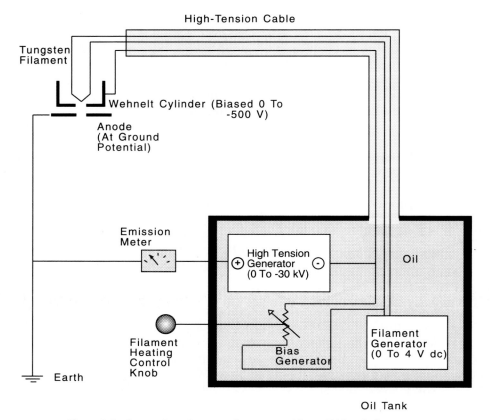

Figure 2-6 Setup of an electron microscope with a self-biased electron gun.

prevent arcing over (shorting) at the voltages that are used. Oil, instead, is used as an insulator. The high-tension generator (the device that establishes the accelerating voltage between the cathode and anode) and the filament generator (which is at the same potential as the accelerating voltage) are isolated in a tank of insulating oil (Fig. 2-6). The high-voltage cable carrying the accelerating voltage to the gun is also often encased in an oil insulating layer.

A **Wehnelt cylinder** (**shield** or **grid cap**) surrounds the filament, with a circular aperture (hole) 1 to 3 mm in diameter at the tip of the filament. The Wehnelt cylinder can be used in one of two ways. The first way is as a diode, where the unbiased Wehnelt cylinder is at the same voltage as the tungsten cathode. The tungsten filament and the Wehnelt cylinder are connected through a pair of balancing resistors directly to the negative terminal of the high-voltage supply. Thus, if the accelerating voltage is 30,000 V, the Wehnelt cylinder and the tungsten cathode will be at −30,000 V, while the anode is held at 0 V (ground). In this system, the electrons for the electron beam are drawn directly from the surface of the hot tungsten cathode. This type of gun is not used in modern electron microscopes, which instead use a triode gun arrangement (Fig. 2-7).

In a triode gun, the Wehnelt cylinder is biased negative in relation to the filament (triode refers to the fact that the Wehnelt cylinder exerts the same kind of control function as does the grid of a triode gas control valve). The Wehnelt cylinder can be biased negatively from 0 to -2500 V relative to the tungsten cathode by means of a bias resistor in the circuit between the filament and the negative lead of the high-voltage supply. The filament is connected to the high-voltage supply through both a bias resistor and a pair of balancing resistors, while the Wehnelt cylinder is attached directly to the negative lead of the high-voltage supply. The bias on the Wehnelt cylinder relative to the tungsten filament is variable depending on the current in the electron beam. When the beam current is zero (no electron beam), there is no bias between the tungsten cathode and the Wehnelt cylinder (both are at the accelerating voltage); the maximum bias is achieved when the tungsten cathode is saturated and the beam current is at a maximum. The bias is generated by that part of the current in the electron beam that strikes the Wehnelt cylinder (the remainder of the beam current passes through the aperture of the Wehnelt cylinder on its way to the anode) and passes to the bias resistor along with the negative lead of the high-tension supply. The resistance to the current passing through the bias resistor produces a voltage drop across the resistor, thereby resulting in a voltage difference between the anode and tungsten cathode. Thus, the bias voltage depends on (1) the accelerating voltage selected, (2) the amount of beam current, and (3) in a self-biasing gun (see p. 26), the amount of resistance set up by the operator. This can be seen more clearly in the following example of the gun at three different filament currents.

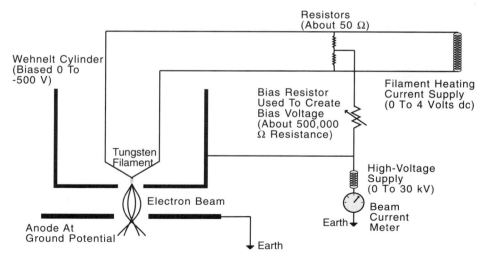

Figure 2-7   A self-biased gun showing the position of the resistors that are used to set up the bias.

1. *Current through the bias resistor is near zero:* When there is no heating current to the tungsten cathode, few electrons are being emitted from the cathode (although there is a small amount of current, called the **dark current**, due to some electrons being pulled off the cold cathode by the accelerating voltage), few electrons are striking the Wehnelt cylinder from the beam, and a minuscule amperage is passing from the negative lead of the high voltage supply to the tungsten cathode through the bias resistor. If essentially no amperes are passing through the bias resistor, then there is no resistance through the bias resistor and the tungsten cathode and the Wehnelt cylinder are at the same voltage.

2. *Current through the bias resistor is 100 microamperes (μA):* The filament heating current has increased, but not to the saturation temperature, so a moderate number of electrons is escaping from the heated tungsten cathode to form a beam of moderate brightness. One hundred microamperes of current travels from the Wehnelt cylinder through the bias resistor between the negative lead of the high-voltage source and the tungsten cathode. If the bias resistor has a resistance of 500,000 ohms (Ω), then the voltage difference between the Wehnelt cylinder and the tungsten cathode will be 50 V, according to the equation

$$\text{voltage} = \text{current} \times \text{resistance}$$

$$= 100 \ \mu A \times 500{,}000 \ \Omega$$

$$= 0.0001 \ A \times 500{,}000 \ \Omega$$

$$= (1.0 \times 10^{-4} \ A)(5 \times 10^{5} \ \Omega)$$

$$= 50 \ V$$

The tungsten cathode is now $+50$ V more positive than the Wehnelt cylinder.

3. *Current through the bias resistor is 300 μA:* At this point the beam current is probably near saturation. The 300 μA of current passing through the bias resistor between the tungsten filament and the negative lead of the high-voltage supply produces a bias voltage of 150 V.

$$\text{voltage} = 300 \ \mu A \times 500{,}000 \ \Omega$$

$$= 0.0003 \ A \times 500{,}000 \ \Omega$$

$$= (3.0 \times 10^{-4} \ A)(5 \times 10^{5} \ \Omega)$$

$$= 15.0 \times 10^{1} \ V$$

$$= 150 \ V$$

The tungsten cathode is $+150$ V more positive than the Wehnelt cylinder. If the accelerating voltage is 80,000 V, the Wehnelt cylinder is at $-80{,}000$ V and the tungsten cathode is at $-79,850$ V. Thus, as the filament heating current is increased, the Wehnelt cylinder becomes more negative than the tungsten cathode by means of the bias created by the increased beam current passing through the bias resistor.

The beam current is usually read indirectly by monitoring that part of the beam current that strikes the Wehnelt cylinder. An ammeter placed in this circuit measures that part of the beam current that drains down the Wehnelt cylinder (Fig. 2-7).

Biased electron guns are preferred in electron microscopy because of the higher brightness and smaller effective source of electrons compared to unbiased guns. In an unbiased gun, the electrons are drawn directly from the tungsten filament, resulting in a larger diffuse crossover point below the Wehnelt cylinder. The diameter of the electron beam is ultimately determined by the diameter of this crossover point. In a biased gun, the electrons are drawn from the relatively small electron cloud in the aperture of the Wehnelt cylinder, resulting in a smaller crossover diameter beneath the Wehnelt cylinder and a small, bright electron beam.

Most modern electron microscopes use a variation of the self-biased gun called the **variable self-biased gun**, where the operator is able to change the value of the resistance in the bias resistor by a potentiometer knob on the instrument panel. This allows the operator to control the beam current by changing the bias between the tungsten cathode and Wehnelt cylinder. The greater the voltage difference between the two is, the less the beam current and the dimmer the beam (at high enough bias, all the electrons are pinched off by the high bias of the Wehnelt cylinder and there is no electron beam). This is useful when the operator is working with specimens that are sensitive to damage, since the electron beam interacts with the specimen, causing the specimen to heat up and become damaged.

Wehnelt Cylinder

In a self-biased gun, the Wehnelt cylinder is biased negative in relation to the tungsten cathode. As a result, the Wehnelt cylinder acts as an electrostatic lens on the electron beam. Since electrostatic lenses are no longer used in other positions in electron microscopes, the action of electrostatic lenses will be discussed here.

Effect of electrostatic lenses on electrons.   A basic rule of physics is that negative particles (such as electrons) are attracted to positive charges and repelled by negative charges. This is the basis of the action of an electrostatic lens.

The simplest electrostatic field is set up by two parallel plates of opposite charge. The setup can be constructed by placing two identical metal plates near each other and connecting the metal plates to the terminals of a battery (Fig. 2-8). Ignoring the inhomogeneity of the electric field at the edges of the charged plates,

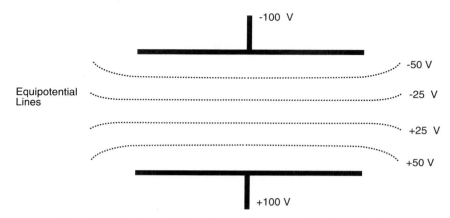

Figure 2-8   Equipotential lines between two oppositely charged plates.

the electrical field can be represented by a series of equipotential planes parallel to the plates. A negative electron entering the electrostatic field parallel to the lines of equipotential electrostatic force is affected by the electrostatic force perpendicular to the initial path toward the plate of positive charge. Thus the negative electron entering the electrostatic field is deflected from its initial path toward the plate of opposite charge, the positive plate (Fig. 2-9).

Wehnelt cylinder as an electrostatic lens.   Lenses in electron microscopes are not parallel plates, but instead are holes in charged surfaces. The Wehnelt cylinder is therefore an electrostatic lens. The equipotential lines of electrostatic force in an electron gun can be drawn as in Fig. 2-10. The equipotential lines of force represent a gradual change in electron potential in space, which allows an illustration of the effects of an electron field on the path of the electron beam. The potential difference between the Wehnelt cylinder and the tungsten cathode is less than the potential difference between the Wehnelt cylinder and the anode. Because of this, the equipotential lines of electrostatic force protrude into the space where

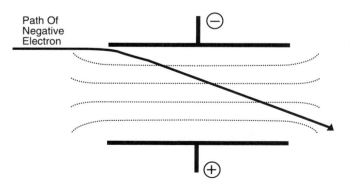

Figure 2-9   Path of an electron as it passes between two oppositely charged plates.

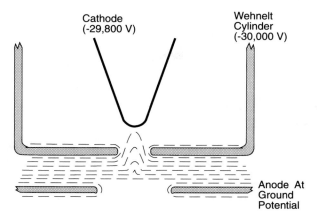

Cathode
(-29,800 V)

Wehnelt
Cylinder
(-30,000 V)

Anode At
Ground
Potential

Figure 2-10   Equipotential lines of electrostatic force in a biased gun.

the electric field is weaker. This is between the tungsten cathode and the Wehnelt cylinder, since the potential difference here is only 200 V, whereas between the Wehnelt cylinder and the anode the potential is 30,000 V. As the electrons of the electron beam leave the cathode, they will attempt to move perpendicular to the equipotential lines of force. The electrons will be deflected toward the axis, with the electrostatic field acting as a **convergent** or **positive** lens (Fig. 2-11).

Operation of the Self-biased Gun

Increasing the heating current to the filament in a self-biased gun causes a characteristic increase in the beam current passing to the anode and ultimately striking the specimen (Fig. 2-12). The bias on the Wehnelt cylinder barely affects the

Figure 2-11   Path of the beam electrons in a self-biased gun.

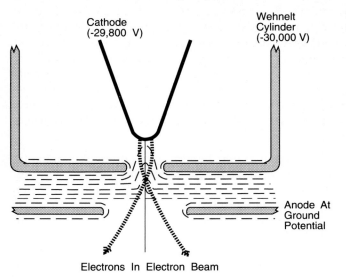

Cathode
(-29,800 V)

Wehnelt
Cylinder
(-30,000 V)

Anode At
Ground
Potential

Electrons In Electron Beam

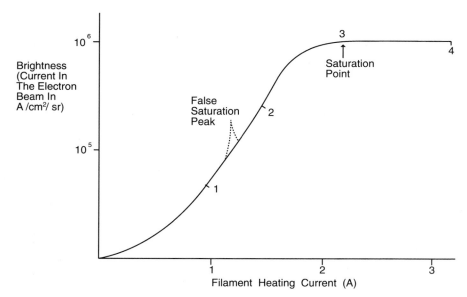

Figure 2-12   Relationship between the brightness of the electron beam and the filament heating current in a self-biasing gun with a tungsten cathode.

number of electrons *emitted* by the filament, but it does influence the number of emitted electrons that *pass through* the aperture of the Wehnelt cylinder to the anode. The configuration of the electron beam at four different places on the curve can be used to illustrate the mechanism of electron beam formation (Fig. 2-12).

1. A small rise in filament heating current results in the tip of the filament being heated. Emission of electrons occurs primarily from the hot filament tip. Since there is little bias on the Wehnelt cylinder at this point (there is little current from the electron beam across the Wehnelt cylinder to the bias resistor), the electrons emitted from the tungsten cathode are attracted directly to (readily "see") the anode and pass through the aperture of the Wehnelt cylinder down the column. The electron beam strikes the specimen as an elongated spot that represents an image of the filament, even to the striations on the surface of the tungsten wire of the cathode (Fig. 2-13).

2. At this filament heating current (point 2 on the curve in Fig. 2-12), the tip of the filament and the area of the rear of the tip are heated to a high temperature. The current of the electron beam has been increased to such a point that the beam current passing along the Wehnelt cylinder to the bias resistor is significant. This produces a moderate bias difference between the tungsten cathode and the Wehnelt cylinder. The relatively negative Wehnelt cylinder repels the electrons emitted from the tungsten cathode. If the Wehnelt cylinder is at $-30,000$ V and the tungsten cathode is at $-29,950$ V, the electrons in the electron beam will be at $-29,950$

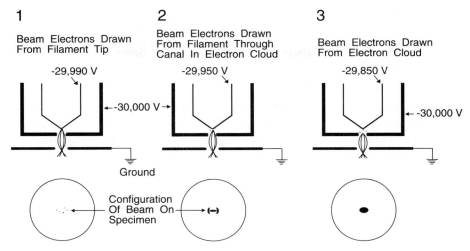

Figure 2-13   Potential of the electron gun at positions 1 to 3 in Fig. 2-12. The configuration of the beam on the specimen is also shown for each point.

V and will be repelled from the $-30,000$ V Wehnelt cylinder. The negative Wehnelt cylinder forces the beam electrons away and produces a cloud of beam electrons (**space charge**) in front of the filament and above the aperture of the Wehnelt cylinder. At this filament heating current, the anode is still powerful enough to extract some electrons *directly* from the tungsten filament through a canal in the negative field of the electron cloud. The electron beam striking the specimen still consists in part of a direct image of the filament, but additional electrons have also been drawn from the area on the side and to the rear of the tip of the filament. The beam configuration is that of a central spot (from the tip of the filament) surrounded by an incomplete ring (from electrons emitted from the sides and rear of the filament tip).

3. Here the filament heating current has reached the **saturation point** (a tungsten filament temperature of about 2700 K or 2427°C), where an increase in filament heating current results in increased electron emission but does not result in an increase in beam current. At this point a relatively large amount of beam current is passing to the Wehnelt cylinder by direct electron bombardment. This current is conducted through the Wehnelt cylinder and the bias resistor to produce a relatively large voltage difference between the Wehnelt cylinder and the tungsten cathode. If the bias voltage is -150 V, then the Wehnelt cylinder is 150 V more negative than the beam electrons coming off the tungsten cathode. The filament is reduced in voltage to $-29,850$ V since the Wehnelt cylinder is stabilized at $-30,000$ V. The electron gun is at **saturation**, with the electrons coming off the hot tungsten cathode being forced by the negative Wehnelt cylinder into a discrete electron cloud at the tip of the tungsten filament and above the aperture of the

Wehnelt cylinder. In this case the space charge of the electron cloud is so strong that the electrons forming the electron beam are only extracted from this electron cloud (instead of directly from the surface of the filament). The space charge of the electron cloud is blocking the access of the extracting force of the anode field to the filament. The removal of only those electrons in the center of the electron cloud at the saturation point produces a very small (0.1-mm diameter), very concentrated spot of electrons striking the specimen. Therefore, the **saturation point** represents the *maximum filament heating current at which the electron microscope should be operated, since it produces the maximum brightness of the beam for the minimum filament heating current* (which translates into the maximum life of the filament).

4. Here the filament heating current has been raised above the point where the beam current striking the specimen is not increased by increasing the filament heating current. Increasing the filament heating *does* increase the number of electrons leaving the filament and therefore the beam current striking the Wehnelt cylinder. However, the increased portion of the beam current passing along the Wehnelt cylinder to the bias resistor makes the Wehnelt cylinder more negative with respect to the tungsten cathode and the electron beam. The increased negative bias of the Wehnelt cylinder forces a greater proportion of the electrons back onto the tungsten filament. Thus there is no increase in the total number of electrons in the electron cloud or in the total number of electrons being extracted from the electron cloud by the anode field. The filament has developed such a large positive charge from electron loss that the number of electrons entering the space charge is equal to those returning to the filament. Therefore, at the saturation point, the space charged electron cloud is complete, and an increase in the filament heating current and bias above the saturation point results in no total increase in the number of electrons that can be extracted from the electron cloud by the anode field. *Above the saturation point, the filament is operating at a higher temperature, which results in the tungsten filament burning out more quickly, without an increased current in the electron beam* (Fig. 2-14).

In summary, the main difference between the unbiased and self-biased electron gun is that in the unbiased gun the electrons are drawn directly from the surface of the hot tungsten filament, whereas in the saturated self-biased gun, the negative Wehnelt cylinder results in the formation of an electron cloud in front of the filament from which the electrons of the beam are drawn by the anode field. The situation in the self-biased gun is more desirable, since the electron cloud is a homogeneous source, resulting in an even beam of electrons. A nonhomogeneous electron beam is produced in an unbiased gun because the electrons are drawn from the surface of the hot tungsten cathode, which has an irregular surface. Moreover, filament life is usually shorter in an unbiased gun because filament heating is not easily monitored or controlled, resulting in frequent overheating of the filament and resultant melting and failure.

The main advantage of working with a self-biasing gun is that the electrons

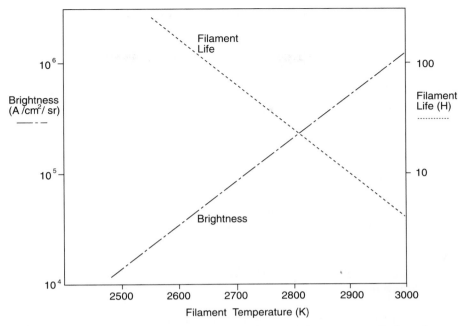

Figure 2-14 Relationship between the temperature of a tungsten cathode and brightness and filament life. Note this is a log scale for filament life.

are concentrated to a small electron source in the electron cloud, which results in a relatively small diameter of the electron beam at the crossover point above the anode. The crossover point acts as a **virtual source** of electrons to the beam, so the smaller the diameter of the crossover point is, the brighter the usable portion of the electron beam. The electron gun makes the electrons diverge from this crossover point, which is only tens of micrometers in diameter, although the electrons come from a region on the filament that is some hundreds of micrometers in extent. This is desirable in electron microscopy, since a small concentrated beam is optimal in viewing specimens.

In the scanning electron microscope, the beam current is measured indirectly by examining a y-modulated scan of a single line over the specimen on the cathode ray tube. A y-modulated scan is generated when the brightness in a scan line is represented by a series of peaks, while the dark areas in the scan line are represented by valleys (in other words, the brightness is modulated in the vertical of y-direction). Therefore, as the filament heating current is increased, the brightness increases and the height of the y-modulated scan line becomes higher. The four points on the saturation curve of filament heating current (Fig. 2-12) are represent in y-modulated scans in Fig. 2-15.

False Saturation Peaks.   As the filament heating current is increased, peripheral parts of the filament reach the saturation emission temperature before other

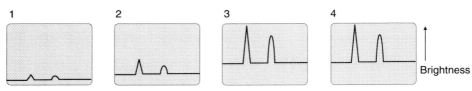

Y-Modulated Line Scans On Cathode Ray Tube Of An SEM

Figure 2-15   Appearance of a *Y*-modulated line scan on the cathode ray tube of a scanning electron microscope as the filament heating current is increased to the four positions illustrated in Fig. 2-12.

areas. This can give rise to what are called **false saturation peaks** as the filament heating current is increased (Fig. 2-12). A false saturation peak can be recognized by a decrease in the brightness of the beam as the filament heating current is increased beyond the peak. In the true saturation peak, there is no increase (or decrease) in brightness as the filament heating current is increased.

Distance between the Tungsten Cathode and the Wehnelt Cylinder

The distance between the tungsten cathode and the anode is fixed in electron microscopes. However, the distance from the tungsten cathode to the Wehnelt cylinder is variable and is adjusted when a new filament is placed in the gun assembly. The distance from the tungsten cathode to the Wehnelt cylinder is usually adjusted by turning the Wehnelt cylinder on a thread at the base of the electron gun.

When the tungsten filament is close to the aperture of the Wehnelt cylinder (Fig. 2-16), the electrons emitted from the hot tungsten cathode can readily "see" the anode, and a large number of electrons are pulled from the accumulated electron cloud in front of the filament to produce a bright beam with a high beam current. A high filament heating current is needed to produce a saturated beam because the negatively biased Wehnelt cylinder is close to the tungsten cathode, producing a strong negative field in the Wehnelt cylinder that makes it difficult for the electrons to escape from the tungsten cathode by thermionic emission. Nevertheless, in this configuration, the electron gun produces a high beam current.

When the tungsten filament is far from the Wehnelt cylinder, the electrons in the electron cloud are relatively far back from the aperture of the Wehnelt cylinder, and it is not easy for these electrons to "see" the anode field. This results in a weak beam with little current. Little filament heating current is required to saturate the filament, since the negative field of the Wehnelt cylinder is relatively far away from the tungsten cathode. Saturation occurs at a lower temperature because the electrons are able to escape more easily from the hot tungsten filament, producing a greater concentration of electrons and a greater beam current at a lower filament temperature. However, the maximum beam current is very

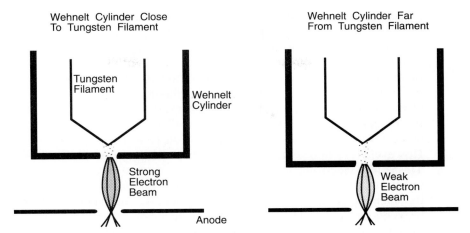

Figure 2-16 Effect of changing the Wehnelt cylinder to cathode distance in an electron gun. The distance from the anode to the tungsten cathode is constant.

low, resulting in a relatively weak beam. The distance between the tungsten cathode and Wehnelt cylinder must be adjusted to a predetermined value each time a tungsten filament is replaced in an electron gun in order to get the desired beam current.

Failure of the Tungsten Filament

During operation of the electron gun, tungsten atoms are slowly evaporated from the hot tungsten filament and deposited on the cooler parts of the tungsten cathode and the Wehnelt cylinder. This is because the operating temperature of the filament is close enough to the melting point that some of the metal is evaporated in the high vacuum. This vaporized tungsten makes the surface of the Wehnelt cylinder bluish-black in color and has to be removed by polishing when the filament is replaced in the electron gun. The filament becomes thinner as the tungsten evaporates from the hot filament, increasing the resistance of the filament to electrical current and reducing the amount of current that is needed to keep the filament at saturation temperature. Therefore, the longer the tungsten filament has been in use, the less filament heating current that is needed to saturate the filament. Eventually, the tungsten filament fails when the filament becomes so thin that local temperatures increase to the point where a fracture (melting) occurs. The hottest part of the filament is usually just to the rear of the tip, and this is usually where the failure occurs.

Failure of a tungsten filament that has been saturated properly will occur after 100 to 150 h if a vacuum of 10^{-8} mbar (10^{-6} Pa) (attainable in modern electron microscopes fitted with an ion getter pump; see Chapter 12) is maintained in the gun chamber. At this vacuum, few electronegative gases (such as oxygen) are present, and failure of the tungsten filament is due mostly to evaporation of tungsten from the filament. At poor vacuums (10^{-3} to 10^{-4} mbar; 10^{-1} to 10^{-2} Pa), a properly saturated tungsten filament will last only 5 to 10 h before failure occurs. In this case, failure is due primarily to oxidation of the hot tungsten atoms in the filament, and not due directly to evaporation of elemental tungsten. Above 400°C (673 K), tungsten is very susceptible to oxidation. At 800°C (1073 K), sublimation of the oxide becomes significant, so at operating temperatures (2427°C, 2700K) the oxides of tungsten are evaporated off as soon as they are formed. Above 600°C (873 K), tungsten also reacts vigorously with water vapor to form oxides. However, tungsten is stable to nitrogen at temperatures above 2300°C (2573 K). This is why electron gun chambers with hot filaments are vented with nitrogen gas instead of atmospheric gas. The tungsten filament takes some time to cool off after the filament heating current has been turned off, and the presence of nitrogen gas prevents oxidation. Therefore, in the presence of oxygen or water vapor, the hot tungsten filament oxidizes rapidly. The tungsten oxide vaporizes quickly at the operating temperatures of the filament, and the filament soon fails.

Examination of the broken tungsten filament can usually determine the cause of failure. A properly saturated filament will eventually fail to produce a break to one side of the tip, with the filament thinning down to the area of the break (Fig. 2-17). A failure in a tungsten filament that has been operated in the *oversaturated condition* at too high a temperature will produce a break with molten balls of tungsten at the break edges (Figs. 2-17 and 2-18). *Poor vacuum conditions* (less than 10^{-4} mbar, 10^{-2} Pa) result in all the hot areas of the tungsten filament combining with oxygen to produce oxides. The oxides evaporate at a relatively lower temperature, so the oxides evaporate as soon as they are formed, resulting in thinning of the whole hot area of the filament (not just the hottest part of the

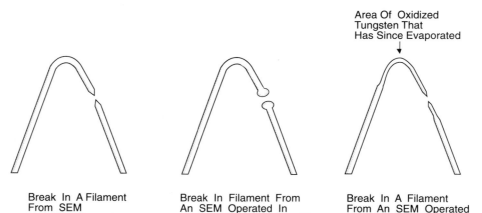

Break In A Filament
From SEM
Operated Properly

Break In Filament From
An SEM Operated In
The Oversaturated Condition

Break In A Filament
From An SEM Operated
With A Poor Vacuum

Figure 2-17   Condition of a failed tungsten filament operated under different conditions.

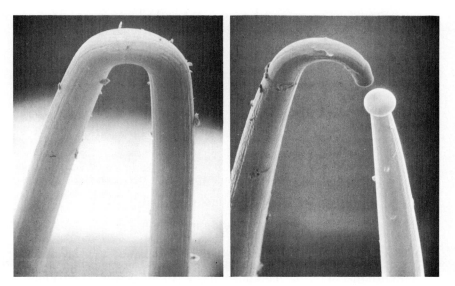

Figure 2-18 Scanning electron micrographs of (a) an unused tungsten filament and (b) a tungsten filament that has failed after being used at oversaturation. The failed filament has a ball of tungsten that was formed when the tungsten melted.

filament behind the tip). Inspection of a failed tungsten filament operated under a poor vacuum shows most of the filament tip to be thin (Fig. 2-17).

Brightness of Illumination

The important property of an electron gun is the number of electrons per unit of solid angle of beam and not the total intensity of the beam. This is because of the small angular aperture in electron microscopes, which produces a very thin beam of electrons. It is important to concentrate the electrons into this small beam. The number of electrons in the beam is usually defined by brightness (β). The **brightness** of the electron beam is the current density per unit of solid angle and is measured in amperes per square centimeter per steradian. In contrast, the **intensity** of the electron beam is expressed in amperes per square centimeter (and is inversely proportional to the magnification). Thus the intensity of the beam represents the electrons striking a specific place, whereas brightness is the average number of electrons in the electron beam over a steradian (sr) (more commonly expressed in millisteradians; msr). See Appendix I for a definition of steradian.

The solid angle of the beam is small (typically 0.3 msr) because of the high magnifications used. The small angle, combined with the small apertures in the electron microscope, restricts the size of the electron beam to that of a thin pencil. This, in turn, reduces the spherical aberration of the lenses (see Chapter 4).

The brightness of the electron beam (in amperes per square centimeter per steradian) is given by the following formula:

$$\beta = \beta_c \frac{eV}{kT}$$

where β = brightness

β_c = current density at the cathode surface

e = electronic charge $(1.59 \times 1^{-19} \text{ C})$

V = accelerating voltage (potential between the cathode and anode)

k = Boltzmann's constant $(8.6 \times 10^{-5} \; eV/K)$

T = absolute temperature in kelvins

From this equation, it is apparent that the brightness (β) of the electron beam is proportional to the accelerating voltage (V). The effect of temperature (T) is more complex. T appears as a denominator of the equation, so, on first glance, it would appear that, according to the equation, the brightness would decrease as the temperature is raised. However, the current at the surface of the cathode (β_c) rises exponentially with temperature; therefore, the brightness actually increases rapidly with the increase in temperature of the tungsten cathode.

For a temperature (T) at the tungsten cathode of 2800 K (2527°C) and a current density at the cathode surface (β_c) of 0.035 A/mm^2, the brightness (β) is about 2 A/mm^2/msr^{-1}.

Effect on the Electron Beam of Biasing the Wehnelt Cylinder

Changing the negative bias on the Wehnelt cylinder affects the total current in the beam, the brightness of the beam, the total beam angle, and the diameter of the beam at crossover. The *total beam current* is decreased as the bias is increased (Fig. 2-19a). This is because the increased negative condition of the Wehnelt cylinder as the bias is increased allows fewer electrons to pass through to the anode. The *total beam angle* also decreases as bias increases (Fig. 2-19b), since the electrons that do escape come from a smaller area at the higher biases. The *brightness* increases as the bias increases up to a point and then the brightness decreases (Fig. 2-19c). The increase in brightness at the lower biases is due to the fact that the total beam angle is affecting the brightness more than the total beam current. Thus increases in the lower values of bias cause more electrons to be squeezed into a smaller beam angle, resulting in a total increase in brightness per millisteradian. Eventually, the point is reached where brightness decreases with increase in bias due to a decrease in total beam current becoming more important than the decrease

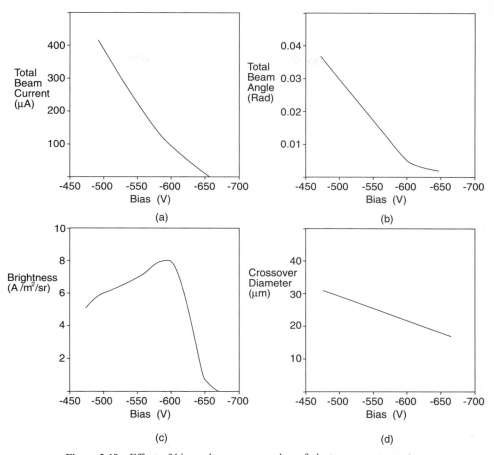

Figure 2-19   Effect of bias voltage on a number of electron gun parameters.

in total beam angle, so the brightness per millisteradian decreases. The last parameter, the *diameter of the beam at crossover*, decreases slightly as the bias is
increased (Fig. 2-19d).

Accelerating Voltage

In most scanning electron microscopes the accelerating voltage can be varied from
1 to 30 kV. Higher accelerating voltages result in better signal collection from the
specimen and an image with less electronic noise since the signal does not have
to be boosted as much. The trade-off is that at higher accelerating voltages the
beam electrons penetrate to deeper depths in the specimen, resulting in the signal
coming from an increasing depth within the specimen. At lower accelerating voltages, more of the signal comes from the surface layers, resulting in a final image

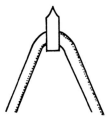

Figure 2-20 A pointed tungsten cathode created by welding a sharpened piece of tungsten to a V-shaped tungsten filament.

that more clearly represents the surface of the specimen. However, less electrons are received by the detector, resulting in less signal. The details of specimen–beam interactions will be discussed later.

Pointed Tungsten Filaments

The conventional V-shaped, hairpin tungsten filament produces a beam of electrons that is focused by the electrostatic field of the Wehnelt cylinder to a crossover of approximately 20 to 50 μm in diameter above the anode. One method of increasing the brightness of the electron beam is to reduce the diameter of the filament tip. The size of the filament tip can be reduced by using a pointed filament that consists of a tungsten wire etched to a fine point 0.1 μm in diameter, which is welded to another piece of tungsten wire (Fig. 2-20). This results in a very fine source of electrons with symmetrical geometry (compared to the asymmetrical geometry of the V-shaped hairpin filament). The pointed filament tip is also removed from the asymmetrical field created by current flowing through the sharp bend at the filament tip. The pointed filament, although requiring a higher filament heating current, offers a gain factor of 10 in gun brightness level, in comparison to the more diffuse source offered by the hairpin filament. Also, the size of the virtual source at the crossover point above the anode is approximately 60% smaller than that of a conventional V-shaped hairpin filament. The accelerating voltage fields are very high near the tip, resulting in field emission of electrons. The basis of the field emission type of gun is discussed later.

The reason that pointed tungsten filaments are not widely used is that they are very fragile, and since the pointed tip receives heat by conduction from the heated filament itself, higher filament heating currents are required, resulting in filament lives that are only 10% to 20% that of conventional filaments.

LANTHANUM HEXABORIDE CATHODE

In 1969, Alec N. Broers (Fig. 2-21) described an electron gun using a lanthanum hexaboride rod as a cathode. Lanthanum hexaboride (LaB_6) cathodes have the advantage of a long life and of producing an electron beam with a brightness five

Figure 2-21   Alec N. Broers, Conrad G. Bremer, and Oliver C. Wells (back to front). The picture was taken in 1973 at the scanning electron microscope at IBM's Thomas J. Watson Research Center. The electron column is at the left foreground (Photograph from *Photo Methods for Industry* 17:24, 1974). (*Photograph courtesy of Professional Photography of America*) Alec N. Broers was born September 17, 1938, in Calcutta, India. He received his B.Sc. in physics from the University of Melbourne in 1959, a B.A. in mechanical science from Cambridge University in 1962, and a Ph.D. in electrical engineering at Cambridge University in 1965. From 1965 to 1984 he worked at IBM's Thomas J. Watson Research Center. In 1984 he moved to Cambridge University in England where he is head of the Electrical Division of the Department of Engineering. Oliver C. Wells was born February 14, 1931. He received his BA, MA, and Ph.D. from Cambridge University in 1950, 1953, and 1957 respectively. From 1957 to 1959 he developed transistor circuits for information storage at the Ericsson Telephone Company, Nottingham, England. From 1959 to 1964 he (with T. E. Everhart) built the first scanning electron microscope at a corporate research laboratory at the Westinghouse Research and Development Center in Pittsburgh, Pennsylvania. From 1964 to 1965 he worked on the deposition of epitaxial silicon films by evaporation in ultra-high vacuum at CBS Laboratories, Stamford, Connecticut. From 1965 to the present time he has been engaged in research in the theory and application of scanning electron microscopy at the IBM Thomas J. Watson Research Center, Yorktown Heights, New York. His publications have included his book *Scanning Electron Microscopy*.

to ten times that of a thermionic tungsten wire operating at saturation temperatures. The disadvantages of lanthanum hexaboride cathodes are that they are relatively expensive (about 50 times more than a tungsten filament cathode) and that they require the maintenance of a high vacuum (10^{-6} mbar, 10^{-4} Pa or better) to prevent poisoning of the lanthanum hexaboride rod. Lanthanum is extremely chemically reactive when hot and readily forms compounds with all elements except carbon and rhenium. When this happens, the surface of the LaB_6 cathode becomes occluded with these compounds, and the LaB_6 ceases to be an effective electron emitter.

It is the low work function of lanthanum hexaboride that enables the production of an intense electron beam. Polycrystalline lanthanum hexaboride has a work function on the order of 2.4 eV, while a single crystal of lanthanum hexaboride can have a work function as low as 2.0 eV (compared to 4.5 eV for tungsten). At 2700 K, each 0.1-eV reduction in the work function increases the electrons leaving the cathode surface about 1.5 times. Therefore, because of the lower work function of LaB_6, many more electrons are emitted from a LaB_6 cathode than from a tungsten wire cathode at the same temperature. Lanthanum hexaboride melts at 2800°K.

A LaB_6 cathode consists of a small rod or single crystal with a tip machined down to a tip of 10 μm. The tip can be formed by carefully grinding the rod on a diamond wheel. The electron gun for LaB_6 is similar to that for tungsten, with the LaB_6 cathode encased in a Wehnelt cylinder and subtended by an anode. The tip of the LaB_6 cathode sits above the aperture of the Wehnelt cylinder (Figs. 2-22

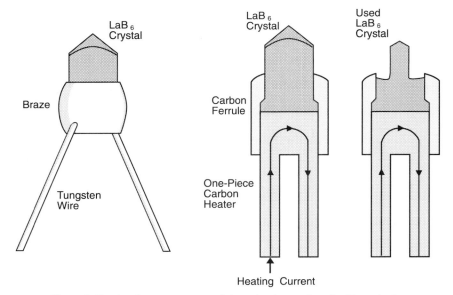

Figure 2-22   An electron gun containing a lanthanum hexaboride cathode.

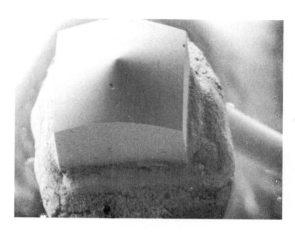

Figure 2-23 Scanning electron micrograph of the tip of a lanthanum hexaboride crystal that has been mounted to form a cathode for an electron gun.

and 2-23). In most modern electron microscopes, the LaB_6 is held in a base made of carbon. The LaB_6 is heated directly by placing a current of 1 to 2 A and 2 to 3 W through the base. Operating the LaB_6 cathode at 1500 K (1227°C) results in approximately the same amount of electron emission that occurs from a tungsten cathode (brightness of 1.0×10^5 A/cm^2/sr), while at 2000 K (1727°C) the brightness is approximately 100 times this value (1.0 and 10^7 A/cm^2/sr).

The LaB_6 cathode is usually operated at around 1850 K (1587°C) in order to reduce evaporation of lanthanum and to prolong life of the cathode. At this temperature the brightness is about ten times that of a thermionic tungsten cathode. The actual brightness is also affected by the size of the LaB_6 tip (the smaller the tip, the greater the brightness), the distance from the Wehnelt cylinder to the LaB_6 cathode tip, and the bias on the Wehnelt cylinder (factors that also affect the brightness of thermionic emission from a tungsten cathode). The increased brightness of the LaB_6 cathode means the beam can be manipulated so that a significantly smaller probe size occurs at the same beam current or so that a larger beam current occurs at the same probe size as for a thermionic tungsten emitter.

Lanthanum hexaboride has the lanthanum atoms contained in a lattice formed by the boron atoms. The small boron atoms form a three-dimensional framework that surrounds the large lanthanum metal atoms. The boron framework is made up of octahedrons, one at each end of a cube. The melting point of LaB_6 is characteristic of the boron structure and not of the lanthanum. At the operating temperatures of the LaB_6 cathode (1500 to 2000°K, 1227° to 1727°C), the lanthanum metal atoms slowly evaporate away. The lanthanum atoms are, however, immediately replaced by diffusion of metal atoms from below the surface through the boron latticework. The boron framework does not evaporate, but remains intact. The diffusion of lanthanum atoms to the surface of the cathode occurs only when there is a vacancy at the surface. This keeps evaporation losses at a minimum and at the same time provides a mechanism for constantly maintaining an active cathode surface. The life of a LaB_6 cathode operated at 10^{-6} mbar (10^{-4} Pa) can be as high as a couple of years in a properly used electron microscope. Ultimately, during the use of the LaB$_6$ cathode, so much lanthanum is removed from the cathode that it is not possible to replace the lanthanum evaporated at the surface, the boron lattice collapses, and electron emission becomes sharply reduced, necessitating the insertion of a new LaB$_6$ cathode.

After being exposed to the atmosphere, it often takes from a few seconds to a few minutes for the LaB$_6$ cathode to activate itself. During this period of time, contaminants on the surface are being removed by evaporation, and it is not possible to achieve electron emission through this contaminating layer. Once this layer is evaporated, electron emission begins. It is important during this period that the temperature of the cathode is not increased to attempt to induce electron emission, as the LaB$_6$ cathode can be destroyed by overheating.

The normal saturation curve of a LaB$_6$ cathode has a significant false saturation peak as the filament heating current is increased. To properly saturate the LaB$_6$ cathode, it is necessary to increase the filament heating current past this false saturation peak, causing a decrease and then and increase in the emission of electrons, until the true saturation peak is reached.

FIELD EMISSION

Appreciable electron emission occurs when a cold metal surface in a vacuum is subjected to an electron accelerating field on the order of 10^8 V/cm^2. This process, called **field emission**, was first reported in 1897 by R. W. Wood and quantified using quantum mechanics by E. Fowler and L. Nordheim in 1928. The Fowler–Nordheim theory of field emission from a cold metal in a vacuum is based on the energy diagram in Fig. 2-24. In thermionic emission, it is necessary to heat the orbital electrons of the metal to raise them to sufficient energy to escape the metal surface. In field emission, the potential energy curve is changed. If a strong electron accelerating field is applied to the metal surface, there is a fall in the potential energy needed for an electron to escape. The farther from the surface of the metal, the greater is the fall in necessary energy (Fig. 2-24). This is represented by a curve with its maximum energy just outside of the surface, followed by a curve sloping to the right (farther from the surface of the metal). The peak potential energy of the curve is still greater than the maximum energy of the orbital electrons. Classically, therefore, it would appear impossible for electrons in the "cold" metal to acquire enough energy to overcome the potential energy barrier and escape from the metal surface. However, the application of quantum mechanics to the situation shows that the potential energy barrier is porous, and electrons can pass through the barrier if a high enough electric field is applied. In an electron accelerating field of 10^8 V/cm^2, electrons are able to **tunnel** through the barrier. In **tunneling**, the electrons of the metal do not acquire additional energy (necessary to go over the top of the potential energy curve), but instead are pulled through the potential energy of the barrier by the electron accelerating field. Field emission is thus quite distinct from thermionic emission. Field emission is substantially

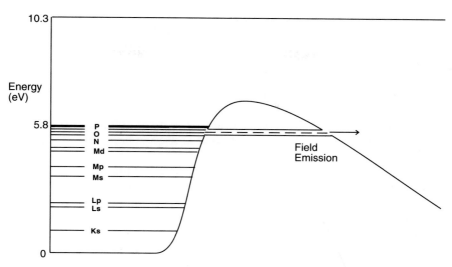

Figure 2-24   Field emission from tungsten. The energy of the highest-orbital electrons of the tungsten atom is insufficient to move over the energy barrier at room temperature. However, when a strong electrical field is placed in front of the metal, the orbital electrons are able to tunnel through the potential energy barrier and escape from the surface of the metal.

independent of temperature (as long as the metal remains cold), and electron emission from the surface of the metal increases with an increase in the electron accelerating field according to an exponential law. Thermionic emission is very sensitive to temperature changes, but relatively insensitive to changes in field strength.

Field Emission in Electron Microscopy.   *The main advantage of using a field emission source in electron microscopy is the large number of electrons that are emitted in a relatively small beam spot.* Current densities are about 10^6 A/cm² for a field emission source, compared to 10 A/cm² for a thermionic tungsten source, an increase in emission of 1 million times. A field emission source can produce a brightness of 0.4 A μm² sr⁻¹. This is particularly important in the scanning electron microscope, where a field emission source allows increased resolution because of the ability to produce a smaller beam spot size that has adequate current. A scanning electron microscope with a thermionic tungsten emitter usually can use only a 10-nm or larger beam spot diameter. This is because reducing the beam spot diameter below 10 nm results in so few electrons in the beam that the beam must rest on each pixel of the specimen for an impractical length of time to produce an image with an acceptable noise level. However, a field emission source produces an electron beam containing so many electrons that the beam spot diameter can be smaller and still have sufficient electrons to form an image. A smaller beam spot diameter results in increased resolution of the image.

Field emission of electrons from metal at room temperatures requires an electron field of 10^8 V/cm². This is far too great an electric field to be practically acquired by the generation of a direct potential between two flat or curved electrodes (Fig. 2-25). The only practical way to generate such a field is to manufacture a sharp edge or fine needle tip and apply an electric field to this with a potential of a few thousand volts. All the positive charges now accumulate on the sharp needle tip, producing a strong electrical field. Field emission sources in electron microscopes use a tungsten cathode with a tip of 100-nm or 0.1-μm diameter (a very small tip; the diameter of the point of a pin in comparison is 100 μm) (Fig. 2-26).

The strength of the field can be calculated as follows. You are using a scanning electron microscope that has a field emission source. The diameter of the tip of the tungsten cathode is 100 nm (radius = 50 nm), and the voltage between the first anode and the tungsten tip is 1000 V. The field at the surface of the 100-nm tungsten tip can be calculated from the following formula:

$$\text{Field} = \frac{\text{volts}}{\text{radius of the cathode tip}}$$

$$F = \frac{V}{R}$$

$$= \frac{1000 \text{ V}}{50 \text{ nm}}$$

$$= \frac{1 \times 10^3 \text{ V}}{0.5 \times 10^{-5} \text{ cm}}$$

$$= \frac{2.0 \times 10^8 \text{ V}}{\text{cm}}$$

The tungsten tip is produced by electrolytically or chemically etching the tip. Tungsten is usually used as a cathode because the electric field produced at the tip (10^8 V/cm) is so high that there is a very large mechanical stress on the cathode, and only very strong materials, such as tungsten, can withstand it without failing. The crystal orientation of the tungsten is also important because the potential energy barrier to electron escape is a function of the crystal orientation of the surface through which the electrons leave. It is necessary to have a single crystal at the emitting tip with a particular orientation in order to obtain the highest electron emission. When solid tungsten is formed by cooling liquid tungsten, the solid tungsten will have a limited number of crystal symmetries. The physical properties of the different crystal symmetries are different. Each crystal will have a number of facets, and one of the physical characteristics that varies from one facet to another is the amount of current that can be extracted by field emission. Tungsten wire, produced by drawing metal through small holes, has an orientation

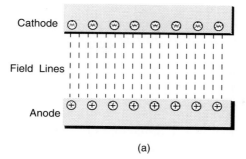

Cathode

Field Lines

Anode

(a)

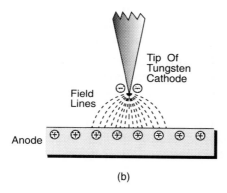

Tip Of
Tungsten
Cathode

Field
Lines

Anode

(b)

Figure 2-25 How a very high electric field can be generated at the tip of a fine point on a tungsten cathode: (a) When a potential is applied across two plates, the charges are spread along the plates and a disperse electric field results. (b) When the cathode is a fine tip, the field lines from the anode concentrate at the tip, resulting in the formation of a very strong electric field at the tip of the cathode.

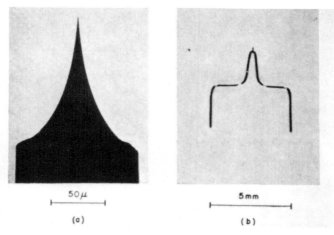

50μ

5mm

(a)   (b)

Figure 2-26 A tungsten cathode with an etched tip suitable for field emission. (a) Shadowgraph showing the configuration of the tip. (b) An etched tungsten tip mounted on a hairpin filament. (*From* A. V. Crewe and others, *Rev. Sci. Instruments* 39:576–584, 1968). (*Photograph courtesy of American Institute of Physics*)

called (110), whereby almost no electrons emerge by field emission in a direction toward the anode of the electron microscope. In choosing tungsten for field emission sources, tungsten metal in the (310) and (111) orientations is selected as cathodes.

To obtain electron emission from the fine tungsten tip, *it is essential that the metal surface be clean and that there be no atoms or molecules of foreign material on the surface*. Even a single foreign atom sitting on the surface will lower the electron emission. A vacuum of 10^{-9} mbar (10^{-7} Pa) is necessary to prevent accumulation of atoms on the surface of the tungsten cathode. Even at this vacuum, a few gas molecules will occasionally land on the tungsten tip and cause fluctuations in the emission current. At 10^{-9} mbar (10^{-7} Pa), tungsten cathodes will operate for thousands of hours before needing to be replaced. If the surface of the tungsten cathode is contaminated by a brief exposure to a poor vacuum, the cathode can be cleaned by rapid heating to around 2000°C (2273 K) for a few seconds. This results in reevaporation of the contaminating atoms.

Construction of a Field Emission Source.   A typical field emission source consists of the tungsten cathode and two anodes (Fig. 2-27). There is no Wehnelt cylinder around the cathode. The tip of the tungsten cathode is about 1 cm from the first anode. There is a voltage difference of a couple of thousand volts between the first anode and the tungsten cathode. The first anode serves to set up the electron accelerating field of 10^8 V/cm on the cathode tip and to strip the electrons from the tungsten cathode. The second anode serves to accelerate the electrons down the column and set up the accelerating voltage. The second anode is at

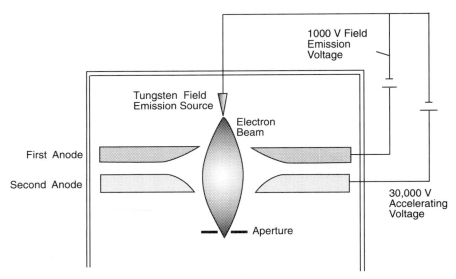

Figure 2-27   Configuration of a field emission gun at the top of an electron microscope column.

ground. In a scanning electron microscope with a field emission source, the second anode is at ground, the first anode might be at $-30,000$ V, while the tungsten cathode is at $-32,000$ V. The two anodes act as electrostatic lenses and contribute to spherical aberration of the image. The peculiar design of the anodes reduces the aberration of the beam. The diameter of the crossover point beneath the second anode is only about 10 nm. This is a very small crossover diameter when compared to the 10-μm crossover diameter for a lanthanum hexaboride source and 50 μm for a thermionic tungsten cathode. Because of the already small-sized, crossover diameter from the field emission source, no condenser lenses are used in the column when a field emission source is used. This is in contrast to the situation when thermionic tungsten or lanthanum hexaboride cathodes are used, where condenser lenses are used to reduce the diameter of the electron beam.

In summary, field emission has three advantages: (1) energy does not need to be supplied to the cold cathode; (2) very large numbers of electrons are emitted; and (3) despite the large number of electrons emitted, the energies of the electrons leaving the cathode are very similar, with a relatively narrow energy spread, thus minimizing chromatic aberration. There are, however, two disadvantages: (1) high electric fields and (2) high vacuums are required.

3

Lenses and Magnetism

PATH OF RADIATION THROUGH LENSES

The function of lenses is to bend rays of light or electrons so that they are deflected in a predictable way from their original paths. In lenses used in light microscopes, the light rays are slowed down as they pass through the lens. Refraction of light occurs as light passes between the media of two refractive indexes. The light rays are slowed down and refracted as they pass from air (or whatever the lens is immersed in) into the glass lens and, again, as they pass from the glass lens into the air. Grinding the lenses into different shapes causes the lenses to bend light in different ways. A beam of parallel rays striking a perfectly convergent lens perpendicular to the central plane of the lens will be bent as the light passes through the lens, to converge to a focal point and then diverge again (Fig. 3-1). The distance from the central plane of the lens to the focal point is called the **focal length** and it is here that the image is in focus.

A strong lens has a thick center and a short focal length, whereas a weak lens has a thin center and a long focal length (Fig. 3-2). This type of lens is called a **positive** or **convergent** lens. A positive lens with a specimen in front of the lens will form a real image of the specimen at the focal point behind the lens. A **negative** or **divergent** lens consists of glass with two concave surfaces (Fig. 3-3). Parallel rays striking the lens in a direction parallel to the axis will diverge as they pass

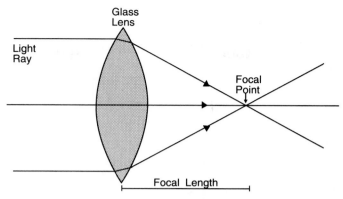

Figure 3-1   Path of light rays through a positive, convergent glass lens.

through the lens and cannot be brought to focus. Both converging and diverging lenses are used in light microscopes, but only converging magnetic lenses are used in electron microscopes. As will be seen, because of the action of magnetic fields on electron beams, it is not possible to construct diverging magnetic lenses. It is possible to construct converging and diverging electrostatic electron lenses (lenses using positive or negative electrical fields); however, electrostatic electron lenses are not used in modern electron microscopes.

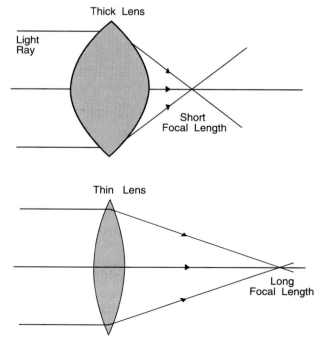

Figure 3-2   Path of light rays through a strong, thick lens and through a weak, thin lens.

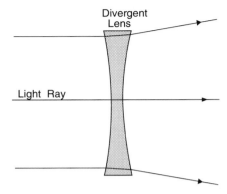

Figure 3-3 Path of light rays through a negative, divergent lens.

Magnetic lenses used in electron microscopes consist of copper wire wound around a central bore. Electrical current passed through the copper wire creates a magnetic field in the bore of the lens that has two effects on electrons passing through the bore (Fig. 3-4). The first effect causes the electrons to spiral as they pass through the bore, and the second effect pushes the electrons toward the center of the bore and is responsible for the focusing action of the lens. Reversing the flow of electrons through the windings causes the electrons to spiral in the opposite direction, but the focusing action is still maintained since the electrons are again pushed toward the center of the bore. A more detailed explanation of the effects of the magnetic field on the electron beam follows later in the chapter.

Magnetic electron lenses differ in a further fundamental way from the glass lenses used in light microscopes, in that the electron beam does not change its velocity as it passes through the magnetic field. In a glass lens the light rays slow in velocity as they strike the glass and are refracted. Also, in a magnetic electron lens, the electrons are affected continuously by the magnetic lines of force of the

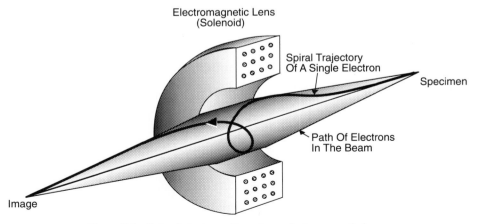

Figure 3-4 Path of electrons through an electromagnetic lens.

lens. Refraction of the electrons is continuous, and there is no sharp interface between the refracting medium (the magnetic field) and the immersion medium (the vacuum within the electron microscope), as there is in a light microscope.

CREATION OF MAGNETIC FIELDS

Magnetic fields are created by electrical currents. Electrons, protons, and neutrons rotate rapidly and in doing so create magnetic fields with characteristic north and south poles, similar to those in a bar magnet (Fig. 3-5). A spinning electron is called a **magnetic dipole** as it has a north and south pole. When free to rotate, the dipole aligns itself with an exterior magnetic field.

It is the spinning of the electrons as they orbit the atomic nucleus that gives materials their magnetic properties. Within an atom, the orbital motion of the electron around the atomic nucleus produces a very weak magnetic field. Most of the strength of the magnetic field is due to the spinning of the electron on its own axis. In most elements, half of the orbiting electrons spin in one direction and the other half in the opposite direction, thereby canceling any magnetic effects. However, in some elements the number of electrons in the outer orbital of the atom spinning in one direction does not equal the number of electrons spinning in the opposite direction. These elements are in the **ferromagnetic group** and include iron, nickel, cobalt, gadolinium, and dysprosium. In the third orbital shell of iron (with 14 electrons), the number of electrons with positive spin does not equal the number of electrons with negative spin. This makes each iron atom act like a small magnet. The unique feature of ferromagnetic materials, however, is that in the solid state whole groups of atoms (ranging in number from 10^{16} to 10^{21} atoms) have

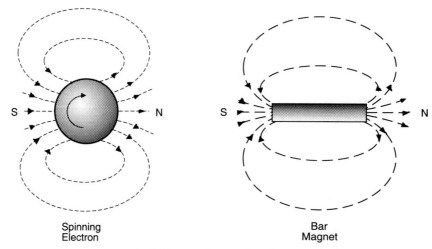

| Spinning | Bar |
| Electron | Magnet |

Figure 3-5 Magnetic fields around a spinning electron and a bar magnet.

Domains Increase
In Size

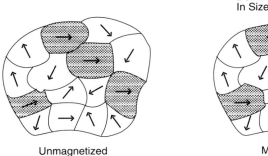
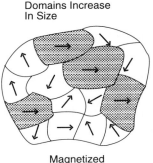

Unmagnetized   Magnetized

Figure 3-6 During magnetization of an unmagnetized piece of a ferromagnetic
substance, the domains that are oriented in the direction of the magnetic field
(stippled domains) grow at the expense of the domains oriented in other directions.

their magnetic fields oriented in the same direction. These groups of atoms form
small volumes, typically from 10^{-6} to 10^{-2} cm³, known as **domains** that are magnetized to saturation (Fig. 3-6). In an ordinary, unmagnetized piece of iron, the
tiny domains face in all directions, and their small magnetic effects cancel out.
But if the material is put in a strong magnetic field, two things happen. Slowly the
domains swing around into the direction of the field. As they come into line, they
also grow larger by reorienting some atoms from adjacent domains, which shrink
in size (Fig. 3-6). When a number of domains have lined up in one direction, the
whole piece of iron becomes a magnet with a north and a south pole.

If a bar magnet is pivoted near the center so that it can move freely in a
horizontal plane, it will set itself up in a north–south direction along the earth's
magnetic field. The south pole of the magnet will be pointed toward the magnetic
north pole of the earth, while the north pole of the magnet will be pointed toward
the south pole of the earth. Thus a north pole is attracted to a south pole, and vice
versa, while like poles (for example, north–north poles) repel each other. Treating
the earth as a magnet, the south pole of the earth's magnet is near the northern
geographical pole, while the north pole of the earth's magnet is near the southern
geographical pole (Fig. 3-7). Since unlike poles attract each other, the north poles
of compasses are attracted toward the northern part of the world where the magnetic south pole is located.

A bar magnet always has two poles, and if an attempt is made to isolate one
of them, by cutting the bar magnet in two, new poles appear at the cut ends of the
two halves of the original bar magnet. The space around a magnet contains a
magnetic field, which is schematically represented by **lines of force (flux)** connecting the north and south poles (Fig. 3-7a through c). The magnetic lines of force
always emerge from the north pole and enter the south pole. The total number of
magnetic lines of force represents the magnetic flux, and the number of lines of
force per unit volume is the flux density.

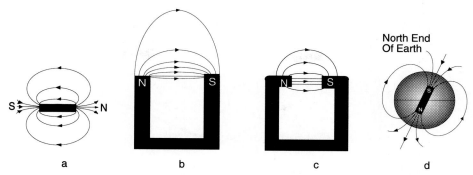

Figure 3-7   Orientation of the magnetic field around different forms of magnets (a–c) and around the earth (d). The magnetic lines of force always go into the south pole and out of the north pole.

MAGNETIC FIELDS CREATED BY AN ELECTRICAL CURRENT

A permanent magnetic field set up in a ferromagnet is not very valuable in electron microscopy because it is not possible to easily vary the magnetic field and therefore manipulate the electron beam. Instead, magnetic lenses in electron microscopes are based on the fact that a conductor carrying an electrical current will create a magnetic field and that the strength of the magnetic field can be varied by changing the current through the electromagnetic lens.

The magnetic effect of an electrical current was discovered by the Danish physicist Hans Christian Oersted in 1819 when he noted, during the course of a demonstration lecture, that a compass needle, which by chance was near a wire connected to a battery, was deflected at right angles to the wire when a current was passed through the wire (Fig. 3-8). When the direction of the current was reversed, the magnetic compass needle reversed its direction, also. Oersted showed that an electric charge always has an electrical field associated with it, and when the charge is in motion, it has a magnetic as well as an electrical field associated with it.

Within a week after Oersted announced his results, Andere Ampere presented the first of a series of papers on the magnetic effects of electrical currents. Ampere demonstrated the configuration of the magnetic field about a conductor passing an electrical current. He did this by passing a copper wire through a piece of horizontal cardboard that had iron filings on its surface (Fig. 3-9). Without a current passing through the wire, the iron filings were oriented randomly on the cardboard. However, when a current was passed through the wire, the iron filings formed a series of concentric circles around the wire as the natural north and south poles of the filings became oriented within the magnetic field. The iron filings align themselves along the lines of magnetic force because the magnetic domains align in the individual filings, causing them to act like small compasses. A compass

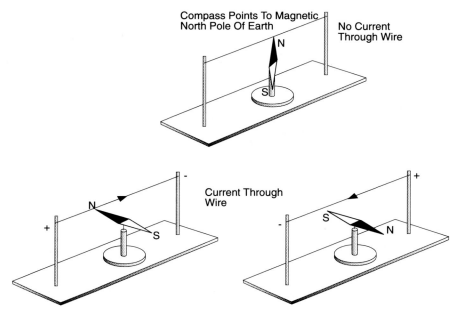

Figure 3-8   The experiments of Oersted involved the induction of a magnetic field by electrical currents. When there is no current through a wire, the north pole of the magnet points to the north pole of the earth. When current is passed through the wire, the magnet orients perpendicular to the wire. Reversal of the electric current causes reversal of the poles of the magnet.

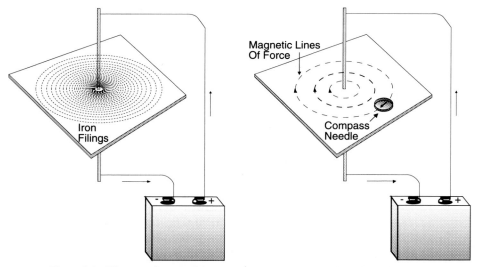

Figure 3-9   The experiments of Ampere demonstrating the orientation of the magnetic field created by electrical current passing through a wire.

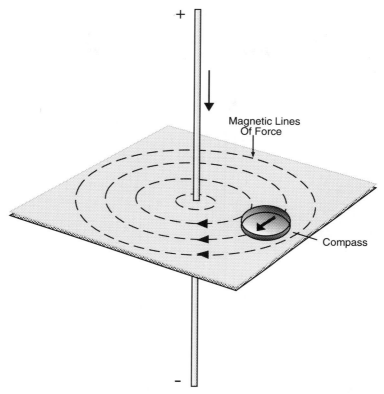

Figure 3-10   The direction of the magnetic lines of force in a magnetic field created
by electrical current passing through a wire are indicated by the magnet.

placed at different points above the iron filings always came to rest tangential to
the circle of iron filings, showing that the lines of magnetic force encircle the
conducting wire (Fig. 3-10). If the direction of the current was reversed, the magnetic field was reversed and the compass needle reversed poles.

The direction of the magnetic field can be remembered by means of the **right-
hand rule** (Fig. 3-11). If the wire is grasped with the thumb extended along the
wire in the direction of the current (from the + pole to the − pole of the battery),
the fingers curl and point in the direction in which the magnetic lines of force
encircle the wire. The magnitude of the magnetic field at any point in the field of
a long straight wire is directly proportional to the current and inversely proportional to the perpendicular direction from the wire.

The relatively weak magnetic field surrounding a wire carrying an electrical
current can be increased by bending the wire into a tight loop. Applying the right-
hand rule, it is apparent that all the lines of magnetic flux pass to the inside of the
loop in the same direction (Fig. 3-12). The side or face of the loop from which the
magnetic lines of force emerge acts as the north pole, and the opposite face, from

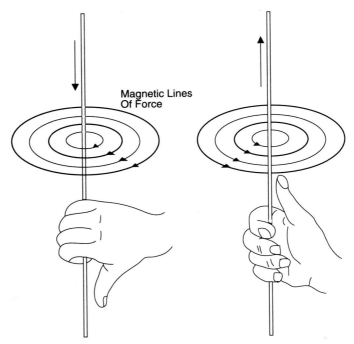

Figure 3-11   The right-hand rule is used to determine the direction of the magnetic lines of force.

which the magnetic lines of force enter, is the south pole. If the current is reversed in the wire, then the poles are reversed in position as the magnetic lines of force circle the wire in the opposite direction. The loop thus acts as a magnet or, to be more precise, an **electromagnet**. The strength of the magnetic field can be increased by (1) increasing the magnitude of the current in the wire loop or (2) producing more loops so that a helical row of loops is built up. A helically arranged row of conducting loops is called a **solenoid** (Fig. 3-13). A magnetic field is established in the center (core) of the solenoid when current is passed through the wire. The

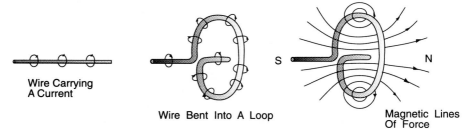

Figure 3-12   Electromagnetic field created by manipulating a wire carrying electrical current into a loop.

magnetic field within the solenoid is fairly uniform except at the ends, and there is a north and south pole. While the magnetic field is increased by increasing the current and the number of coils, the strength of the solenoid is decreased by increasing the length of the solenoid providing no new coils are added. If the solenoid is long, the magnetic lines of force will fan out widely as they come back

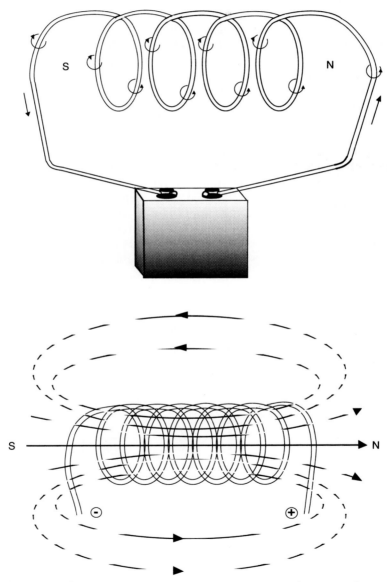

Figure 3-13 Creation of a solenoid by bending a wire carrying current into a number of loops.

around to enter the other end of the solenoid. Thus the magnetic lines of force are widely spaced outside the solenoid, while being close together inside the core of the solenoid. This is an indication that the magnetic field outside the solenoid is many times weaker than inside. A tightly wound solenoid will concentrate the magnetic lines of force (and therefore the strength of the electromagnet) more than a loosely wound solenoid. In an electromagnetic lens in an electron microscope, the wire windings are actually on top of one another.

The strength of the magnetic field in the core of the solenoid electromagnet will also vary according to the **relative permeability** of the substance in the core of the solenoid. Air has a relative permeability (μ) of 1, and an electromagnet with air in its core has a relatively weak magnetic field. Most other substances (other than ferromagnetic substances) have a relative permeability approximately equal to air. A vacuum also has a relative permeability near to that of air. However, **ferromagnetic substances** such as nickel, cobalt, iron, and some of their alloys have very high permeabilities. The relative permeabilities of nickel and cobalt are on the order of 40 to 50, that of soft iron is 3000, and that of permalloy (an alloy of iron and nickel) is over half a million. Thus a strong electromagnet (Fig. 3-14) can be produced by increasing the flux density of the magnetic lines of force with a ferromagnetic substance. The high permeability (and the resulting strong magnetic field) of ferromagnetic substances is due to the magnetic field of the solenoid orienting microscopic domains within the ferromagnetic substance so that all of these tiny magnets add their magnetic fields to the induced field of the solenoid.

The increase in strength of the magnetic field of a ferromagnetic substance

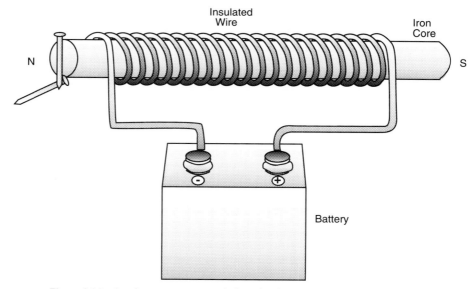

Figure 3-14   An electromagnet, consisting of an insulated wire carrying a current wound around an iron core.

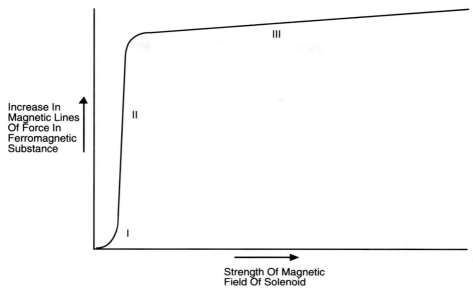

Figure 3-15   Graph of the three phases of increase in the magnetic lines of force
in a ferromagnetic substance induced by the magnetic field of a solenoid.

in a solenoid is not linear (Fig. 3-15). As the magnetic induction of the solenoid
increases, there are three different effects on the magnetic lines of force within
the ferromagnetic substance. Initially (Fig. 3-15, I), as the magnetic induction of
the solenoid increases, only a few of the tiny domains in the ferromagnetic substance reorient to the induced field of the solenoid, resulting in little contribution
of the ferromagnetic material to the total magnetic field density. At higher
strengths of magnetic induction by the solenoid, there is extensive reorienting of
the tiny domains in the ferromagnetic material, resulting in a rapid rise in the total
magnetic field (Fig. 3-15, II). When almost all the domains in the ferromagnetic
material have been reoriented (Fig. 3-15, III), the total magnetic field will rise
slowly with an increase in the induction by the solenoid. When the situation is
approached where all the domains in the ferromagnetic material have been reoriented, a further increase in the induction by the solenoid will result in no further
increase in the contribution of the ferromagnetic substance to the total magnetic
field. The ferromagnetic substance has then reached saturation.

PATH OF AN ELECTRON THROUGH AN ELECTROMAGNET

Electromagnets are used as lenses in electron microscopes. By varying the current
through the electromagnet, it is possible to vary the magnetic field in the core of
the electromagnet and therefore to manipulate an electron beam passing through
the core.

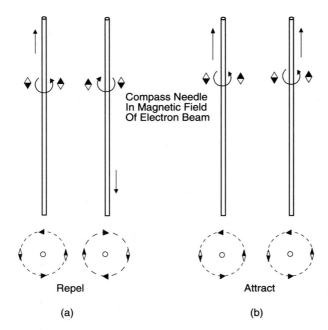

Compass Needle
In Magnetic Field
Of Electron Beam

Repel

(a)

Attract

(b)

Figure 3-16 Adjacent wires carrying electric current in opposite directions **attract** each other, while wires carrying current in the same direction **repel** each other.

To understand how an electron behaves as it passes through an electromagnet, it is best to go to one of Ampere's original findings. Soon after Oersted discovered that a compass needle adjacent to a wire carrying a current experiences a force, Ampere showed that two currents in parallel wires would attract each another if the currents were in the same direction, but would repel each other if they were in the opposite direction (Fig. 3-16).

Qualitatively, this makes sense if we consider the direction of the magnetic lines of force of the current. The magnetic lines of force point in the direction of the south pole. Adding small current needles as in Fig. 3-16 shows that, when currents are traveling in opposite directions along two wires (Fig. 3-16a), the magnetic lines of force are traveling in opposite directions. This means that the magnetic lines of force on the side of each wire toward the adjacent wire are in the same direction; for example, the north poles of each magnetic field are aligned. Since a north pole of one magnet repels a north pole of a second magnet, one wire is repelled from the adjacent wire. When currents are traveling in the same direction along two wires, the opposite is true (Fig. 3-16b), and the two wires attract each other.

If the two wires are running perpendicular to one another in the same plane (and therefore intersect in the same volume of a space, a practical impossibility), the wires would not be affected by each other's magnetic fields, no matter in which direction the current is traveling. This can be seen by looking at the magnetic field of each wire (Fig. 3-17). Since the magnetic field of one wire is perpendicular to the magnetic field of the second wire, there is no net attraction or repulsion of the

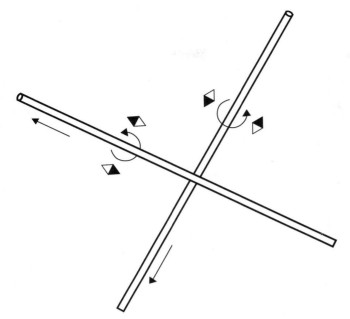

Figure 3-17 Two wires perpendicular to each other have no net attraction or repulsion of each other.

second wire by the first wire. Similarly, a stream of electrons traveling perpendicular to a second stream of electrons will produce a magnetic field that results in no net attraction or repulsion of either electron stream.

VECTOR FORCES IN THE CORE OF AN ELECTROMAGNETIC LENS

The magnetic field of a solenoid can be broken down into three vector forces, along the x and y directions in the radial direction and along the z axis through the center of the electromagnetic lens (Fig. 3-18). Because the magnetic lines of force in the radial direction are symmetrical, the vectors in the x and y directions are combined into one vector, the r vector, when discussing the vector forces affecting an electron beam passing through an electromagnetic lens. As an electron passes through the core of an electromagnetic lens, the electron is subjected to two vector forces at any particular moment in time as it passes through the magnetic lines of force (Fig. 3-19). The first vector force (H_z) is parallel to the core or z axis of the electromagnetic lens, while the second vector force (H_r) is parallel to the radius of the electromagnetic lens.

Before explaining the path of an electron in the magnetic field of the electromagnetic lens, it is necessary to remember that the direction of current in a conductor wire is opposite to the direction of movement of the electrons carrying the

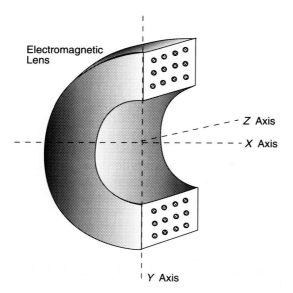

Figure 3-18 The three axes of a solenoid, *X, Y,* and *Z*. The *x* and *y* vector forces in an electron magnetic lens are the same, so they are usually referred to as the single *r* (for radial) vector.

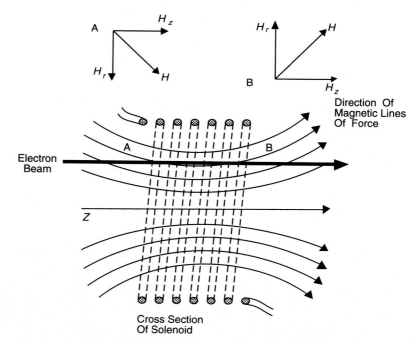

Figure 3-19 Two vector forces, H_z and H_r, created at two points, **A** and **B**, by the electron beam passing through the magnetic lines of force of a solenoid.

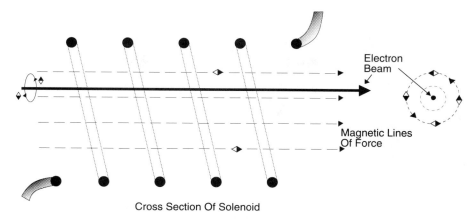

Figure 3-20 An electron beam moving parallel to the magnetic lines of force is not affected by the magnetic lines of force. This is because the magnetic lines of force are perpendicular to the lines of force of the electron beam.

electrical charge in the conductor. Thus an electron beam traveling down the column of an electron microscope actually has the current traveling in the opposite direction, according to the conventional right-hand rule.

To explain the vector force along the z axis (H_z), it is easiest to assume that the magnetic lines of force emanating from the solenoid (electromagnetic lens) are parallel to one another and that the electron beam is passing through the solenoid parallel to the z axis (Fig. 3-20). The north–south faces of a compass needle placed in the solenoid will be parallel to the magnetic lines of force of the solenoid because of the flow of electrons through the wires that make up the solenoid. However, the north–south faces of a compass needle placed along the magnetic lines of force of the electrons moving through the solenoid will be perpendicular to the north–south faces of compass needles in the magnetic field of the solenoid. The magnetic effects of the north and south poles on the spinning electrons in the electron beam cancel out in the magnetic field of the solenoid, and no vector force (H_z) is produced on an electron moving parallel to the z axis of a solenoid with parallel magnetic lines of force (an infinitely long solenoid). Therefore, an electron beam moving parallel to the z axis of an electromagnetic lens has no force applied against it, and there is no reduction in its velocity as it passes through the core of the lens.

However, if the electron beam is moving radial to the z axis and perpendicular to the magnetic lines of force of the solenoid, a significant force (H_r) is applied against the electron beam (Fig. 3-21). In actual fact, electrons never move perpendicular to the magnetic lines of force through a solenoid making up an electromagnetic lens, because the lens is solid in this direction and does not permit the passage of electrons. However, explaining the passage of electrons perpendicular to the magnetic lines of force gives an illustration of one force responsible for the manipu-

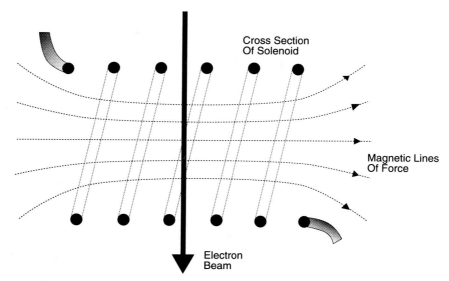

Figure 3-21   Orientation of an electron beam moving perpendicular to the H_r vector lines of force in a solenoid. In an electron microscope, the electron beam cannot move in this direction because the windings of the electromagnetic lens are solid in this area.

lation of the electron beam as it passes through an electromagnetic lens. The direction of force on a moving stream of electrons is determined by the right-hand-thumb rule (Fig. 3-22). In the **right-hand-thumb rule**, *the right hand is positioned in such a way that the index finger points in the direction of the magnetic lines of force, the middle finger, ring finger, and little finger point in the direction the electron beam is traveling, and the thumb points in the direction of the force on the electrons in the electron beam.* Thus, when an electron beam passes through a magnetic field perpendicular to the magnetic lines of force, the electrons are

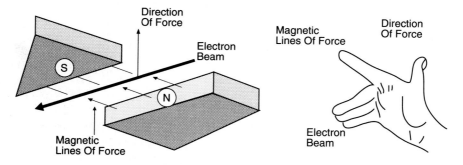

Figure 3-22   Right-hand-thumb rule for determining the direction of force on an electron beam in a magnetic field.

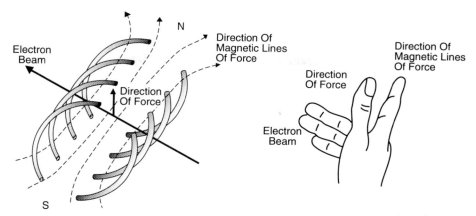

Figure 3-23 Right-hand-thumb rule applied to an electron beam passing through the coils of a solenoid. The direction of force is perpendicular to the electron beam and the direction of the magnetic lines of force.

exposed to a force that moves the electrons in a direction perpendicular to the direction of the electron beam and perpendicular to the magnetic lines of force.

In a solenoid representing an electromagnetic lens, the forces on an electron beam moving perpendicular to the magnetic field are as illustrated in Fig. 3-23. Applying the right-hand-thumb rule, the electrons are subjected to a force that moves the electrons in a direction perpendicular to the magnetic lines of force and perpendicular to the electron beam, resulting in the electrons being pushed toward the windings of the solenoid.

Thus a short solenoid representing an electromagnetic lens has two potential forces applied against it, H_z parallel to the z axis and H_r radial (perpendicular) to the z axis. The electromagnetic lens is symmetrical in the radial direction, so H_r or the radial force is the same in the y and x direction axes, thereby reducing three potential forces on the electron passing through the lens to the two forces, H_z and H_r.

In an electromagnetic lens, the magnetic lines of force are not parallel to the z axis because the lens is a short solenoid with curving magnetic lines of force (Fig. 3-24). Thus, if an electron enters the lens parallel to the z axis, the magnetic field at any point of the electron beam can be divided into two forces, one directed parallel to the z axis of the lens (H_z) and one directed radial to the z axis (H_r). The component parallel to the z axis (H_z) reaches its highest value as the electron passes through the center of the lens and has its lowest value farthest from the center of the lens (Fig. 3-24b). The component perpendicular to the z axis and radial to the lens (H_r) likewise has a low value when the beam electrons are large distances from the radial axis. As the electrons approach the core of the lens, the value of H_r increases to a point where the alignment of the magnetic lines of force becomes nearly parallel to the z axis. At this time the vector component H_r

decreases in value until it reaches zero, when the beam electrons pass through the radial axis. H_r then changes in value as it passes out through the other end of the lens.

These two forces, H_z and H_r, are responsible for two different actions on the electrons, **spiraling** and **focusing**, as they pass through the lens. For electrons entering the lens parallel to the z axis, these two actions are due to the following two steps:

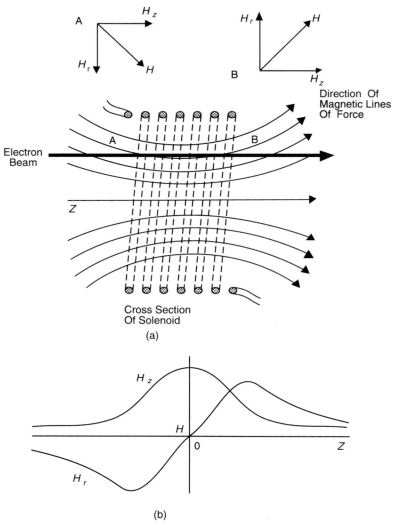

Figure 3-24 (a) Vector forces at points **A** and **B** on an electron beam passing through a solenoid. (b) Relative strength of the vector forces H_z and H_r at different distances from the center of the solenoid.

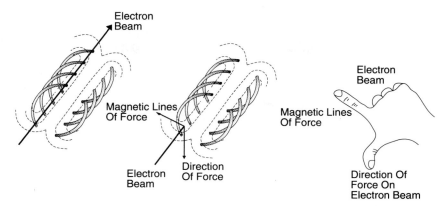

Figure 3-25 As the electrons in the electron beam enter the electromagnetic lens, the electrons are subjected to vector force H_r, which forces the electrons perpendicular to the magnetic lines of force.

1. Initially, vector force H_z has no effect on the electrons since the electrons are moving parallel to the z axis. However, vector force H_r, perpendicular to H_z, pushes the electrons into a **spiral path** through the lens (Fig. 3-25).

2. Once the electrons start to spiral through the electronmagnetic lens, the electrons are no longer moving parallel to vector force H_z. Vector force H_z now forces the electrons toward the core of the lens and sets up the **focusing** action of the lens.

Spiralling. If the electron beam enters the electromagnetic lens parallel to the z axis, then H_z, or the vector force parallel to the z axis, will initially have no effect on the electron beam, no matter where it enters the core of the lens (for example, in the center of the core or near the windings). This is because the electron beam is moving parallel to the vector force H_z. The vector force H_z will not affect the electron beam unless the electron beam is deflected so that it is *not* moving parallel to the z axis (which happens after vector force H_r has deflected the beam into a spiral path). However, the vector force H_r, radial to the z axis, will force the beam electrons in a direction perpendicular to the direction of the beam electrons movement and perpendicular to the magnetic lines of force according to the right-hand-thumb rule. If the beam electrons enter the magnetic field at point A, they will be subjected to the force H_r. Vector force H_r will force the electrons in the beam into a circular (helical) path as they move through the electromagnetic lens. This is because the electrons in the beam have a velocity component (they are moving rapidly down the column). As shown in Fig. 3-24, the radial force H_r changes value in the center of the electromagnetic lens. Past the center of the lens, the force (H_r) on the electrons in the beam reverses. Thus, in the second half of the lens, the radial acceleration of the beam electrons is in the opposite direction, and a force is produced between H_r and the velocity of the electron. This

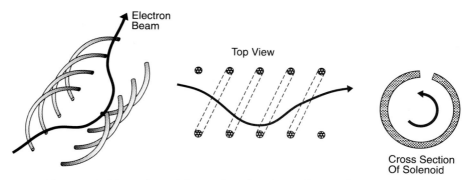

Figure 3-26   Three views of the rotation of an electron beam as it passes through a solenoid.

force counters the rotational velocity of the electron imposed by the first half of the lens. In the second half of the lens, the force H_r causes the rotational alignment to be reversed gradually until the electron has resumed a straight path when leaving the field (the electron beam as a whole has been rotated through the electromagnetic lens, although the rotation is considerably less than 90°) (Fig. 3-26).

Thus, the *radial force (H_r) is responsible for the electron beam following a helical path as it travels through the electromagnetic lens*. However, this force has not yet pushed the electron beam closer to the axis of the electromagnetic lens, a phenomenon that is *necessary if the lens acts to magnify or focus an image*.

Focusing.   The focusing and magnification action of an electromagnetic lens is due to another factor that has been set up by forcing the electron beam into a helical path through the lens. Once the electrons are traveling through the lens in a helical path, the *electrons are no longer moving parallel to the z axis and are now affected by the vector force (H_z)*, which is parallel to the z axis. The helically moving beam electrons are now subjected to the vector component H_z, which pushes the electrons in a direction perpendicular to the magnetic lines of force and perpendicular to the direction that the beam electrons are moving, according to the right-hand-thumb rule. The right-hand-thumb rule pushes the beam electrons toward the center of the core of the electromagnetic lens (Figs. 3-27 and 3-28). The force is away from the windings of the electromagnetic lens since this is the direction of the weaker magnetic lines of force. Thus, as the electron beam passes through the electromagnetic lens, the electron beam is compressed toward the z axis (center of the lens).

In summary, as an electron beam passes through an electromagnetic lens parallel to the z axis, the electron beam is affected by the right-hand-thumb rule, which causes the electrons to rotate through the lens (as the electrons encounter the vector component H_r radial to the lens), which in turn causes the electron beam to become compressed toward the z axis (as the beam electrons moving in a helical path are now affected by the vector component H_z parallel to the z axis).

It should be noted that it does not matter which way the electrical current passes through the electromagnetic lens; the electron beam will still be rotated (although in the opposite direction) and compressed toward the center of the magnetic lens.

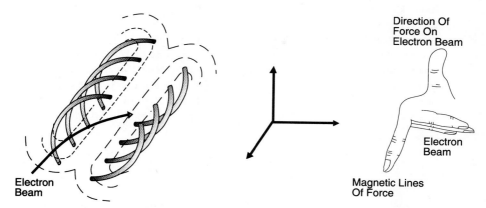

Figure 3-27 The spiraling movement of the electron beam created by the vector force H_r causes the electrons to move so that they are no longer parallel to the z axis. This means that the electrons are now subjected to vector H_z forces pushing the electrons toward the core of the lens and setting up the focusing action of the electromagnetic lens.

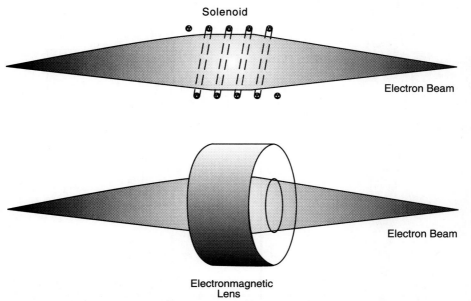

Figure 3-28 Focusing action of a solenoid (electromagnetic lens) on an electron beam.

DESIGN OF ELECTROMAGNETIC LENSES

In 1926 and 1927, H. Busch published a number of classical papers in which he proved both mathematically and experimentally that a short solenoid acted as a lens. He showed that the mathematical formulas of light optics could be applied to electron optics. He demonstrated that a diverging electron beam originating from a point and passing through an electromagnetic lens could be brought to a second point (the focal point) on the optical axis of the electron beam. Thus the image could be reconstructed, point by point, after the electron beam had passed through the electromagnetic lens. However, a simple solenoid had a very weak magnetic field in its core because the magnetic field was spread out for a long distance along its axis (Figs. 3-29 and 3-32). This resulted in the electron beam

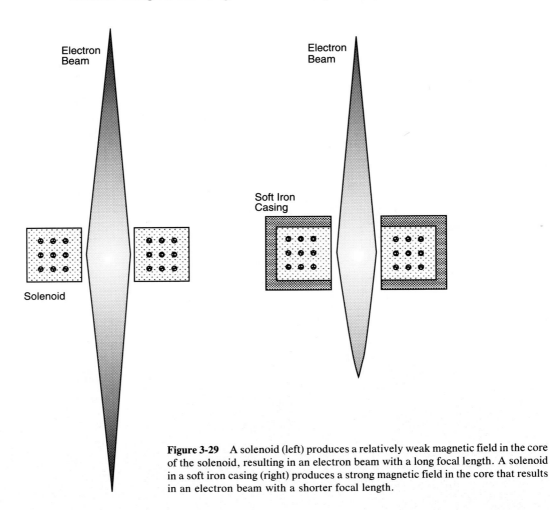

Figure 3-29   A solenoid (left) produces a relatively weak magnetic field in the core of the solenoid, resulting in an electron beam with a long focal length. A solenoid in a soft iron casing (right) produces a strong magnetic field in the core that results in an electron beam with a shorter focal length.

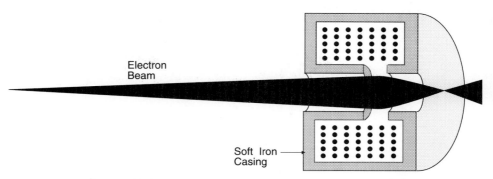

Electron
Beam

Soft Iron
Casing

Figure 3-30 A solenoid encased in soft iron, except for a small gap facing the core, produces a strong magnetic field in the gap, which results in an electron beam with a short focal length.

traveling some distance before reaching crossover, a situation that is not acceptable in the construction of an electron microscope, since focal lengths of a few millimeters are required.

For electromagnetic lenses to be effective, it is necessary that they have a powerful magnetic field that deflects the electrons strongly, resulting in a short focal length. In 1927, D. Gabor encased the windings of the solenoid in a soft iron casing (shroud) on all sides, except on the side facing the core of the solenoid (Fig. 3-29). The magnetic lines of force, which normally were widely spread outside an unshrouded solenoid, were concentrated in the iron casing. This was because the magnetic lines of force of the solenoid resulted in parallel orientation of the magnetic domains in the soft iron. This concentrated the magnetic lines of force in the soft iron casing, except along the unshrouded wall facing the core. This resulted in an increased magnetic field within the core and a reduced focal length of the electron beam passing through the core.

Ernst Ruska and Max Knoll increased the concentration of the magnetic field in 1931 by surrounding the windings with soft iron except for a narrow gap facing the core (Fig. 3-30). The magnetic field was concentrated at the gap in the casing. The strength of the magnetic field was increased to the point where focal lengths in the range of centimeters were obtained.

In the last major development, Ruska in 1934 used soft iron pole pieces over the gap in the soft iron casing of the windings (Fig. 3-31). Two pole pieces were placed, one above each other, in the bore of the electromagnetic lens, each separated from the other by a brass ring. This separated the two pole pieces by a gap of only a few millimeters. The bore of the pole pieces (through which the electron beam passed) was also only a few millimeters.

The focal length of electromagnetic lenses with pole pieces corresponds roughly to the diameter of the bore of the pole pieces (Fig. 3-32). The highly concentrated magnetic field at the gap of the pole pieces allows focal lengths as short as 1 mm before the magnetic field of the soft iron of the pole pieces is

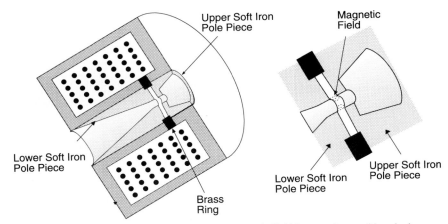

Figure 3-31   A lens with a very strong magnetic field is manufactured by placing
two soft iron pole pieces, separated by a brass ring, in the core of the lens.

saturated, although generally focal lengths in the range of 2 to 3 mm are used. The
strength of the magnetic field is limited by the magnetic saturation of the pole
piece iron. The saturated parts of the pole pieces will not contribute further to the
field strength of the lens when the magnetic flux of the solenoid is increased above
the magnetic saturation of the pole piece iron.

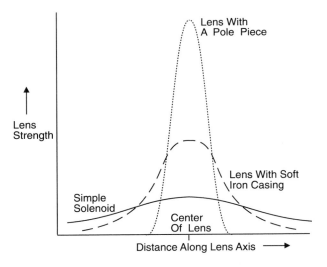

Figure 3-32   How the strength of an
electromagnetic lens can be increased
by casing the solenoid in soft iron and
using pole pieces.

4

Lens Aberrations

Lenses used in electron microscopes, like their counterparts in light microscopes, possess a number of inherent defects. Electromagnetic lenses have additional lens aberrations due to the rotation of the beam electrons as they pass through the electromagnetic lens. The three significant lens aberrations in electron lenses used in scanning electron microscopes are **spherical aberration, chromatic aberration**, and **astigmatism**. Other potential lens aberrations, such as coma and curvature of the field, are not significant in electron microscopes and are not included in the following discussion.

SPHERICAL ABERRATION

Previously, in discussing electron optics, it was assumed that light rays and electrons can be focused to a single focal point by a lens. In actual fact, this does not occur because of a number of characteristics of light rays and electrons as they pass through lenses.

In **spherical aberration**, the light rays or electrons that pass through the *center* of the lens converge at a focal point that is farther from the lens when compared to the focal point of those light rays or electrons that pass closer to the *edge* of the lens. This is because the peripheral light rays and electrons are bent more than the light rays and electrons that pass near the center of the lens (Fig. 4-1). In electromagnetic lenses, this occurs because the strength of the magnetic field decreases as the distance from the windings of the electromagnetic lens increases (Fig. 4-2). Thus electrons passing close to the windings of the lens are subjected to a stronger magnetic field and are bent more than the electrons passing near the center of the lens.

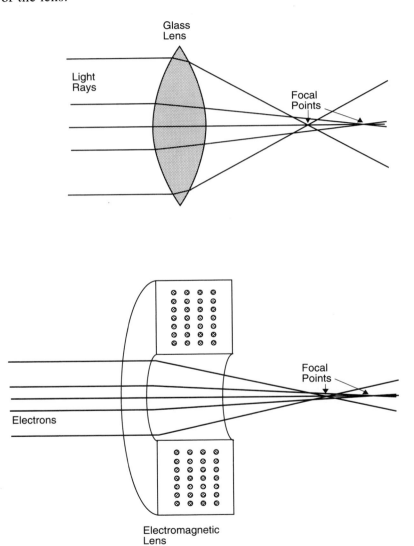

Figure 4-1 Spherical aberration arises when light rays or electrons passing through a lens are brought to different focal points. The differential bending of the light rays or electrons is due to varying lens strength.

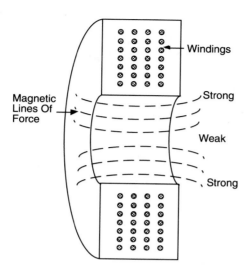

Magnetic Lines Of Force

Windings

Strong

Weak

Strong

Figure 4-2 The strength of the magnetic field in an electromagnetic lens is strongest near the windings and weakest near the core.

Maximum resolution is obtained in a scanning electron microscope when the electron beam is brought to a focal point of minimal diameter. Spherical aberration results in a continuous series of focal points, depending on where the electrons initially enter the lens (Fig. 4-3). This results in an electron beam with the minimum beam diameter (**disc of least confusion** or **circle of minimum confusion**) at a position somewhere between the focal point of the electrons passing through the periphery of the lens and those passing near the center of the lens (Figs. 4-3 and 4-4). The shape of the whole electron beam converging to a disc of least confusion is called a **caustic curve**. The disc of least confusion is the position where the best image is obtained. An image formed by a disc of least confusion will obviously be of lower resolution than an image formed by all the electrons passing through a single focal point.

The term spherical aberration is derived from glass lenses, whose designers usually manufacture lenses that have a spherical surface on each side of the glass lens. Unfortunately, a spherical glass surface does not refract an oncoming wavefront to a perfect focal point, hence the term spherical aberration. Spherical aberration can be corrected in glass lenses in a couple of ways: (1) Adding a divergent lens of a different refractive index to the convergent lens, resulting in the light rays passing to nearly a single focal point after passing through both lenses. While this works with glass lenses, it does not work with electromagnetic lenses, since electromagnetic lenses are always convergent and never divergent. (2) Computers can formulate a simple glass lens so that one surface is ground to an aspheric curve instead of a spherical curve; this parabolic structure is efficient for converging the light rays to one focal length.

The inability to correct for spherical aberration is the principal factor limiting the resolving power of the scanning electron microscope. Since spherical aberration is proportional to the third power of the aperture angle, the most common

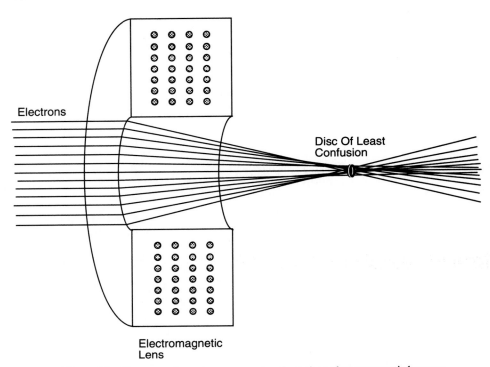

Figure 4-3   Electrons of equal energy passing through an electromagnetic lens are brought to different focal points. The closer the electrons are to the windings, the more the electrons are deflected by the magnetic field and the closer the focal point is to the lens.

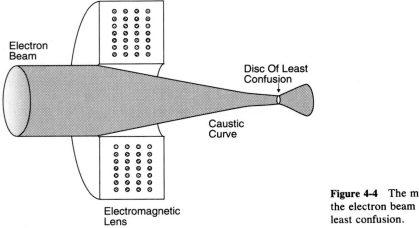

Figure 4-4   The minimum diameter of the electron beam occurs at the disc of least confusion.

way of limiting spherical aberration in an electron microscope is by introducing an aperture in the path of the electron beam in front of the lens (Fig. 4-5). The aperture strip intercepts those electrons that would pass through the periphery of the lens and allows only electrons to pass through near the center of the lens. This results in a smaller disc of least confusion and better resolution of the specimen.

Spherical aberration is also reduced as the focal length of the lens is decreased. This is because there is a smaller disc of least confusion at shorter focal lengths (Figs. 4-6 and 4-7). This is accomplished by increasing the lens current strength, and therefore the magnetic field, to reduce the focal length. Spherical

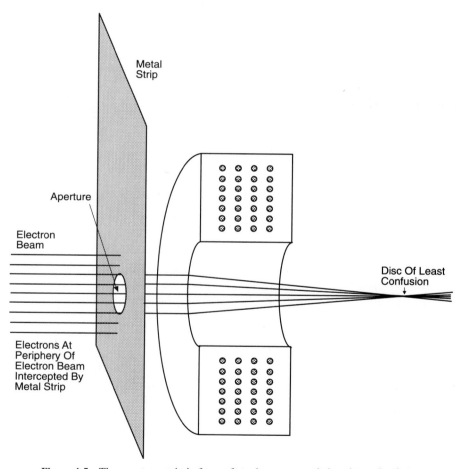

Figure 4-5 The aperture strip in front of an electromagnetic lens is used to intercept electrons in the peripheral areas of the beam. This results in a smaller-diameter electron beam that has a smaller disc of least confusion and less spherical aberration.

aberration decreases rapidly as the focal length is reduced and reaches a near constant value at shorter focal lengths.

Spherical aberration is roughly inversely proportional to the square of the accelerating voltage. Therefore, increasing the accelerating voltage will reduce the spherical aberration, although again there is a trade-off. In the scanning electron microscope, increased accelerating voltage results in a wider specimen–beam interactive zone in the bulk specimen, which results in a wider escape zone for secondary and backscattered electrons, resulting in a loss of resolution.

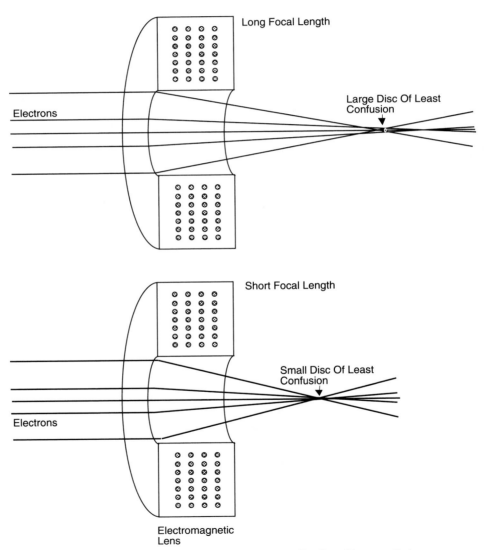

Figure 4-6   A short focal length results in a smaller disc of least confusion.

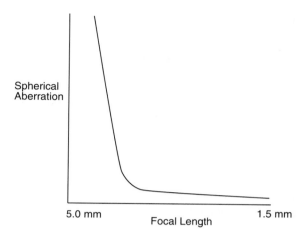

Spherical Aberration

5.0 mm

Focal Length

1.5 mm

Figure 4-7 Decreasing the focal length of a lens results in a decrease in spherical aberration.

CHROMATIC ABERRATION

Chromatic aberration arises when lenses deflect light rays or electrons of different wavelengths to different degrees. When light rays pass through a glass lens, the short wavelengths (blue-greens) are deflected more than the long wavelengths (reds). This results in different wavelengths of light being brought to different focal lengths (Fig. 4-8). If a viewing screen is placed at the focal length of the blue-light rays (A), the image will consist of a blue dot with a reddish halo, while at the focal point of the red light (B) there would be a red dot with a bluish halo. Nowhere is there a truly focused image of a small white dot and, instead, it is necessary to focus at the smallest dot image at the disc of least confusion.

In the light microscope, chromatic aberration can be corrected in two ways:

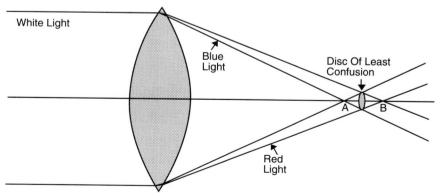

White Light

Blue Light

Disc Of Least Confusion

A   B

Red Light

Figure 4-8 Chromatic aberration in a glass lens results when shorter-wavelength light rays are bent more than longer wavelengths as the light rays pass through the glass lens.

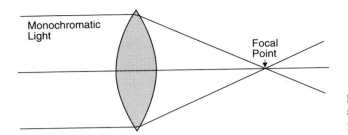

Figure 4-9 Monochromatic light rays are brought to a single focal point by a perfectly constructed glass lens.

(1) by using monochromatic light (light of a single color with a narrow range of wavelengths) so that the light rays are brought to a single focal point (Fig. 4-9); (2) by using glass lenses of different refractive indexes. For example, a converging lens made from crown glass can be combined with a diverging lens from flint glass (Fig. 4-10). The crown glass (alkaline–lime silicate optical glass) deviates the light rays twice as much as the flint glass (glass containing lead oxide, formerly made by using calcined flint as a source of glass) so the composite system is still a converging lens. However, the chromatic effects of each lens are the *same*, so the differential deflection of light rays by the converging lens is canceled by the opposite differential bending of the diverging lens, resulting in correction for chromatic aberration. An **achromatic** glass lens is one that is largely corrected for chromatic aberration. An **apochromatic** glass lens is one that is corrected for both chromatic and spherical aberration.

Chromatic aberration in electron lenses occurs when electrons of different velocities (and therefore wavelengths) enter the lenses at the same place and are bent to different focal points (Fig. 4-11). The greater the velocity of the electron is, the less time it spends in the magnetic or electrostatic field of the lens, the less it is bent as it passes through the lens, and the farther the focal point will be from the lens. Thus fast, short-wavelength electrons are deviated less by an electron lens than are slow electrons of longer wavelength. This effect works in the direc-

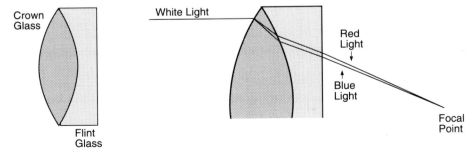

Figure 4-10 Chromatic aberration in a glass lens can be minimized by combining a converging and diverging lens of the correct strengths so that the light rays are brought to a single focal point.

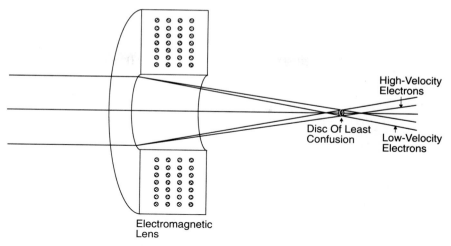

High-Velocity
Electrons

Disc Of Least
Confusion

Low-Velocity
Electrons

Electromagnetic
Lens

Figure 4-11   Chromatic aberration in an electromagnetic lens. The higher-velocity electrons are deflected less by the magnetic field than the lower-velocity electrons, resulting in a series of focal points for the different-velocity electrons.

tion opposite to that in the light microscope, where the short wavelengths of light are bent more by the glass lens.

In the **scanning electron microscope**, chromatic aberration results in a loss of resolution because the minimum beam spot size (which ultimately determines resolution) is represented by the disc of least confusion. Elimination of chromatic aberration would mean that the minimum beam spot size would be at a single focal point resulting in a smaller beam spot size and better resolution.

Causes of Chromatic Aberration in Electron Lenses.   Ideally, chromatic aberration in the electron microscope could be eliminated by producing a monochromatic beam of electrons, where all the electrons have the same velocity. However, this is not possible due to three factors: (1) variation in the emission velocity of the electrons as they leave the electron gun; (2) variation in the accelerating voltage; and (3) variation in lens currents.

1. *Variations in the emission velocity of the beam electrons:* The initial velocity of the electrons when they are emitted from the thermionic electrode (tungsten filament or lanthanum hexaboride rod) can vary. The velocity at which the electrons leave the thermionic electrode depends on the thermal motion of the electrons in the metal, and the velocity is increased by heating the metal. However, the energy of the emitted electrons is usually less than 1 eV (Fig. 4-12). When compared to the accelerating voltage of 30,000 V, the 1-V contribution due to thermionic emission is negligible and is not an important factor in chromatic aberration.

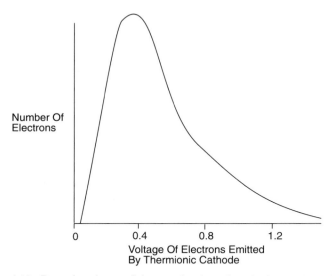

Figure 4-12 Range in voltages of electrons leaving a thermionic tungsten cathode.

2. *Fluctuation in the accelerating voltage:* This is produced by variations in the high-voltage supply and can be minimized by stabilizing circuits so that the voltage fluctuation is less than 10^{-5} or 1/100,000 of the voltage. A dirty electron gun resulting in discharges can also vary the accelerating voltage, leading to chromatic aberration.

3. *Lens current fluctuations:* Variation in the current to the objective lens will change the focal length and cause chromatic aberration. A fluctuation in the lens current (in amperes) will cause a change in the focal point that is exactly equivalent to twice the change in accelerating voltage.

ASTIGMATISM

A **stigma** is a point. **Astigmatism** is the inability of a lens to bring rays of light or electrons to a point. *Astigmatism is the most important factor limiting resolution in the electron optical system comprising the objective lens.* Fortunately, astigmatism in electron lenses is relatively easy to correct with a stigmator.

In the electron microscope, astigmatism results when the electron beam is not perfectly circular in cross section, but is instead elliptical (Fig. 4-13). An electron beam that is elliptical in cross section results when the short axis of the cross section is subjected to a stronger electrostatic or magnetic field in the lens than is the long axis of the electron beam. The stronger the field is, the more the

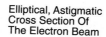

Circular Cross Section
Of The Electron Beam

Elliptical, Astigmatic
Cross Section Of
The Electron Beam

Figure 4-13 An astigmatic image is formed from an electron beam that is not circular in cross section.

electrons in that axis of the electron beam are compressed and the more elliptical is the electron beam in cross section (Fig. 4-14).

The electrons passing through the lens that are subjected to the stronger magnetic field come to crossover closer to the electromagnetic lens than the electrons that are subjected to the weaker magnetic field (Fig. 4-15). This results in the inability of the lens to bring the electron beam to a single focal point, and instead a series of focal points is created, similar to the situation in spherical and chromatic aberration. The best focus is obtained at the disc of least confusion. The resulting image has less resolution and appears blurred.

Astigmatism can be recognized in both the transmission and scanning electron microscopes by an elongated orientation of the image details in one direction just above and below focus. The details are elongated in one direction above focus, lose their elongation at the blurred image at focus, and below focus become elongated at 90° to the direction seen above focus (Figs. 4-16 and 4-17). This is caused by the image being formed at the focal point of one axis of the electron beam above focus and the second axis (at right angles to the first) below focus.

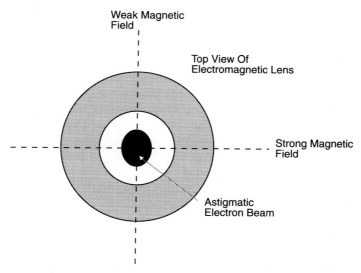

Figure 4-14 An electromagnetic lens with a symmetrical magnetic field will produce an astigmatic electron beam that is elliptical in cross section.

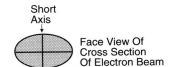

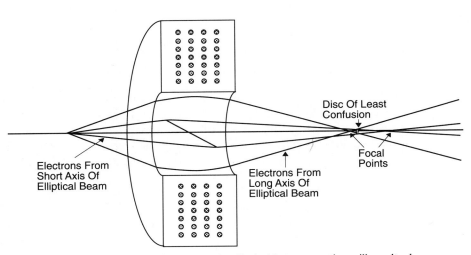

Figure 4-15 An electron beam that is elliptical in cross section will result when the electron beam passes through an electromagnetic lens that has an asymmetric magnetic field. This results in the formation of a continuous series of focal points past the lens.

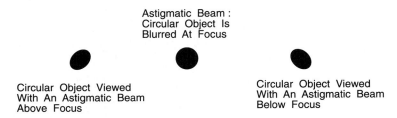

Figure 4-16 A circular object appears elongated in an astigmatic image.

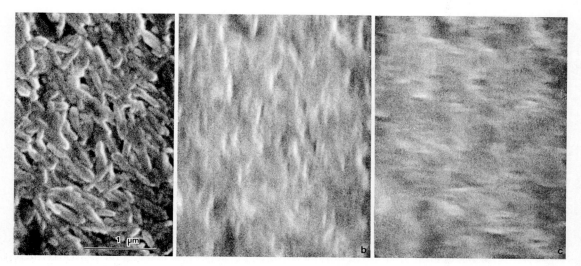

Figure 4-17 Scanning electron micrographs of the same area of magnetic tape taken at an accelerating voltage of 30,000 V. (a) An in-focus micrograph showing no astigmatism. (b) and (c) Images taken with an astigmatic electron beam showing how the orientation of details of the specimen changes 90° as the operator goes from overfocus to underfocus. (b) An astigmatic overfocused image showing the orientation of details primarily in the vertical plane. (c) An underfocused astigmatic image showing the orientation of details primarily in the horizontal direction.

Causes of Astigmatism

There are two common causes of astigmatism in electron microscopes: (1) electron lenses that are manufactured so that the field strength is not perfectly symmetrical and (2) dirty apertures or lenses.

  1. *Nonsymmetrical fields in electron lenses:* It is not possible in practice to make a lens that has perfectly symmetrical field strengths. Because of the imperfections in grinding the lens bore, asymmetry in the windings, and inhomogeneities in the iron that make up the pole pieces, all electron lenses have a focal length in one direction that is slightly different from the focal length formed by electrons at 90° to the first direction. Fortunately, if the astigmatism of the lens is not bad, it is fairly easy to correct using a correcting device called a stigmator, which uses a magnetic or electrostatic field to force the electron beam back to a circular cross section.

  2. *Dirty apertures or lenses*: Astigmatism becomes worse in an electron microscope as dirt builds up in the column, lenses, and apertures. The nonconduct-

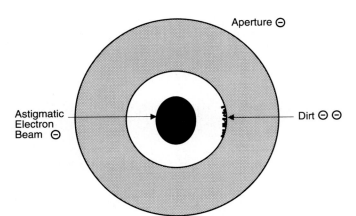

Figure 4-18 A dirty aperture charges, resulting in a symmetrical electrostatic field that produces an astigmatic electron beam.

ing dirt charges (becomes negative because it is bombarded by negative electrons of the column that are not drained away to ground) and distorts the magnetic field of the lens. Aperture strips have small holes of different sizes (10 to 400 μm) drilled in metal strips or discs of metals, such as molybdenum, gold, or platinum. These apertures are placed in the path of the electron beam to eliminate random electrons. The buildup of dirt on one side of the aperture causes the aperture to act as an astigmatic electrostatic electron lens (Fig. 4-18). Because the dirt is nonconducting, the electrons in the dirt are not drained away to ground, and that area of the aperture becomes more highly negative than the remainder of the aperture. This results in the negative electron beam being preferentially repelled from the dirty area of the aperture. An electron beam that is astigmatic with an elliptical cross section results. The same effect results when dirt builds up in the bore of an electron lens.

Stigmators

A stigmator is a device that corrects the astigmatism of the electron beam by superimposing on the lens magnetic field a second magnetic or electrostatic field of asymmetric and variable magnitude, which can be positioned so as to oppose and cancel the existing lens asymmetry. Stated more simply, a field is introduced to the lens that causes the electron beam to become circular. A stigmator is essentially a weak lens of variable **azimuth** (direction) and **amplitude** (strength). All electron microscopes have stigmators in the objective lens and some have them in the condenser lens. Stigmators can be either **electrostatic** or **magnetic**.

Electrostatic Stigmators.   One form of an electrostatic stigmator consists of eight short cylindrical rods with the long axes of the rods parallel to the optic axis (Fig. 4-19). The electron beam passes through the center of the eight rods. Those rods closest to the long axis of the elliptical cross section of the electron beam are made negative with respect to earth, thus compressing the negative electron beam in this axis. Those rods closest to the short axis of the elliptical cross section of

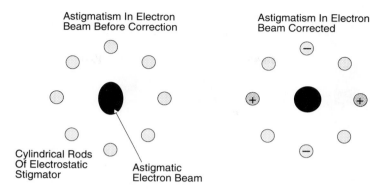

Astigmatism In Electron
Beam Before Correction

Astigmatism In Electron
Beam Corrected

Cylindrical Rods
Of Electrostatic
Stigmator

Astigmatic
Electron Beam

Figure 4-19 An electrostatic stigmator used to correct astigmatism of the electron beam by applying an electrostatic field to push and pull the electron beam into a circular cross section.

the electron beam are made positive, thus expanding the negative electron beam in this direction. Correct adjustment of the magnitude of the positive and negative charge on each pair of rods results in a circular cross section of the electron beam and elimination of astigmatism.

Magnetic Stigmators. This was the first type of stigmator, originally described by J. Hillier and E. Ramburg in 1947. Refinements of their original stigmator produced a number of different designs. One of these designs uses four soft iron slugs mounted on a circular carriage that can be rotated around the electron lens (Figs. 4-20, 4-21, and 4-22). The soft iron slugs are magnetized by the magnetic

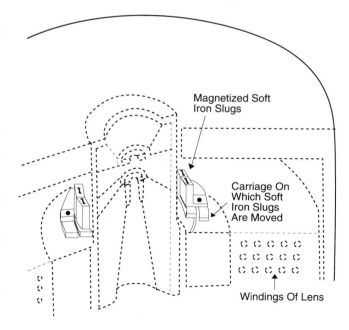

Magnetized Soft
Iron Slugs

Carriage On
Which Soft
Iron Slugs
Are Moved

Windings Of Lens

Figure 4-20 Position of soft iron slugs used to correct the astigmatism of the electron beam in an older type of electron microscope.

Asymmetric Magnetic Field   Symmetric Magnetic Field

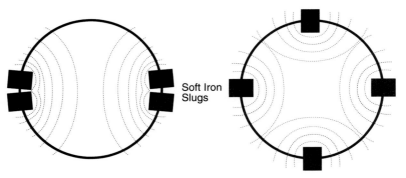

Soft Iron
Slugs

Figure 4-21   The older type of stigmator had magnetized soft iron slugs that could
be physically moved to produce magnetic fields of different symmetries that could
be used to correct the astigmatism of the electron beam.

Uncorrected Astigmatic
Electron Beam, Soft
Iron Slugs Equidistant

Magnetized Soft Iron Slugs
Moved So That Astigmatism
is 90° To Original Astigmatism

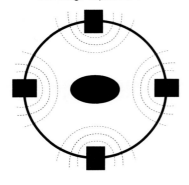

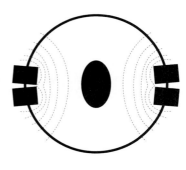

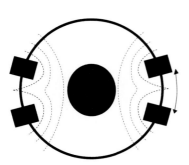

Magnetized Soft Iron Slugs
Moved Apart, Making The
Electron Beam More Circular
In Cross Section

Soft Iron Slugs At Position
Where The Asymmetrical Magnetic
Field From The Slugs Cancels
That Of The Lenses And The Beam
Is Circular In Cross Section

Figure 4-22   Sequence of events used to correct for astigmatism with a stigmator
that has movable soft iron slugs.

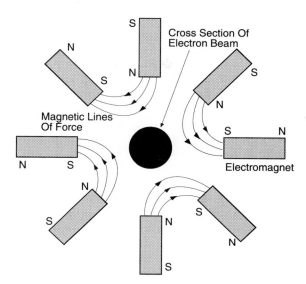

Figure 4-23 Octupole electromagnetic stigmator showing the position of the electromagnets and their magnetic fields.

field from the lens windings and concentrate the magnetic field in this area. Because of the high magnetic permeability of the soft iron, the magnetic field of the soft iron is strong. The four iron slugs are mounted in two pairs. When the slugs are moved so that pairs of slugs are opposite each other, the magnetic field is strong and asymmetric. When the soft iron slugs are moved so that they are equidistant on the rings, the magnetic field is symmetrical. Such a stigmator requires two steps to compensate for the stigmatism of the electron beam (Fig. 4-22). First, the azimuth of the astigmatic field is determined with the soft iron pieces equidistant from one another. In this position, the magnetic field of the stigmator is symmetrical and is not affecting the astigmatism of the electron beam. Once the azimuth has been determined, the pairs of soft iron slugs are brought together to produce a compensating magnetic field that is at right angles to the astigmatism of the electron beam. Unless the astigmatism is very bad (which usually means that the column or apertures must be cleaned), the symmetric magnetic field introduced by the soft iron slugs will compress the astigmatic beam in a direction 90° to the original astigmatism. Once the azimuth of the stigmatism has been determined, the strength of the compensation is adjusted by moving the two pairs of soft iron slugs until the electron beam is circular in cross section.

Most modern electron microscopes use electromagnetic stigmators to correct for astigmatism of the electron beam. **Octupole stigmators** consisting of eight electromagnets are the most common type. The electromagnets are arranged in four pairs, each member of a pair having oppositely directed north–south poles (Fig. 4-23). As the electrons pass down the column, they are affected by the electromagnetic field according to the right-hand-thumb rule. The beam electrons are either pushed toward the optic axis or pulled toward the periphery of the lens, depending on the direction of the magnetic field (Fig. 4-24). Correction of the astigmatism of the electron beam is accomplished by changing the current through the windings of the electromagnets, thereby changing the strength of the magnetic field and expanding or compressing the electron beam in the area of the electromagnets (Fig. 4-25).

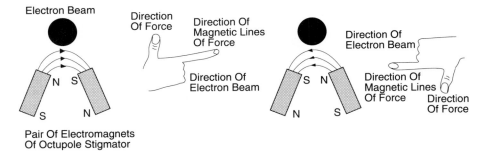

Figure 4-24   The magnetic field of the pair of electromagnetic stigmator elements deflects electrons either toward or away from the optic axis, depending on the direction of the magnetic lines of force.

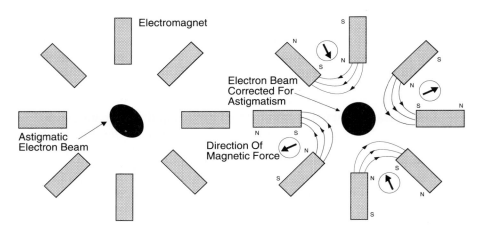

Figure 4-25   Use of an electromagnetic stigmator to correct for astigmatism of an electron beam.

5

Assembled Column of the Scanning Electron Microscope

The assembled column of the scanning electron microscope consists of the following parts (Fig. 5-1):

1. An **electron gun** at the top of the column that produces electrons for the beam.

2. An **anode** under the electron gun. The voltage difference between the electron-emitting cathode in the gun and the anode determines the accelerating voltage of the electrons in the beam.

3. **Beam (shift and tilt) deflection coils** that are used to align the electron beam over the optic axis of the column.

4. **Lenses** that determine the size and intensity of the electron beam and focus the electron beam on the specimen.

5. **Apertures** that remove stray electrons and reduce the beam spot size to minimize spherical aberration.

6. **Scanning coils** that move the focused electron beam over the specimen in a raster pattern of lines.

7. **Stigmator** elements in the objective lens that correct the astigmatism of the beam.

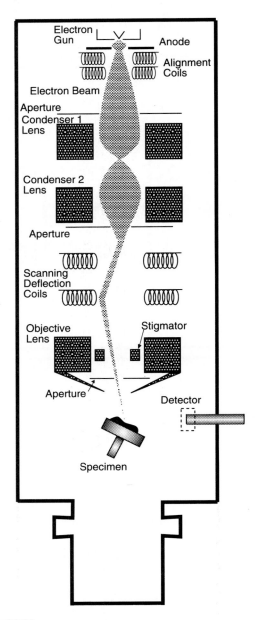

Figure 5-1 Organization of the column of a scanning electron microscope.

ELECTRON GUN

Electron guns using field emission or thermionic emission with either a tungsten or lanthanum hexaboride cathode can be used in the scanning electron microscope. If field emission or thermionic emission from a lanthanum hexaboride cathode is used, the electron gun has a vacuum isolation valve in the column under the gun to keep the gun chamber at high vacuum while the specimen chamber is vented. A column using thermionic emission from a tungsten filament may or may not have a vacuum isolation valve under the gun.

ANODE

The voltage difference between the anode and the cathode usually can be varied by the operator from 0 to 30,000 V in a scanning electron microscope. This voltage difference defines the accelerating voltage of the electrons in the beam. The higher the accelerating voltage is, the more electrons that will be received by the detector, resulting in a clearer image (less noisy) on the cathode ray tube. However, the higher the accelerating voltage is, the greater the penetration of the electrons into the specimen, resulting in a larger specimen–beam interaction volume and poorer resolution of the specimen. The chosen accelerating voltage is a balance between generating sufficient electrons to obtain an image on the cathode ray tube, while having a small enough specimen beam interaction volume to obtain resolution.

Changing the accelerating voltage will result in a change in focus. The higher the accelerating voltage is, the less the electrons are affected by the magnetic field of the lenses and the further the crossover point is from the bottom of the lens. Since optimal focus in the scanning electron microscope is defined as the electron beam striking the specimen at the crossover point of the beam, changing the position of the crossover point by changing the accelerating voltage will affect the focus of the image.

BEAM (SHIFT AND TILT) DEFLECTOR COILS

The maximum number of beam electrons striking the specimen occurs when the filament in the electron gun is directly over the optic axis of the column. In practice, it is often difficult to precisely center the filament mechanically over the optic axis. In addition, during operation of the scanning electron microscope, the tungsten filament will often shift slightly in position due to the heating of the filament (if a thermionic cathode is used). To correct for slight misalignment of the filament and the electron gun, two pairs of deflector coils are placed in the column to align the electron beam directly over the optic axis to obtain maximum illumination of the specimen. Each deflector coil has half of the coil on one side of the column and the other half on the opposite side of the column (some electron microscopists refer to each half of the deflector coil as a coil). The two halves are joined by the coil wire that passes around the column (Fig. 5-2). The electron beam passes down the optic axis of the column between the two halves of the deflector coil. When electrical current is passed through the deflector coil, each half of the deflector coil acts as a solenoid, setting up a magnetic field. The magnetic lines of

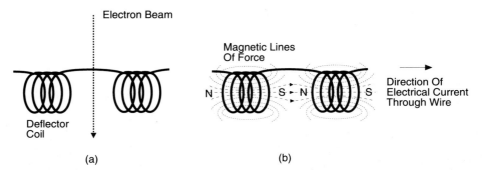

Figure 5-2 Deflector coils (for convenience the coils are drawn as circular coils, although in many cases they are manufactured with a square profile). (a) A deflector coil consists of two coil halves joined by the wire that passes around the column. The electron beam passes between the two coil halves. (b) A magnetic field is generated when an electrical current is passed through the wire of a deflector coil.

force travel in a single direction between the two halves of the deflection coils (Fig. 5-3).

The electron beam is deflected according to the right-hand rule as the electrons in the beam travel through the magnetic lines of force in the center of the column (Fig. 5-3). As the electrons in the beam enter the magnetic field, the electrons are forced in a direction perpendicular to the magnetic lines of force and the direction of the electron beam. If the magnetic flux density is uniform, the path of the deflected electrons is circular through the deflector coil. On leaving the deflection field of the coils, however, the electrons travel in a straight line. The stronger the electrical current through the deflection coils, the stronger is the magnetic field and the more the electrons in the beam are deflected. Reversing the flow of electrons through the coil reverses the direction of the magnetic lines of force, resulting in a reversal of the force on the electrons in the beam (to 180° of the original force) causing the electrons to circle in the opposite direction.

In the deflector system of an electron microscope, there are two pairs of deflector coils. Each pair of deflector coils has one coil above the other and perpendicular to the other (Fig. 5-4). Changing the current through the two pairs of deflector coils allows the operator to center the beam to rectify any misalignment of the electron gun over the optic axis of the column (gun shift) or to tilt the beam to rectify misalignment of the cathode within the Wehnelt cylinder (gun tilt).

Gun Shift. If the electron gun is not centered over the optic axis of the column, the aperture of the probe-forming (objective) lens could intercept the electron beam, allowing only a part or none of the electron beam to strike the specimen, resulting in a cathode ray tube with the specimen image partially or totally obscured by the final aperture. The operator is able to use the deflector coils to electronically correct for misalignment of the electron gun. **Shifting** the

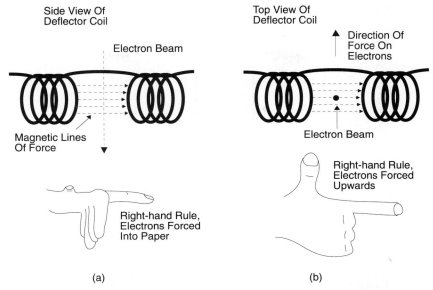

Figure 5-3 (a) Side view and (b) top view of deflector coils showing the direction of force on electrons traveling through the magnetic lines of force generated by the deflector coils. The electrons are subjected to a force perpendicular to the magnetic lines of force and the direction of the electron beam.

gun image over the optic axis is carried out by having the upper and lower deflector coil pair deflect the electron beam to the same angle. Thus, if the upper deflector coil pair causes a 1° inclination in the electron beam, the lower deflector coil pair will also cause a 1° inclination in the beam, but in a direction opposite to the inclination induced by the upper deflector coil pair. Such an adjustment results in the electron beam still traveling down the column parallel to the optic axis, but in

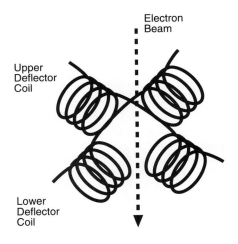

Figure 5-4 In one pair of deflector coils, one coil is positioned one above the other, and perpendicular to the other, in the electron microscope column. In the column, two pairs of deflector coils make up the deflection system.

a different place (Fig. 5-5). By changing the degree and direction of the deflection of the beam, the electron beam can be centered down the optic axis of the column.

Gun Tilt.   The deflector coils can be used to correct a misaligned beam caused by a cathode that is not properly centered above the aperture of the Wehnelt cylinder. If the electron-emitting cathode is not mechanically centered over the center of the Wehnelt aperture, the electron beam that is accelerated down the column will be generated from an electron cloud that is partially obscured by the

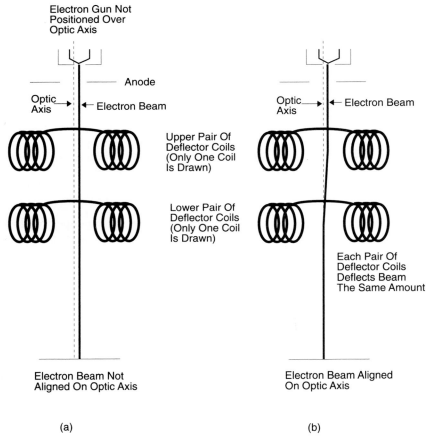

(a)   (b)

Figure 5-5   Gun shift. The set of deflector coils is used to align the electron beam with the optic axis of the column. Note that only one deflector coil of each pair is drawn. The electron gun is not precisely positioned over the optic axis. (a) Without any current through the deflector coils, the electron beam does not follow the optic axis of the column. (b) When the proper amount of current is passed through the deflector coils, the beam becomes aligned with the optic axis. The electron beam is bent the same amount by each deflector coil, so the electron beam is parallel to the optic axis.

Wehnelt cylinder. This will result in an electron beam with less than the maximum electrons in the beam. In a properly saturated cathode, this will result in a relatively dim image of the specimen on the viewing CRT. If the cathode is unsaturated, the electron beam will strike the specimen with a dark image of the cathode of center in the electron beam (Fig. 5-6).

The deflector coils are used to bring the portion of the beam containing the maximum number of electrons onto the optic axis to ensure optimal illumination

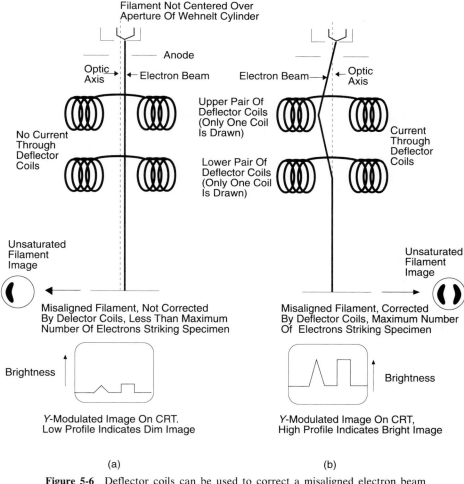

(a)   (b)

Figure 5-6 Deflector coils can be used to correct a misaligned electron beam caused by the cathode not being in the center of the Wehnelt cylinder aperture. (a) No current is passing through the deflector coils and the beam is misaligned. (b) Current is passing through the deflector coils, so the electron beam travels parallel to the optic axis after passing through the lower set of deflector coils.

of the specimen (Fig. 5-6). The upper deflector coils take the area of the electron beam containing the maximum current under the anode and deflect this area of the electron beam onto the optic axis at the position of the lower deflector coils. The lower deflector coils align the electron beam down the optic axis. Thus, in gun tilt, the upper deflector coils deflect the electron beam at a greater angle than the lower deflector coils.

LENSES AND APERTURES

A scanning electron microscope will normally have two condenser lenses and one objective (probe-forming) lens. Each lens usually has one aperture associated with the lens. The lenses and apertures have two functions in the scanning electron microscope:

1. *Demagnification of the electron beam to produce the final probe diameter:* The ultimate resolving power of the scanning electron microscope is determined by the smallest beam diameter that strikes the specimen with sufficient beam current to generate an image. The diameter of the electron beam is determined by the condenser lenses and the objective lens in association with the final (objective or probe-forming) aperture.

2. *Focus:* The objective (probe-forming) lens is used to bring the electron beam to crossover at the specimen. Focus in the scanning electron microscope is defined as the electron beam crossing over at the specimen.

Demagnification of the Electron Beam to Produce the Final Probe Diameter

The two condenser lenses and the objective lens are used to demagnify the electron beam to form the final diameter of the electron beam as it strikes the specimen. The electron beam has a diameter of approximately 50 μm at the crossover under the electron gun when a thermionic tungsten cathode is used. The lenses and apertures are used to demagnify the diameter of the electron beam so that it is approximately 10 nm (0.01 μm) when it strikes the specimen, a demagnification of 500 times. For high-resolution examination of specimens, the electron beam can be further demagnified to 3 to 5 nm. Beneath this size there are insufficient electrons in the beam to generate enough signal to form an image on the viewing cathode ray tube.

The electron beam is produced by the electron gun and is accelerated down the column of the scanning electron microscope. Initially, the electron beam encounters the first aperture of the condenser lens, which is usually around 100 μm in diameter. The maximum number of X-rays, potentially harmful to the operator, are produced when the relatively high current electron beam encounters the first condenser aperture. It is in this area where the maximum shielding to absorb X-

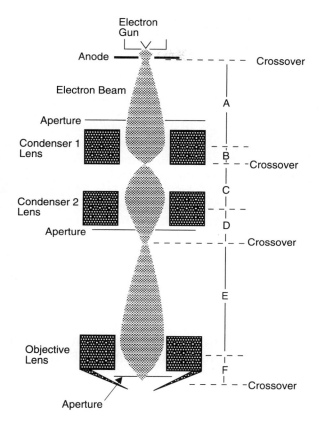

Figure 5-7 Path of the electron beam through the lenses. The relative size and angles of the electron beam have been exaggerated for illustration purposes. The demagnification through the condenser 1 lens is A/B, the demagnification through the condenser 2 lens is C/D, and the demagnification through the objective lens is E/F.

rays should be positioned. As the electron beam passes down the column, it has less current and produces fewer X-rays on interaction with parts of the column. All the apertures in the condenser lenses are "splash" apertures that are normally larger than the diameter of the electron beam at crossover and function primarily to intercept stray electrons.

A modern scanning electron microscope usually has two condenser lenses. The first condenser lens is strong and reduces the diameter of the electron beam at crossover beneath the lens to about 1 μm. The second condenser lens is used to project the next crossover point to a position that will determine the number of electrons that will pass through the objective lens and objective aperture.

The amount of demagnification that is produced by the first condenser lens is determined by the ratio of the distance from the crossover point of the electron beam under the electron gun to the first condenser lens (A in Fig. 5-7), compared to the distance from the first condenser lens to the crossover point of the electron beam under the first condenser lens (B in Fig. 5-7). Thus, in Fig. 5-7, the demagnification caused by the first condenser lens is A/B. For example, if A is 100 mm and B is 2 mm, then the demagnification produced by the first condenser lens is 50. A demagnification of 50 will take an electron beam that is 50 μm in diameter at the crossover point under the electron gun to a diameter of 1 μm at the crossover point under the first condenser lens. The stronger the current is through the coils of the first condenser lens, the closer the electron beam undergoes crossover to the lens and the greater is the demagnification.

The demagnification caused by the second condenser lens can be calculated in a similar manner. In Fig. 5-7, the demagnification caused by the second condenser lens is represented by C/D. Normally, this demagnification is around 1.

The objective (probe-forming) lens accepts the electron beam from the condenser lens system. The objective lens functions to further demagnify the beam and to focus the electron beam to crossover on the specimen. The definition of focus in the scanning electron microscope is the objective lens bringing the electron beam to minimal spot size (crossover) on the specimen. The demagnification produced by the objective lens is equal to the distance from the crossover point of the electron beam above the objective lens to the center of the objective lens (E in Fig. 5-7), divided by the distance from the center of the objective lens, to the crossover point of the electron beam under the objective lens (F in Fig. 5-7). Thus, in Fig. 5-7, the demagnification produced by the objective lens is E/F.

Working Distance

The amount of demagnification produced by the objective lens and the amount of current (number of electrons) in the electron beam are affected by the working distance. The **working distance** is defined as the distance from the bottom of the objective lens pole piece to the top of the specimen (Fig. 5-8). The working distance can be varied by the operator from 5 mm to around 30 mm on most scanning electron microscopes. Crossover of the electron beam occurs farther from the objective lens as the working distance is increased (in order to keep the specimen in focus) (Fig. 5-9).

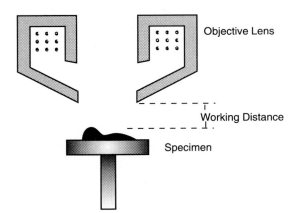

Objective Lens

Working Distance

Specimen

Figure 5-8 The working distance in a scanning electron microscope is the distance from the bottom of the lower objective lens pole piece to the top of the specimen.

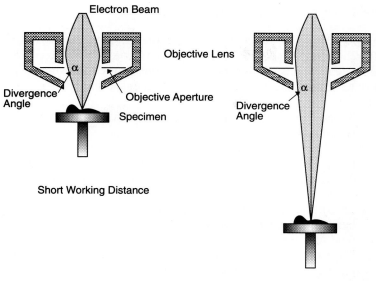

Figure 5-9 Effect of working distance on the divergence angle of the beam. A short working distance results in a large divergence angle (α) and shallow depth of field. A long working distance results in a small divergence angle (α) and deep depth of field. The angles and size of the electron beam have been exaggerated for illustration purposes.

Changing the working distance varies two parameters:

1. *Demagnification of the electron beam decreases as the working distance increases.* In Fig. 5-7, demagnification of the electron beam is equal to E/F. The greater the working distance is, the greater is F and the less the demagnification of the electron beam and the larger the beam spot. The size of the beam spot striking the specimen determines the resolution of the scanning electron microscope. The larger the beam spot is, the poorer the resolution. Therefore, to attain maximum resolution, a short working distance should be used to obtain minimal beam spot size.

2. *Depth of field (focus) increases as the working distance increases.* As the working distance increases, the divergence angle (α in Fig. 5-9) decreases and the depth of field increases. The depth of field is also affected by the diameter of the objective lens aperture. The smaller the aperture is, the greater the depth of field. Depth of field will be covered in more detail later when pixel diameters are discussed.

Design of the Objective Lens

The objective (probe-forming) lens is designed asymmetrically and is therefore different from the axially symmetrical construction of the condenser lenses. The objective lens must be designed to account for the following difficulties (which are not encountered in the design of the condenser lenses):

1. Secondary electrons, produced on interaction of the electron beam with the specimen, have energies of 50 eV or less. The secondary electrons (along with backscattered electrons) are collected by the Everhart–Thornley detector and are used to form the image of the specimen. The Everhart–Thornley detector collects the secondary electrons by attracting them with a relatively weak positive electrostatic field. The specimen is relatively close to the bottom of the objective lens (5 to 30 mm, the working distance). If the objective lens creates even a moderate magnetic field near the specimen, the magnetic field will interfere with the collection of secondary electrons by the detector. Therefore, the objective lens must be designed so that it creates a minimal magnetic field near the specimen.

2. Many scanning electron microscopes are used to perform X-ray microanalysis of specimens. The design of the final lens must allow unimpeded passage of the X-rays to the detector.

3. The center (bore) of the objective lens must be large enough to accommodate the stigmator elements, objective (beam-limiting) aperture, and, often, scanning coils. The electron beam is deflected a considerable distance from the optic axis of the column by the scanning coil, and the bore of the objective lens must be large enough to allow the electron beam to pass.

To accommodate these considerations, the objective lens has an asymmetrical design (Fig. 5-10). The outer pole piece slopes down under the lens toward the specimen. The diameter of the outer sloping pole piece is much smaller than the inner pole piece. This is because the inner pole piece has the stigmator, beam-limiting aperture, and, often, the scanning coils inside it. The small bore of the

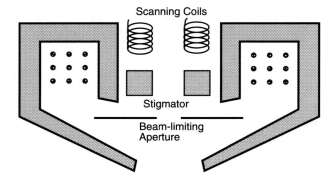

Scanning Coils

Stigmator

Beam-limiting Aperture

Figure 5-10 Structure of an objective (probe-forming lens) in the column of a scanning electron microscope. The outer pole piece extends inward toward the specimen.

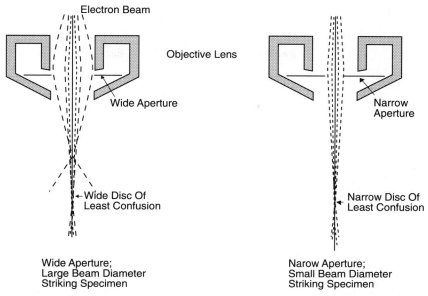

Figure 5-11 Effect of using different-sized apertures in the objective (probe-forming) lens on the size of the electron beam striking the specimen. A wide aperture results in a large beam spot because of the large disc of least confusion produced by spherical aberration. A narrow aperture results in a small beam spot because of the reduction in spherical aberration. The angles and size of the electron beam have been exaggerated for illustration purposes.

outer sloping pole piece restricts the magnetic field in this area. This results in a minimal magnetic field near the specimen and minimal interference in collection of the secondary electrons by the detector. Also, because of the sloping design of the outer pole piece, there is minimal physical interference of X-rays reaching the detector.

The ultimate minimal size of the electron beam (probe) striking the specimen, and therefore the ultimate resolving power of the scanning electron microscope, is determined by spherical aberration of the objective lens. The spherical aberration caused by the objective lens is in turn determined by the diameter of the objective lens aperture. The larger the objective aperture is, the greater the spherical aberration, the larger the disc of least confusion, and the larger the diameter of the electron beam (Fig. 5-11).

Demagnification and the Amount of Current in the Electron Beam

The condenser lenses in combination with the aperture in the probe-forming (objective) lens determines the current in the final probe spot striking the specimen. The stronger the current is through the coils of the condenser lenses, the closer

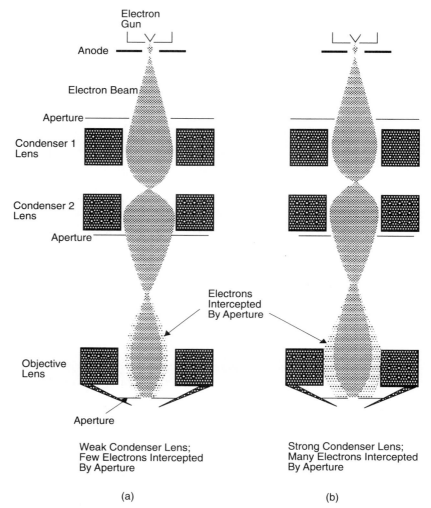

Electron
Gun

Anode

Electron Beam

Aperture

Condenser 1
Lens

Condenser 2
Lens

Aperture

Electrons
Intercepted
By Aperture

Objective
Lens

Aperture

Weak Condenser Lens;
Few Electrons Intercepted
By Aperture

Strong Condenser Lens;
Many Electrons Intercepted
By Aperture

(a)   (b)

Figure 5-12 Amount of the electron beam that is intercepted by the objective aperture when the condenser lenses have a weak magnetic field (a) compared to the situation when the condenser lenses have a strong magnetic field (b). Weak condenser lenses result in an electron beam crossover point that is relatively far from the lens, a large beam spot, and a large number of electrons (current) in the beam spot. Strong condenser lenses result in an electron beam crossover point that is near the lens, a small beam spot, and few electrons in the electron beam. The size and angles of the electron beam have been exaggerated for illustration purposes.

the electron beam crossover points are to the lens, the greater the demagnification produced by the lenses, and the less electrons (current) there will be in the final probe as it strikes the specimen. This can be seen in the two illustrations in Fig. 5-12. In Fig. 5-12a, little current is passing through the coils of the condenser lenses and relatively little demagnification occurs, resulting in a large final probe size that contains a relatively large number of the electrons from the electron beam above the condenser lens. However, if the electron beam is strongly demagnified by the condenser lenses to produce a small probe size, many of the electrons in the beam are intercepted by the apertures, resulting in a small final probe with minimal current (Fig. 5-12b). Here, the current through the coils of the condenser lenses has been increased, resulting in increased lens strength, shorter crossover distances, increased demagnification, and more of the electrons in the beam being intercepted by the apertures because of much wider cones of electrons under each lens. The stippled areas of the electron beam in Fig. 5-12 represent the areas of intercepted electrons. *The electrons (current) that enter the next lens is equal to the current in the electron beam multiplied by the result the square of the semiangle of the electron beam accepted by the next lens* (α_a), *divided by the semiangle of the electron beam produced by the first lens* (α_i). In Fig. 5-13, this is represented by $(\alpha_a/\alpha_i)^2$ times the current in the electron beam passing through the first lens.

   It is important to realize that the current (number of electrons) in the electron beam decreases as the final spot (probe) size of the electron beam decreases (Fig. 5-14). *The current in the electron beam increases to the 8/3 power as the probe diameter is increased.* A scanning electron microscope with a thermionic tungsten cathode generating an emission current of 4.1 A/cm^2 at 2820 K with a final probe size of 10 nm will have a current of about 5×10^{-10} A as the electron beam strikes the specimen. Increasing the probe size to 100 nm will result in a probe current of

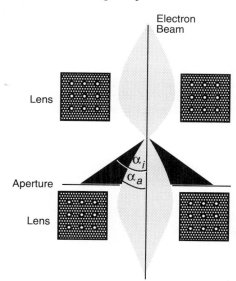

Figure 5-13 Angle produced by all the electrons that pass through one lens (α_i) compared to the angle produced by those electrons that are accepted by the next lens (α_a).

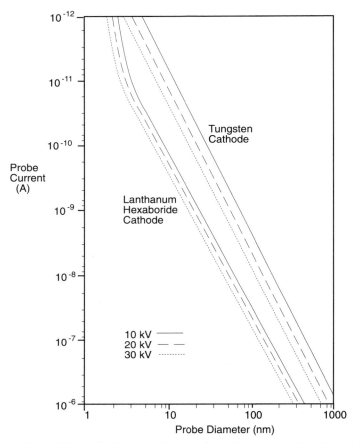

Figure 5-14 Relationship between the size of the electron beam (probe) and the current in the electron beam for a thermionic tungsten filament and a lanthanum hexaboride rod operating at 10, 20, and 30 kV. The calculations are based on an emission current density at the cathode of 4.1 A/cm² for the thermionic tungsten cathode and 25 A/cm² for the lanthanum hexaboride cathode. (*Adatped from* J. I. Goldstein and others, *Scanning Electron Microscopy and X-ray Microanalysis*, Plenum Press, New York, 1981)

approximately 1×10^{-8} A. The minimum probe current necessary to produce an image with an Everhart–Thornley detector in a scanning electron microscope is 10^{-12} A. Beneath this current, too few secondary and backscattered electrons are produced on interaction of the probe with the specimen to form a satisfactory image. The necessary beam current of 10^{-12} A can be produced with a 5-nm electron beam if thermionic emission from a tungsten filament is used or with a 3-nm beam if a lanthanum hexaboride rod is used. Therefore, the only way of increasing the ultimate resolving ability of a scanning electron microscope to the limits of spherical aberration is to decrease the size of the electron beam striking the specimen, while still maintaining the necessary current (number of electrons) in the electron beam. This can only be done by increasing the emission from the cathode in the electron gun, with a field emission source providing the maximum electron emission.

In conclusion, if the operator minimizes the size of the final probe (electron beam) striking the specimen to attempt to increase resolving power at high magnification, there will be a considerable loss in the number of electrons (current) in the probe. The less electrons that are in the final probe, the less backscattered and secondary electrons produced and received by the detector and the more the signal will have to be boosted electronically, resulting in a noisier or snowy image on the cathode ray tube. The operator, therefore, has to strike a balance between minimum beam (probe) spot size and probe current to obtain an optimal image.

FOCUSING

In the light microscope, the image is in focus at the focal length where the focal point occurs (Fig. 5-15). Focus in the light microscope may be obtained in two ways:

1. *Varying the curvature of the converging objective lens:* This changes the focal length, which changes the place where the image is focused (Fig. 5-16). In actual fact, it is not practical to change the curvature or refractive index of lens to adjust the focal plane, so the second method is used in light microscopes.

2. *Focusing by varying the distance between the objective lens and the specimen:* In light microscopes, the distance from the objective lens to the ocular or eye is fixed. For illustration purposes, the distance can be thought of as being equal to the focal length of the lens. To focus the image, the distance

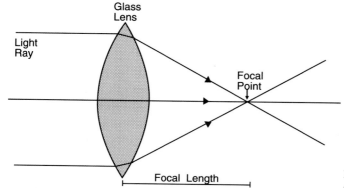

Figure 5-15 Focusing action of a convergent glass lens.

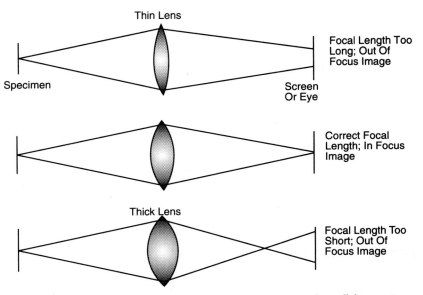

Figure 5-16 Converging glass lenses of different curvatures focus light rays to different focal lengths.

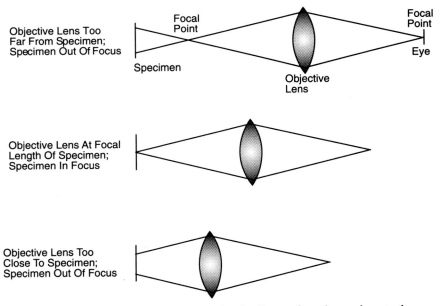

Figure 5-17 In a light microscope, varying the distance from the specimen to the objective lens allows the observer to bring the specimen into focus.

from the specimen to the objective lens is varied. When the specimen is at the front focal point of the lens (a lens has a focal point in front of the lens and a second focal point behind the lens), the specimen will be in focus (Fig. 5-17).

In the electron microscope, it is not possible to move the position of the lenses, so it is necessary to change the focal length of the objective lens to focus the specimen. This is done by varying the current through the windings of the electromagnet that makes up the objective lens. This varies the magnetic field in the bore of the lens and therefore the focal length (Fig. 5-18). Changing the strength of the magnetic field in an electron microscope is thus equivalent to a change in the refractive index or curvature of a glass lens in a light microscope.

Instead of using electromagnets, it is possible to use permanent magnets as lenses in electron microscopes. With permanent magnets, the focal length can be varied by changing the gap length between the upper and lower pole pieces, thereby changing the magnetic field in the bore of the lens. It is also possible to

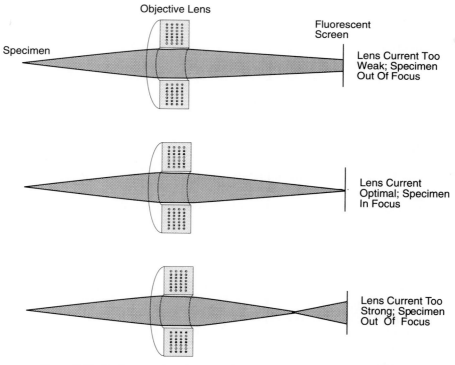

Figure 5-18 In the scanning electron microscope, varying the electrical current to the objective lens changes the magnetic field and allows the operator to bring the electron beam to crossover on the specimen. When the electron beam crosses over at the specimen, the specimen is in focus.

change the position of the specimen in the column and therefore the focus, similar to the way focus is obtained in the light microscope. Most modern transmission electron microscopes have a device for varying the position of the specimen in order to maintain the eucentric position of the specimen for the taking of stereoscopic photographs at different tilt positions. Even though the device is present, its main function is not that of focusing the specimen.

APERTURES

An **aperture** *is a hole in a nonmagnetic refractory metal (a metal resistant to corrosion, especially at high temperature), such as gold, silver, platinum, or molybdenum.* Strictly speaking, the aperture is just the hole in the metal. However, the term aperture is commonly used by electron microscopists when referring to the metal piece and the hole or holes. Aperture holes are placed in the center of the electron beam so that only the desired part of the electron beam is passed through the aperture, with the rest of the electrons being absorbed by the metal of the aperture. In both the transmission electron microscope and the scanning electron microscope, apertures are used to define the size of the beam spot, the number of electrons in the beam, the aperture angle of the beam, and the depth of field of the microscope.

Types of apertures.   There are two basic types of apertures: (1) *thick aperture strips* or *discs* that require periodic cleaning, and (2) *thin* or *foil apertures* that are self-cleaning.

Thick Aperture Strips or Discs.   These apertures are manufactured by boring holes in strips or discs of platinum or molybdenum. The size of the holes varies from 100 to 500 μm for condenser lens apertures and 10 to 60 μm for objective lens apertures. Platinum and molybdenum are used since they are readily machinable to give openings that are smooth and very small. Also, molybdenum and platinum have a high electrical conductivity, so the beam electrons that strike the aperture are readily drained to the ground. These metals also dissipate the heat produced by interaction of the electron beam with the apertures. In the transmission electron microscope, the aperture strips or discs (Fig. 5-19) are placed in a tray that can be moved mechanically in the column by means of *x*- and *y*-direction micrometer knobs mounted on the outside of the column. The mechanical mechanism also has a click-stop mechanism for selecting the size of the aperture hole to be placed in the path of the electron beam. In the scanning electron microscope, the apertures are often fixed in position in the column, although in some scanning electron microscopes the final or objective aperture consists of a number of aperture holes in a movable tray.

Contamination of aperture holes is a problem with thick aperture strips or discs. **Contamination** *is the buildup of a layer of material that does not readily*

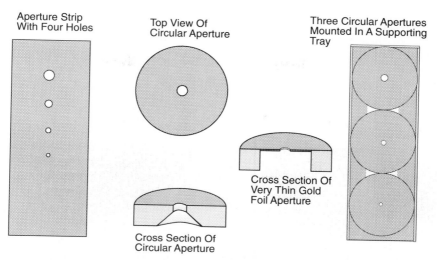

Aperture Strip
With Four Holes

Top View Of
Circular Aperture

Three Circular Apertures
Mounted In A Supporting
Tray

Cross Section Of
Very Thin Gold
Foil Aperture

Cross Section Of
Circular Aperture

Figure 5-19   Aperture strips and discs.

conduct electrons on the edge of an aperture. When in the electron beam, the contaminating layer on the aperture accumulates electrons, becomes charged, and causes the electron beam to become astigmatic, thereby affecting the quality of the specimen image. The contamination on the aperture is the result of the interaction of gases with the electron beam. The problem gases are primarily of organic origin and include oil vapor from the oil diffusion pump, vacuum grease, and O-rings and gases produced on interaction of the electron beam with the specimen, which causes heating of the specimen. All these gases move about within the column. As they move randomly into the path of the electron beam, orbital electrons are knocked out of the gas molecules by the collision with the high-energy beam electrons. This produces positively ionized gas molecules. These ionized gases are swept downward in the beam by the beam electrons. The ionized gas molecules strike the apertures in the path of the electron beam. If there are any negative charges on the apertures (due to previous contamination on the aperture), the ionized gases are attracted to the negative areas and attach to the aperture.

Once attached to the aperture, the gases are carbonized by the electron beam and form a nonconducting waxy layer at the aperture hole. The nonconducting layer of contamination causes the aperture to be transformed into an astigmatic electrostatic lens since electrons accumulate on the nonconducting contamination (Fig. 5-20). This results in a buildup of negative charge at the edge of the aperture hole. Astigmation of the electron beam results because the contaminating layer is not distributed uniformly around the edge of the aperture hole. This causes one part of the aperture to have a more negative charge than other parts of the aperture hole. An *uncontaminated* aperture hole in the path of the electron beam will not act as an electrostatic lens since the electrons striking the aperture metal are

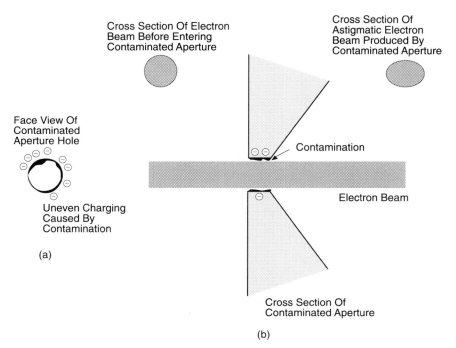

Figure 5-20 Contamination of an aperture. (a) Face view of a contaminated aperture showing the uneven deposition of contamination around the hole. In the electron beam the contamination charges unevenly, resulting in an uneven distribution of negative charges. (b) Effect of a contaminated aperture on an electron beam passing through the aperture. The aperture removes those stray electrons not aimed directly at the aperture. Those electrons that pass through the aperture are affected by the uneven charge on the aperture due to the contamination. This results in an uneven electric field on the beam and an astigmatic beam that is elliptical in cross section after it has passed through the aperture.

immediately drained to ground. The aperture metal is therefore at ground, as is the rest of the column under the electron gun. However, the buildup of an unevenly charged layer of contamination on the aperture hole causes the aperture to become an astigmatic electrostatic lens. As the electron beam passes through the aperture, it is compressed unevenly, and an astigmatic electron beam that is elliptical in cross section results. The smaller the aperture hole is, the more quickly the contamination will obscure the aperture hole and cause astigmatism. Therefore, the smaller the aperture hole, the more rapidly the aperture hole will become unusable due to contamination.

Contamination also occurs on other parts of the column, although it usually does not affect the electron beam as much as contamination on the apertures. Pole pieces of lenses, because they are relatively narrow, are probably the second most important areas as far as contamination is concerned. Many electron microscopes have a removable bronze tube lining the lenses that can be removed for easy cleaning.

Contaminated thick apertures can be cleaned a number of times. Platinum and molybdenum apertures are commonly cleaned by placing them on a tungsten boat strung between two electrodes in a vacuum. Electrical current is passed between the two electrodes, causing the tungsten boat to heat up. The temperature of the boat is brought to where the apertures glow yellowish, at which time most of the contamination is evaporated off the apertures. Platinum apertures can also be cleaned by placing them in a platinum crucible and heating them in a propane torch to evaporate the contamination. These methods have two disadvantages. Not all of the contamination is volatile, and the nonvolatile components burn into the metal aperture. Also, in the case of molybdenum, pointed crystals will grow into the holes, which after a few cleanings make the apertures useless.

A better method consists of heating the apertures to redness in a flame for a minute and dropping the red-hot aperture into 1 ml of concentrated hydrofluoric acid in a small propylene beaker in a hood. The acid penetrates into the tiny aperture holes and attacks the contamination. After 1 min, the acid is diluted with 5 ml of distilled water, left for 5 min, and decanted off. Five percent ammonia solution is added, boiled for 5 min, and replaced with distilled water. The apertures are dried. Molybdenum apertures are slightly attacked by this method, while platinum apertures are not harmed.

Very Thin Foil Apertures. Very thin foil apertures consist of a 0.5-μm-thick gold foil with an appropriate-sized hole in the gold. The gold foil is mounted on a support such as a copper ring or phosphor bronze ring (Fig. 5-19). The gold foil is a poor conductor of heat and heats to a high temperature when the high-energy electron beam bombards the foil. Therefore, little contamination is deposited on the apertures because they are so hot. Foil apertures can be used for long periods of time without being cleaned or replaced.

Very thin foil apertures are manufactured by evaporating gold onto a thin plastic film with a hole in it. The foil is mounted in a supporting ring for insertion into the electron microscope column. Another method for manufacturing foil apertures is to evaporate gold over spheres of dextran on a collodion or Formvar membrane. The dextran and membrane are removed with a solvent, leaving a metal sheet with holes the size of the dextran spheres.

OBJECTIVE STIGMATOR

Stigmator elements are in the bore of the objective lens to correct for astigmatism of the beam. The principles of astigmatism correction have been previously discussed in the section on lens aberrations.

SCANNING COILS

Scanning coils are used to manipulate the focused electron beam over the specimen in a series of consecutive lines to form a rectangular **raster pattern**. The number of lines in a raster can normally be varied from 250 to 4000 by the operator. The length of time spent on each line can also be varied, commonly from 2 to 512 ms per line.

There are two scanning coils (Fig. 5-21) located either in or directly above the objective lens. One scanning coil manipulates the electron beam in the *y* direction, while the other scanning coil manipulates the beam in the *x* direction. The scanning coils manipulate the electron beam so that it is "elbowed" through the scanning coils, passing near the center of the objective (final or probe-forming) aperture and striking the specimen.

The scanning coils are magnetic and work on the same principles as the deflection coils (these principles have already been discussed in the section on deflection coils). It is possible to use either electrostatic (consisting of positively or negatively charged plates) or magnetic scanning coils. Scanning electron microscopes use magnetic scanning coils for the following reasons:

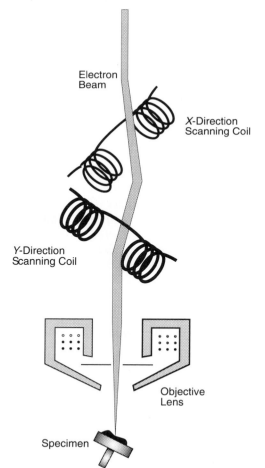

Figure 5-21   Scanning coils and objective lens. The electron beam is deflected by the scanning coils so that it is elbowed and manipulated to pass near the center of the objective lens aperture to strike the specimen.

1. It is essential that the image that is reconstructed on the viewing and photographic cathode ray tubes (CRTs) be of the highest quality. CRTs using an electrostatic deflection system produce poorer quality images because of problems with the internal electron beam becoming defocused (larger) at the periphery of the CRT. Therefore, CRTs with magnetic deflection systems are used in scanning electron microscopes. In the scanning electron microscope, it is desirable to use a single scan generator to drive either the x or y scanning coil in the column and in the CRT. Therefore, if the deflection coils in the CRT are magnetic, then they should also be magnetic in the column of the scanning electron microscope.

2. Electrostatic scanning coils require higher voltages to deflect the electron beam to the same degree when compared to magnetic scanning coils.

A minor disadvantage of magnetic scanning coils is the more time required (in the range of tens of milliseconds) to deflect the electron beam. This is because a change in position of the electron beam requires a change in position of the magnetic field, which in turn requires a change in current through the scanning coils.

In the construction of the scanning electron microscope, one scan generator is used for the x (horizontal) direction and a second for the y (vertical) direction. The currents required to deflect the electron beam in the column are considerably smaller than those required to deflect the electron beam in the CRTs. A suitable network is therefore inserted in the circuit from the scan generator to attenuate the current going to the scanning coils in the column, with no attentuation in the current to the CRTs.

The scanning coils move the electron beam over the specimen to produce the rectangular raster patter (Fig. 5-22). The x-direction scanning coil moves the electron beam relatively slowly to produce a horizontal line scan. Once at the end of the line (Fig. 5-23a), the internal electron beam in the cathode ray tube is **blanked** by the equivalent of the Wehnelt cylinder becoming highly biased so that it pinches

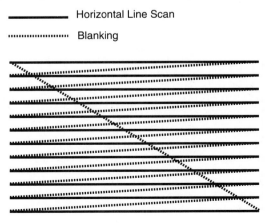

——————— Horizontal Line Scan

·················· Blanking

Figure 5-22 Raster pattern formed by the electron beam as it strikes the specimen. The electron beam passes horizontally over the specimen until it reaches the right side of the line. The electron beam is then quickly brought back to the left side and dropped one line by the scanning coils. During this period, the beam is blanked by the cathode ray tube. At the end of the last line of the raster pattern, the electron beam is quickly returned to the position to again begin rastering the specimen. The actual raster pattern contains many more raster lines than indicated in this drawing.

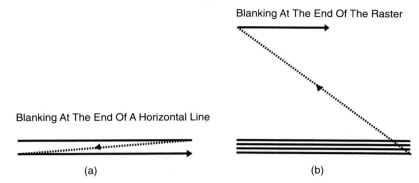

(a)   (b)

Figure 5-23   Blanking. (a) Blanking between horizontal lines in the raster pattern. (b) Blanking at the end of the raster pattern so that a new raster pattern can begin.

off the electron beam in the cathode ray tube, and the electron beam does not reach the phosphor on the viewing surface. The electron beam in the column of the scanning electron microscope is quickly swept back to the left starting position by a rapid change in the magnetic field of the x-direction scanning coil. A small change in the magnetic field of the y-direction scanning coil brings the electron beam to the beginning of the next horizontal scan line. These rapid movements of the electron beam on the specimen are not recorded on the CRT since the beam is blanked during this period. Next, blanking of the beam in the cathode ray tube ceases, and the relatively slow scan of the next line in the raster begins on the specimen and is recorded on the face of the CRT.

The remainder of the raster patter is produced in like fashion. When the end of the last line in the raster is reached, the beam is again blanked in the CRT, there is a major change in the magnetic fields in the scanning coils (Fig. 5-24), and the electron beam in the column of the scanning electron microscope returns to the beginning of the first line of the raster patter (Fig. 5-23b). The next raster of the specimen now begins.

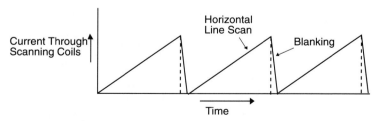

Figure 5-24   Sawtooth pattern of current to the x direction scanning coil. Increasing the current results in an increased magnetic field, causing greater deflection of the electron beam as it moves from left to right across the specimen. At the right edge of the line, the current is returned to base level, and the electron beam quickly sweeps to the left during the blanking period. At this time the current to the y scanning coil changes enough to drop the electron beam to the next scan line.

6

Electron Beam–Specimen Interactions

TYPES OF SIGNALS PRODUCED

A number of different types of signals are produced in the scanning electron microscope as the energy of the electron beam is transformed on interaction with the specimen. The two basic types of emitted signals that are used to view the specimens are (1) electrons and (2) electromagnetic radiation.

Electrons.   Electrons are wave particles that have a mass of 9.11×10^{-28} g and a charge of 10^{-19} C. Even though they have mass, they also behave as waves, with their wavelength (λ) determined by their velocity:

$$\lambda = \frac{h}{mv}$$

where h is Planck's constant, m is the mass of the electron, and v is the velocity of the electron. Electrons used to image the specimen in the scanning electron microscope are classified according to their energy. **Backscattered electrons** are electrons that escape from the specimen with an energy of more than 50 eV (an **electron volt** [eV] is the kinetic energy an electron would acquire in being accelerated through a potential difference of 1 V). **Secondary electrons** escape from the

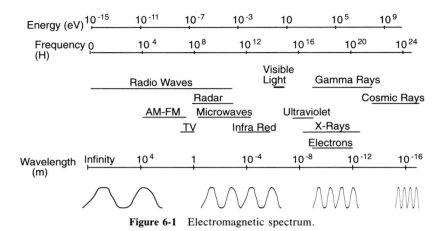

Figure 6-1 Electromagnetic spectrum.

specimen surface with an energy of less than 50 eV, while **Auger electrons** are electrons created after the absorption of X-rays.

Electromagnetic Radiation.   Electromagnetic waves are related patterns of electric and magnetic force. Electromagnetic waves are generated when beam electrons interact with the nuclei or shell electrons of the specimen atoms and lose energy. A quantum of electromagnetic energy is the **photon**. Unlike electrons, electromagnetic radiation has mass only as long as it is moving. An ordinary particle in motion can be characterized by its momentum, its mass times velocity. The photon has velocity, so while it is moving it has mass. However, the photon has no rest mass; if a photon comes to rest by being absorbed, it ceases to exist and its energy is converted to some other form of energy. Electromagnetic rays travel through space at the speed of light, 186,272 miles (299,792 km) per second. No material medium (for example, water or air) is needed for the transmission of electromagnetic waves.

The **electromagnetic spectrum** (Fig. 6-1) is divided into cosmic rays, gamma rays, X-rays, and ultraviolet, visible, infrared, and radio waves, depending on the energy of the radiation. The energy of the radiation is related to the wavelength and the frequency of the radiation. Imaging in scanning electron microscopes can utilize the X-ray, ultraviolet, visible, and infrared portions of the electromagnetic spectrum.

ATOMIC STRUCTURE IN METALS AND INSULATORS

The structure of atoms will be reviewed before discussing the generation of electrons and electromagnetic radiation on interaction of the specimen with the electron beam. In an atom, the outer electrons are in either of two bands, the lower **valence band** or the upper **conduction band** (Fig. 6-2). The valence band corre-

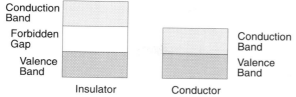

Figure 6-2 In an insulator there is a relatively large energy difference (the forbidden gap) between the valence and conduction band. In a conductor there is no forbidden gap between the valence and conduction bands.

sponds to the ground state of the outer electrons. Normally, the electrons reside in the valence band, where they are held tightly to the individual atoms. There is a forbidden gap between the valence band and conduction band that contains energy levels that no electrons are allowed to have. A valence electron can acquire energy and jump to the conduction band, but it cannot reside at any level between the valence and conduction band. In an **insulator** the valence band is full, the conduction band is essentially empty, and the forbidden gap between them is large; so there are no free states for the electron to move into, unless given sufficient energy (for example, by collision with a high-energy electron) to jump up to the higher conduction band. Once in the conduction band, the electron is no longer bound to a particular atom, and the electron can move about in the lattice (Fig. 6-3). If the gap between the bands is large, the material is an insulator, since few

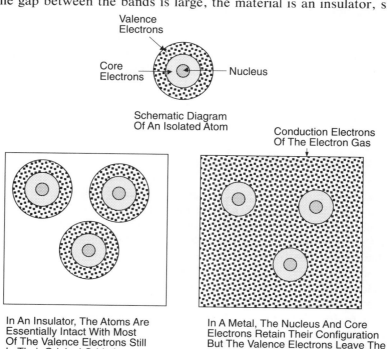

Figure 6-3 Distribution of orbital electrons in an insulator and a metal. In an insulator the atoms are essentially intact, with most of the valence electrons still in their orbitals. In a metal the nucleus and core electrons retain their configuration, but the valence electrons leave the atom to form the electron gas.

TABLE 6-1 NUMBER OF ELECTRONS
IN THE ELECTRON GAS OF SOME METALS

Li	$4.70 \times 10^{22}/cm^3$	Fe	$17.0 \times 10^{22}/cm^3$
Na	2.65	Mn	16.5
K	1.40	Zn	13.2
Cu	8.47	Al	18.1
Ag	5.86	Ga	15.4
Au	5.90	In	11.5
Mg	8.61	Tl	10.5
Ca	4.61	Pb	13.2
Ba	3.15		

electrons have sufficient energy to reach the conduction band. In a good **conductor**, however, there is no gap, the two bands can overlap, or there is one band, and the electrons are free to move and carry an electrical current.

In a **metal** (an element that easily loses electrons) **conductor**, the core electrons remain bound to the atomic nucleus when isolated atoms condense into a solid, but the valence electrons move into the conduction band and are free to wander far away from the parent atom (Fig. 6-3). In the metallic context, they are called conduction electrons and make up the **electron gas** between the metal atoms. The number of electrons in the electron gas is not directly related to atomic number. The number of electrons in the electron gas varies from $0.91 \times 10^{22}/cm^3$ for cesium up to $24.7 \times 10^{22}/cm^3$ for beryllium (Table 6-1). These densities are typically a thousand times greater than those of a classical gas at normal temperature and pressure.

ELASTIC VERSUS INELASTIC SCATTERING

The column of the scanning electron microscope directs the electron beam onto the specimen in a series of raster lines. The size of the nuclei and electrons of the specimen atoms is on the order of 10^{-6} to 10^{-7} nm, while the distances between atoms is around 0.1 nm. Because of these distances, it seems highly improbable that a beam electron would make a direct hit on a nucleus or electron of a specimen atom. However, the nuclei and electrons have large coulomb (electric charge) fields associated with them, and it is this coulomb field that usually interacts with the beam electron. The collision of the electron beam (usually with an energy of 1 to 30 keV) with the specimen results in a large variety of specimen–beam interactions. These interactions can be divided into two general groups, elastic and inelastic. **Elastic events** result when a beam electron interacts with the electrical field of the *nucleus* of a specimen atom, resulting in a change in the direction of the beam electron *without a significant change in the energy of the beam electron* (Fig. 6-4). If the elastically scattered beam electron is deflected back out of the specimen, the electron is called a backscattered electron. *A* **backscattered**

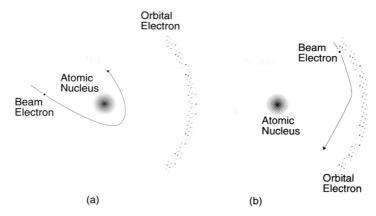

Figure 6-4 (a) A backscattered electron is formed when an electron from the electron beam interacts with the atomic nucleus of a specimen atom. (b) A secondary electron is formed when an electron from the electron beam interacts with an orbital electron of a specimen atom.

electron *is defined as an electron escaping the surface of the specimen with an energy of more than 50 eV, while a* **secondary electron** *is defined as an electron escaping the specimen surface with an energy of less than 50 eV.* **Inelastic events** result when a beam electron interacts with the *electric field of electrons* (primarily conduction band electrons) of specimen atoms, resulting in a transfer of energy to the specimen atom. The interaction of the beam electron and the conduction band electron results in the transfer of a few electron volts of energy to the conduction band electrons. The increased energy of the conduction-band electrons can result in the expulsion of an electron from the atom as a secondary or Auger electron.

Elastic events result in little or no transfer of energy because the nucleus is of far greater mass than the beam electron. The effect is similar to dropping a small rubber ball on a large steel ball. The rubber ball rebounds at an angle, but to almost the same height that it was dropped. The large steel ball remains essentially motionless. *Inelastic events* result in energy transfer because the beam electron and the orbital electron of the specimen atom are of about the same mass. The effect is similar to a fast-moving rubber ball (the beam electron) colliding with a slow-moving rubber ball (the orbital electron in the specimen atom). The fast-moving rubber ball slows on collision and imparts an increased velocity to the slow-moving rubber ball, and they both change direction.

YIELD OF SECONDARY AND BACKSCATTERED ELECTRONS IN METALS AND INSULATORS

The yield of secondary and backscattered electrons is lower in conducting metals and higher in insulators. The reason insulators produce more backscattered and secondary electrons than conducting metals is as follows. In conducting metals the generated electrons can lose energy to electrons in the conduction band, result-ing in the production of fewer electrons emitted from the specimen surface. Also, the electrons travel relatively short distances (a couple of nanometers) in conduct-ing metals before undergoing collisions because of the large numbers of electrons in the conduction band. This makes it more difficult for generated electrons to reach the specimen surface and escape as secondary or backscattered electrons. This results in a yield of around unity at the accelerating voltage of the beam electrons, which produces maximum emission of secondary and backscattered electrons. Specimens that are insulators have a large forbidden zone between the valence and conduction bands. Once the energy of the backscattered or secondary electrons is less than the forbidden gap, the electrons cannot lose energy to other electrons and can only give up energy in relatively infrequent interactions with the lattice to form phonons. Also, the distance an electron travels between collisions in an insulator is greater, making it easier for an electron to travel to the surface of the specimen and escape. The combination of greater production of secondary electrons along with the greater distance between collisions results in more elec-trons escaping from the surface of the specimen.

INFLUENCE OF ATOMIC NUMBER OF THE SPECIMEN ON ELASTIC AND INELASTIC EVENTS

In a specimen of **low atomic number**, such as carbon, the beam electron undergoes *three times as many inelastic collisions as elastic collisions* as the beam electron penetrates into the specimen. The contrary is true for a **high atomic number** speci-men, such as gold, where the beam electron sustains *four times as many elastic collisions as inelastic*. This point is important in regard to the **depth of penetration** of the beam electrons into the specimen. The angle at which the beam is deflected during inelastic scattering is very small (less than 0.001 to 0.002 rad) compared to the much greater deflection of the beam that occurs during elastic scattering. Small deflections result from inelastic collisions caused by a high-energy electron interacting with one or more electrons in a single atom. Due to the abundance of orbital electrons, a number of electron–electron encounters occur at random orientations and tend to cancel, resulting in minimal deflection of the incident high-energy electron. However, larger deflections of the incident electron result from elastic events, since an encounter between an incident electron with a nucleus of a specimen atom is single and unbalanced, resulting in a larger deflection of the incident electron. This is the reason that beam electrons penetrate deeper into a specimen composed of low atomic number elements. In a specimen composed of higher atomic number elements, there is more elastic scattering of the beam elec-trons over greater angles of deflection, resulting in a shallower penetration of the beam electrons into the specimen.

DIMENSIONS OF THE SPECIMEN INTERACTION VOLUME

The shape of the specimen–beam interaction volume depends on (1) the energy of the beam electrons and (2) the atomic number of the specimen.

Effect of the Energy of the Electron Beam. The depth of penetration of the beam electrons into a particular specimen is determined by the energy of the beam electron; the higher the energy of the beam electron is, the greater the penetration (Fig. 6-5). The number of inelastic events increases as the energy of the beam electrons increases, resulting in less deflection of the electron beam on scattering and deeper penetration of the electrons into the solid. At higher energies, the electrons can penetrate to greater depths since they retain a larger fraction of their initial energy after traveling a fixed distance. The *shape* of the specimen–beam interaction volume, however, does not change significantly with a change in beam energy. The electrons travel equally farther in the lateral and depth directions with increases in beam energy.

Effect of Atomic Number of the Specimen. In *low atomic number specimens* (where the beam electrons undergo three times as many inelastic as elastic collisions), the deflection of the beam electrons is small and a pear-shaped (light-bulb–shaped) specimen–beam interaction volume results (Fig. 6-6). When the high-energy beam electrons penetrate into the specimen, most of the interactions are inelastic; the beam electrons penetrate with little deflection, resulting in the formation of the neck of the pear-shaped interaction volume. Once the penetrating electrons lose energy, elastic scattering becomes more probable at lower energies, greater deflection occurs with the increased elastic scattering, and the electrons deviate considerably from their original direction. The lateral scattering results in the formation of the lower bulbous region of the pear-shaped interaction volume. At 30 kV, the electron beam will typically penetrate 10 μm into a low atomic number specimen such carbon.

In *high atomic number specimens*, a greater number of elastic collisions occur when the electron beam initially strikes the specimen, resulting in greater

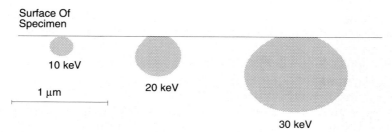

Figure 6-5 Shape and size of the specimen–beam interaction volume in a specimen of iron with an electron beam at 10-, 20-, and 30-kV accelerating voltages.

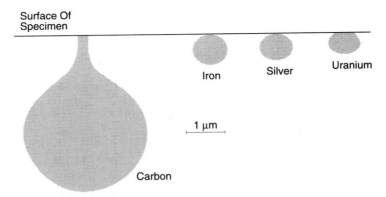

Figure 6-6 Shape and size of the specimen–beam interaction volume in specimens of different atomic number. Accelerating voltage = 20 keV.

deflection of the electrons in the specimen. The specimen–beam interaction zone is without the neck characteristic of low atomic number specimens (Fig. 6-6). Also, the beam electrons do not penetrate to as great a depth as they do in specimens of low atomic number, because the mean free path between collisions is shorter in higher atomic number specimens and there are more elastic collisions. The volume of a solid occupied by a single atom does not vary greatly with atomic number. The shorter mean free path is caused by more protons, neutrons, and electrons packed into a volume of a solid, resulting in a greater chance that the high-velocity electrons will collide with a subatomic particle. At 30 kV the mean free path between elastic collisions in aluminum is 52.8 nm, for copper it is 13.1, and for gold it is 5 nm. A 30-kV electron beam will typically penetrate only 0.5 μm into a high atomic number specimen such as uranium.

DIRECTION OF ELECTRONS ESCAPING FROM A SPECIMEN

The maximum number of electrons generated by elastic and inelastic events escapes from the surface of the specimen in a direction parallel to the electron beam (Fig. 6-7). Progressively fewer electrons escape from the specimen surface in directions from 0° to 90° of the electron beam orientation. The minimum number of electrons escape from the specimen surface in a direction slightly less than 90° from the direction of the electron beam.

Tilting the specimen (Figs. 6-8 and 6-9) results in a symmetric specimen–beam interaction volume that is shallower in respect to the specimen surface, but that is elongated in direction of the optic axis. This shape results when the beam electrons are scattered in a forward direction, nearer to the surface. Since the detector in a scanning electron microscope is fixed at about 60° to 80° to the specimen, tilting the specimen toward the detector results in increased collection of backscattered electrons.

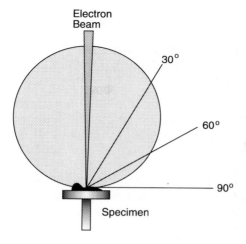

Figure 6-7 Probability of escape of backscattered electrons from the surface of a specimen as a function of the path angle of the backscattered electron. The distance that the angular line travels through the stippled area represents the relative numbers of backscattered electrons escaping the surface of the specimen in that direction. The maximum number of backscattered electrons escape the specimen surface in a direction parallel to the electron beam.

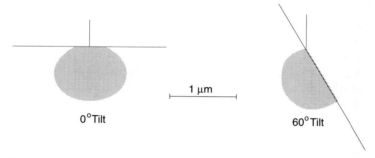

Figure 6-8 Specimen–beam interaction volume in iron at 0° and 60° tilt of the specimen.

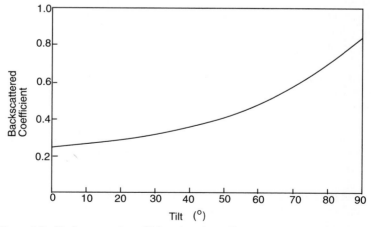

Figure 6-9 Backscattered coefficient (number of backscattered electrons escaping the specimen surface per beam electron) as a function of tilt in a specimen composed of iron.

127

ELASTIC SCATTERING PRODUCES BACKSCATTERED ELECTRONS

Elastic scattering of a beam electron occurs when the beam electron interacts with the electrical (coulomb) field of the nucleus of a specimen atom (Fig. 6-4). The direction of the beam electron is changed, but its velocity and energy are virtually unchanged, less than 1 eV of energy is transferred from the beam electron to the specimen. The amount of energy transferred is negligible when compared to the energy of the beam electron. The angle at which the beam electron deviates from its original path on interaction with the nucleus of the specimen atom can vary from 0° up to 180°, with normal values around 5°. If the deflection in the first collision of the beam electron with the nucleus of the specimen atom is 90° or more, the beam electron will leave the specimen as a backscattered electron. If the deflection is less than 90°, the beam electron continues into the specimen, making more elastic and/or inelastic collisions until the energy of the electron is spent or until the combined deflection angles of the multiple collisions result in the electron being ejected from the specimen as a backscattered electron. An electron beam striking a copper specimen has about 70% of the electrons expending all their energy within the specimen, while the remaining 30% of the electrons is scattered out of the specimen as backscattered electrons. A beam electron that has penetrated 100 nm into a carbon specimen, or 20 nm into a gold specimen, will have a 36.8% chance of being elastically scattered once, 18.4% chance of being elastically scattered twice, 8.0% chance of undergoing multiple elastic collisions, and a 36.8% chance of not being elastically scattered at all.

By definition, an electron leaving the specimen with an energy of more than 50 eV is a backscattered electron, while an electron leaving the specimen with an energy of 50 eV or less is a secondary electron. In actual fact, few electrons leave the specimen with an energy around 50 eV. Most of the backscattered electrons have an energy slightly less than the beam electrons, while most secondary electrons have energies between 0 and 30 eV, with an average value around 2 to 5 eV (Fig. 6-10). The backscattered coefficient is defined as the number of backscattered electrons escaping the specimen per incident electron. With the exception of an infrequent high-energy secondary electron (more than 50 eV) formed by inelastic electron–electron collision, backscattered electrons originate from beam electrons undergoing elastic collisions.

Effect of Atomic Number of the Specimen on the Backscattered Coefficient.   The backscattered coefficient shows a gradual increase with increase in the atomic number of the specimen (Fig. 6-11).

Effect of Beam Energy on the Backscattered Coefficient.   There is a small variation in the backscattered coefficient as the energy of the beam electrons

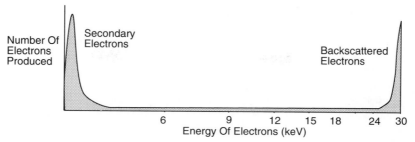

Figure 6-10 Energy distribution of the secondary and backscattered electrons produced when an electron beam with an accelerating voltage of 30 keV reacts with a specimen. Most of the electrons that are produced have an energy of a few electron volts (secondary electrons) or an energy just beneath the accelerating voltage of the electron beam (backscattered electrons).

(accelerating voltage) increases (Fig. 6-12). For low atomic number specimens, the backscattered coefficient decreases gradually with increase in the energy of the beam electrons. This trend decreases as the atomic number increases until at silver (Ag, Z = 47), the backscattering coefficient is independent of beam energy. As the atomic number of the specimen increases past silver, the backscatter coefficient increases slowly with increased beam energy.

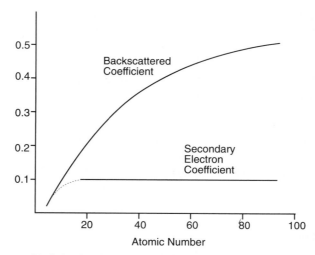

Figure 6-11 Variation in the backscattered coefficient (number of backscattered electrons escaping the specimen surface per beam electron) and the secondary electron coefficient (number of secondary electrons escaping the specimen surface per beam electron) with atomic number using an electron beam with an accelerating voltage of 30 keV.

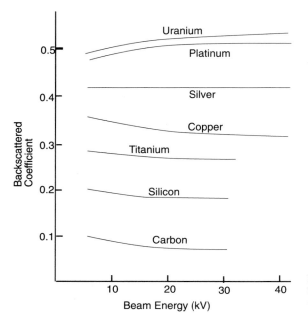

Figure 6-12 Variation of the backscattered coefficient with beam voltage for some elements.

INELASTIC SCATTERING PRODUCES SECONDARY ELECTRONS, AUGER ELECTRONS, X-RAYS, CATHODOLUMINESCENCE, AND HEAT

Inelastic scattering results when a beam electron or a backscattered electron produced from earlier collisions of the beam electron with specimen atoms encounters the coulomb field (electric charge field) of an orbital electron of a specimen atom. Energy is transferred from the beam electron to the specimen atom, and the energy of the beam electron decreases by an amount equal to the transfer of energy to the specimen atom. The energy transfer can occur by a number of possible processes. From the standpoint of scanning electron microscopy, the five processes of energy transfer of interest in inelastic scattering are (1) **secondary electrons**, (2) **X-rays**, (3) **Auger electrons**, (4) **cathodoluminescence**, and (5) **heat**.

Secondary Electrons

Secondary electrons are electrons that are ejected from the valence and conduction bands of the specimen atoms with an energy of less than 50 eV. Some backscattered electrons are included in this energy range, but their numbers are so few that they make up a negligible contribution to the electrons in these energy ranges. If any of these electrons reach the surface of the specimen with an energy greater than the work function of the specimen, the electrons will escape from the surface of the specimen. These low-energy secondary electrons escape from a relatively shallow depth (1 to 10 nm) in the specimen. *Secondary electron emission is, therefore, very sensitive to the topography of the sample and is the type of signal usually used to image the specimen in the scanning electron microscope.* The secondary electrons that are generated deeper than about 10 nm into the specimen do not have sufficient energy to escape from the specimen.

The yield (secondary electron coefficient) of secondary electron emission is defined as the number of secondary electrons emitted per incident beam electron. The first secondary electron is formed within 10^{-13} to 10^{-14} seconds following the primary collision. The initiating electron loses a discrete amount of energy, but still generally possesses sufficient energy to permit additional secondaries to be formed. The last secondary electron is formed within 10^{-11} seconds following the initiation of secondary electron production.

Escape Depth of Secondary Electrons. A secondary electron can escape the bulk of the sample only if it possesses sufficient kinetic energy to surmount the surface energy barrier (work function), usually between 3 and 6 eV. Secondary electrons in aluminum, for example, have an escape depth of 5 to 8 nm. The escape depth of a secondary electron from a specimen depends on (1) how deep the secondary electron is generated within the specimen, and (2) the mean free path of the secondary electron within the specimen. Generally, the maximum depth of emission of secondary electrons is about five times the mean free path of the secondary electron (although this figure will vary somewhat depending on the energy of the secondary electron; the greater the energy is, the deeper the maximum escape depth). The mean free path of secondary electrons in conducting metals is about 1 nm, while it can be up to 10 nm in insulators. This means that the maximum escape depth for conducting metals is about 5 nm, and for insulators it is up to 50 nm. The greater mean free path of secondary electrons in insulators is due to the fact that inelastic collisions of secondary electrons take place mostly with conduction band electrons, which are abundant in metals and scarce in insulators. The escape depth for secondary electrons is about one-one hundredth that of backscattered electrons.

Both Beam Electrons and Backscattered Electrons Generate Secondary Electrons. Secondary electrons escaping the surface of the specimen can be generated by the electrons in the beam as they enter the specimen or by backscattered electrons as they exit the specimen (Fig. 6-13). Generally, about three or four times more secondary electrons are generated (per beam electron) by backscattered electrons as are generated by the beam electron itself. More electrons are generated by backscattered electrons than by the beam electron for two reasons: (1) Incident beam electrons strike the specimen perpendicular to the specimen surface and have the shortest possible route through the escape volume of the secondary electrons. Backscattered electrons, however, usually pass through the secondary electron escape volume at a shallow angle, resulting in a longer path length through the escape volume. The longer path length in the secondary electron escape vol-

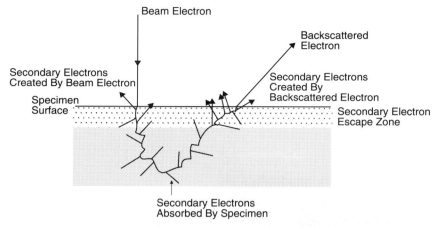

Figure 6-13   Relative number of secondary electrons produced as the beam electron enters the specimen and as the same electron leaves the specimen as a backscattered electron.

ume results in the generation of more secondary electrons within the escape volume. (2) The generation of secondary electrons is more efficient when the incident electron is of lower energy. Thus, the lower-energy backscattered electrons are more efficient at generating secondary electrons than the higher-energy beam electrons.

The total number of backscattered and secondary electrons generated on interaction of the electron beam with the specimen at a particular accelerating voltage depends on (1) the number of secondary electrons generated by *backscattered electrons* interacting with the specimen atoms, and (2) the number of backscattered and secondary electrons generated by interaction of the *electron beam* with the specimen atoms. The number of *secondary electrons generated by backscattered electrons compared to those generated by the electron beam* is about three to four to one, and this ratio is relatively stable in specimens of different atomic number. However, as the atomic number of a specimen increases, there is a *great increase in the number of backscattered electrons produced per beam electron*. Thus, in low atomic number specimens, few backscattered electrons are generated, so most of the secondary electron signal comes from secondary electrons generated by beam electrons. In high atomic number specimens, many more backscattered electrons are produced, so most of the secondary electrons come from interaction of the specimen with backscattered electrons.

Effect of Specimen Atomic Number on Secondary Electron Yield.   The total number of secondary electrons per incident beam electron that escapes from the surface of the specimen (secondary electron coefficient or yield) is relatively insensitive to the atomic number of the specimen (Fig. 6-11). Most elements have about one secondary electron escaping from the surface of the specimen per 10 beam electrons at 20 keV accelerating voltage, although this number can be as high as 2:10 for some metals such as gold. The number of secondary electrons per incident beam electron depends primarily on the characteristics of the specimen that allow the secondary electrons to escape from the specimen (number of outer shell electrons, atomic radius, and density) and not on the total number of electrons that are generated by the incident beam electron. This behavior contrasts with the number of backscattered electrons produced per beam electron, which increases continuously with the atomic number of the specimen.

As is the case with the backscattered electron coefficient, tilting the specimen results in an increase in the yield of secondary electrons from the specimen. The electron beam has a longer path in the secondary electron escape volume in a tilted specimen. This results in more secondary electrons generated in the escape volume and an increased yield of secondary electrons.

Effect of the Energy of the Beam Electrons on Secondary Electron Yield. The emission yield of secondary electrons reaches a maximum when the beam electron has an energy of a couple hundred electron volts (Fig. 6-14). When the beam electrons have less energy than a couple hundred electron volts, the beam electrons do not penetrate to the bottom of the escape depth of secondary electrons in the specimen. The yield of secondary electrons therefore increases with increasing beam electron energy up to a couple hundred electron volts, at which time the beam is penetrating to the maximum escape depth of the secondary electrons. The secondary electron yield peaks at this point at slightly above unity for metals. Above approximately 200 volts, the penetration of the beam electrons is significantly greater than the escape depth of the secondary electrons; and as the beam electron energy increases, the production of secondary electrons near the surface drops off, and fewer electrons escape from the surface of the specimen. At an accelerating voltage of 20 kV, the secondary electron yield for metals drops to about 0.1. It would therefore seem that it would be advantageous to examine the specimen at accelerating voltage of approximately 200 volts. Unfortunately,

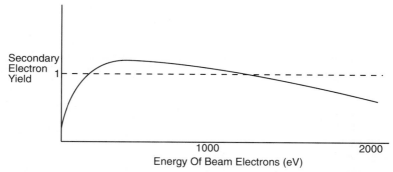

Figure 6-14 Relationship between the yield of secondary electrons and the energy of the electron beam (accelerating voltage).

because of the characteristics of electron emission in a scanning electron microscope, an electron beam of approximately 200 volts has very few electrons in it,
resulting in a weak signal from the specimen. To be able to generate a sufficient
signal to examine the specimen, it is usually necessary to boost the accelerating
voltage up to at least 5 keV in order to have enough electrons in the beam to
generate sufficient signal to obtain a good image.

X-Rays

Before considering the generation of X-rays, Auger electrons, and cathodoluminescence, the basic structure of the atom will be reviewed.

   The Bohr model of the atom has the nucleus at the center with the electrons
occupying orbits (shells) around the nucleus (Fig. 6-15). Bohr's electron shells are
designated as K, L, M, N, or O, starting with the shell closest to the nucleus.
Electrons in the K shell contain the least amount of energy (Fig. 6-15). The electrons must acquire increasing amounts of energy to occupy the shells farther from
the nucleus. Thus the electrons in the K shell have the least amount of energy,
are held most tightly by the nucleus, and require the most amount of energy to be
ejected from the atom.

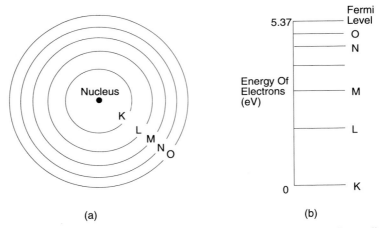

(a)   (b)

Figure 6-15   (a) Bohr model of the atom. Each shell above the K shell actually
has several energy levels (subshells) that are not shown. (b) Energy levels in the
lithium atom.

TABLE 6-3   ENERGY LEVELS OF ORBITAL ELECTRONS

Principal energy level	1	2		→ 3		
Letter designation	K	L		→ M		
X-ray nomenclature	K	L_I	L_{II}	L_{III}	M_I	M_{II}	M_{III}
Number of electrons in subshell	2	2	2	4	2	2	4

TABLE 6-2 CAPACITIES OF THE ELECTRON SHELLS

Principal energy level	1	2	3	4	5
Letter designation	K	L	M	N	O
Capacity ($2n^2$)	2	8	18	32	50

According to the Bohr model, a stable atom contains some minimum amount of energy and is in the ground state. A **ground state** atom has the electrons filling the shells closest to the nucleus, with the electrons remaining indefinitely in these shells. A stable atom can absorb energy and move an electron to a higher shell. When this happens, the amount of energy that is absorbed is equal to the energy difference between the two shells. After the absorption of energy and elevation of one or more electrons to higher levels, the atom is in an **excited state**. An atom in an excited state has a vacancy in a lower shell and is unstable; it achieves stability when its electrons return to their original shells and emit energy as photons of light or X-rays. The energy of the emitted light or X-ray photons is approximately the same as the energy difference between the shell energy of the electron in the excited state and the shell energy of the electron in the ground state.

The energy-level diagram of the lithium atom is presented in Fig. 6-15. The ground state atom has two electrons in the K shell and one electron in the L shell. If a ground state atom absorbs energy, the electrons absorb energy and can reside at higher shells in the excited state atom. If the electrons acquire over 5.37 V of energy, they can escape the lithium atom and are added to the continuum electrons of the matter.

Like the Bohr model, modern atomic theory considers an atom to have a number of shells where the probability of finding an electron is high. Each shell is known as a **principal energy level** and is designated by a positive whole number, 1, 2, 3, 4, or 5, corresponding to the letter designations K, L, M, N, and O of the Bohr model. The higher the principal energy level is, the higher the energy of the electron in the shell, the farther the shell is from the nucleus, and the less tightly the electron is held by the binding force of the nucleus. Each shell can accommodate only a certain number of electrons; this number is equal to $2n^2$, where n is the principal energy level (Table 6-2). Thus, when $n = 1$, there are two electrons, when $n = 2$, there are eight electrons, and so on. There are subshells also within each shell. The number of electrons in each subshell plus the X-ray nomenclature of the subshells is given in Table 6-3.

TABLE 6-3 (continued)

		→ 4						
		→ N						
M_{IV}	M_V	N_I	N_{II}	N_{III}	N_{IV}	N_V	N_{VI}	N_{VII}
4	6	2	2	4	4	6	6	8

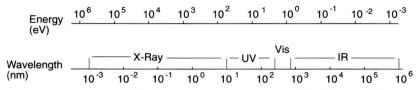

Figure 6-16 Portion of the electromagnetic spectrum containing X-rays and ultraviolet, visible, and infrared radiation.

X-ray generation.   X-rays are electromagnetic radiation between 10^{-3} and 10-nm wavelength (Fig. 6-16). Two basic types of X-rays are produced on inelastic interaction of the electron beam with the specimen atoms in the scanning electron microscope. (1) **Characteristic X-rays** result when the primary electron ejects *electrons* from the inner shells of the specimen atoms. (2) **Continuum X-rays** result when the primary electron interacts with the *nucleus* of the specimen atoms.

Characteristic X-rays.   A vacancy can be created in an inner shell of a specimen atom when a high-energy beam electron, backscattered electron, or high-energy secondary electron strikes an electron of an inner shell, knocking the electron out of the atom (Fig. 6-17). The specimen atom has a vacancy in the inner

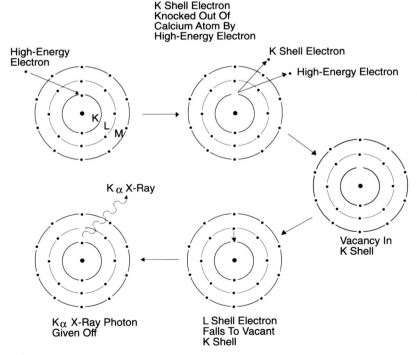

Figure 6-17 Process that results in the generation of a K_α X-ray in a calcium atom.

shell and is ionized and in an unstable state. An electron from a higher shell falls to the vacant inner shell. The specimen now has an amount of excess energy equal to the difference between the two shells. This energy is commonly given off as an X-ray with an energy equal to or slightly less than the energy difference between the two shell electrons in the specimen atom. For example (Fig. 6-17), an atom of calcium (atomic number = 20) has two electrons in the K orbital, eight in the L orbital, and ten in the M orbital. A high-energy beam electron colliding with a K shell electron in the calcium atom will eject the K shell electron from the calcium atom. The vacancy that is created in the K shell can be filled by an electron from the L shell falling to the vacant K shell. A Ca K_α X-ray, with an energy of 3.691 eV (K_{α_1}) or 3.689 eV (K_{α_2}) will be emitted from the specimen atom. The energy of the X-ray is determined by which of the two K shell electrons is knocked out of the specimen atom by the high-energy electron. Differences in electron spin cause slight differences in energy between electrons in the same shell, resulting in two characteristic X-ray energies for K series X-rays. There are two distinct α energies (α_1 and α_2), up to four β energies (β_1, β_2, β_3, and β_4), and many γ energies. If the energy difference caused by the difference in spin within a shell is very small (less than 0.01%), then the subscript denoting the electron spin is often deleted. If the difference is as great as 1% or 2% and the energies are averaged, the average energy is written as $K_{\alpha_{av}}$.

If the atomic number of the specimen atom is high enough so that there are electrons in the M shell, there is also the possibility that an electron could drop from the M shell to the vacant K shell (Fig. 6-18). This would result in the production of a Ca K_β X-ray with an energy of 4.012 eV.

Thus, when an electron falls one shell into a vacant shell, an α (alpha) X-ray is produced; when the electron falls two shells, a β (beta) X-ray is produced; and when the electron falls three shells (if the specimen atom is of a high enough atomic number), a γ (gamma) X-ray is produced. The probability of an electron falling one shell is greater than the probability of falling two shells. In low atomic number atoms the height of the K_α X-ray peaks is about five times that of the K_β X-ray peaks (Fig. 6-19), since the probability of the electrons falling from the M

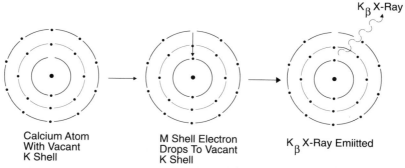

Calcium Atom
With Vacant
K Shell

M Shell Electron
Drops To Vacant
K Shell

K_β X-Ray Emiitted

Figure 6-18 Process that results in the generation of a K_β X-ray in a calcium atom.

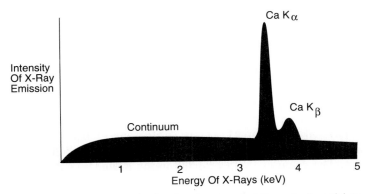

Figure 6-19 X-ray spectrum showing the characteristic peaks for calcium.

shell to the vacant K shell is more remote than the possibility of an electron falling from a L shell to a vacant K shell.

Electrons in shells higher than the K shell can also be ejected from the specimen atoms, resulting in the production of characteristic X-rays. For example, in a specimen atom composed of calcium, an L shell electron can be ejected by a high-energy electron (Fig. 6-20). The vacancy in the L shell can be filled by an

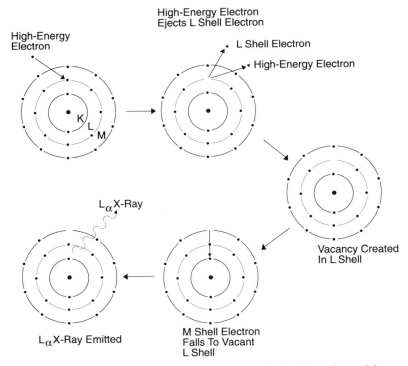

Figure 6-20 Process that results in the generation of a L_α X-ray in a calcium atom.

138

TABLE 6-4 THE TYPES OF X-RAYS THAT ARE PRODUCED ON EJECTION OF AN ELECTRON FROM AN ATOM

		Orbital from which electron is ejected								
		K	L_I	L_{II}	L_{III}	M_I	M_{II}	M_{III}	M_{IV}	M_V
	L_I									
	L_{II}	(K_{α_1})								
	L_{III}	(K_{α_2})								
	M_I			L_n	L_I					
	M_{II}	K_{β_3}	L_{β_4}							
Orbital	M_{III}	(K_{β_1})	L_{β_3}							
That	M_{IV}	$K_{\beta_{10}}$		L_{β_1}	(L_{α_2})					
Provides	M_V	K_{β_9}			(L_{α_1})					
Electron	N_I			L_{γ_5}	L_{β_6}					
	N_{II}	K_{β_2}	L_{γ_2}							
	N_{III}	K_{β_2}								
	N_{IV}			L_{γ_1}	$L_{\beta_{15}}$			$M\gamma_2$		
	N_V				L_{β_2}			M_{γ_1}		
	N_{VI}			L_v					M_{β_1}	(M_{α_2})
	N_{VII}			L_v						(M_{α_1})

electron dropping from an M shell. A Ca L_α X-ray with an energy of 0.341 eV can be emitted from the calcium atom with the energy approximately equal to the difference in energy between the L and M shell of calcium.

Thus the nomenclature of X-rays has the shell from which the electron was emitted first (K, L, or M), followed by the number of shells that the second electron falls to fill the vacant orbital (α, β, or γ). The types of X-rays are presented in Table 6-4 along with the electron orbital transitions that produce the X-rays. X-rays that commonly occur are circled in this table.

Certain rules apply to the energy of X-rays:

1. γ X-rays are more energetic than β X-rays, which are always more energetic than α X-rays (for example, Ca K_α X-rays = 3.690 eV, while Ca K_β X-rays = 4.012 eV).

2. K series X-rays have more energy than L series (for example, Ca K_α X-rays = 3.690 eV, while Ca L_α X-rays = 0.341 eV), and L series X-rays have more energy than M series X-rays.

Measuring either the energy or wavelength of the X-rays enables the investigator to determine the elements from which the X-rays were emitted, since no two elements have exactly the same energy difference between electron shells. The X-rays generated in this manner are called **characteristic X-rays**, since they can be used to characterize the elements from which they were emitted.

Normally, in the electron microscope the energies of the X-rays are measured by energy dispersive spectroscopy, with the energies of the X-rays displayed along a horizontal axis and the intensity of the X-ray production on the vertical axis (Fig. 6-19). It is also possible to measure the wavelength of the X-rays in wavelength dispersive spectroscopy, since the energy of the X-ray photons is related to the wavelength according the Planck's law:

$$E = \frac{hc}{\lambda}$$

where

E = energy of the X-ray photon

h = Planck's constant = 6.6262×10^{-34} J $\times$ s

λ = wavelength of the X-ray photon

c = speed of light = 3.0×10^8 m/s

Continuum X-rays.   The X-ray continuum represents the background on which the characteristic X-ray peaks are imposed (Fig. 6-19). The X-ray continuum results when the incident beam electrons or backscattered electrons interact with the coulomb (electrical) field of the *nucleus* of the specimen atom. On interaction with the nucleus, the primary electron loses energy that can be given off as continuum X-rays. Since the primary electron can give up any amount of energy in the interaction, the energy distribution is continuous and is not characteristic of the atomic number of the specimen atoms from which it was produced. The energy of the continuum X-rays reflects only how close the primary electron comes to hitting the nucleus of the specimen atom (Fig. 6-21). The closer the primary electron comes to the actual nucleus of the specimen atom, the stronger is the interaction between the primary electron and the coulomb field of the nucleus, the more energy is lost by the primary electron, and the more energetic is the X-ray photon that is emitted.

The X-ray continuum is greatest at the lower X-ray energies and tails off toward the energy of the accelerating voltage of the beam electrons (Figs. 6-19 and 6-22). The lower X-ray energies result when the primary electron makes a wide miss of the atomic nucleus through the coulomb field, losing a small amount of energy and generating a low-energy X-ray (Fig. 6-21). The highest-energy X-rays result when a beam electron (at the accelerating voltage, commonly 20 keV when X-ray microanalysis is being performed) makes a direct hit on the nucleus of a specimen atom. This results in the primary electron losing all its energy, with the emission of an X-ray photon with an energy approximately equal to that of the beam electron. There is the highest possibility of wide misses of the atomic nucleus

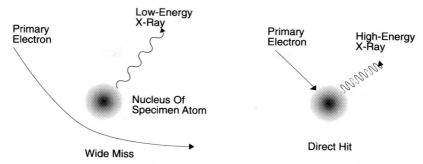

Figure 6-21 A low-energy continuum X-ray results when a primary electron (beam electron, backscattered electron, or high-energy secondary electron) makes a wide miss of the nucleus of the specimen atom and loses a relatively small amount of energy on interacting with the coulomb field of the nucleus. A high-energy X-ray results when the primary electron makes a direct hit on the nucleus of the specimen atom.

by the primary electrons, with decreasing probability as the primary electron approaches a direct hit of the nucleus of the specimen atom. This results in the characteristic X-ray continuum with more X-rays at the lower energies, with the number of X-rays tailing off toward the accelerating voltage of the beam electrons (Figs. 6-19 and 6-22). Most acquired X-ray spectra show a dropping off of the X-ray continuum at the lowest energies (Fig. 6-22). This is an artifact of the detection process. Most X-ray detectors used on electron microscopes have the sensing element of the detector separated from the specimen by the window made of a thin polymer or beryllium. The material that makes up the window absorbs the lowest-energy X-rays, resulting in their absence in the spectrum.

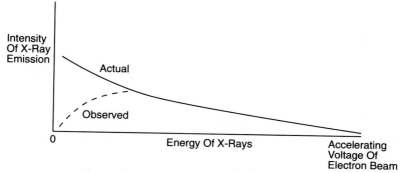

Figure 6-22 Graph of the intensity of the continuum radiation as a function of X-ray energy. The fall-off in the observed X-rays at low energies is due to the absorption of the low-energy X-rays by the window between the detector and the specimen.

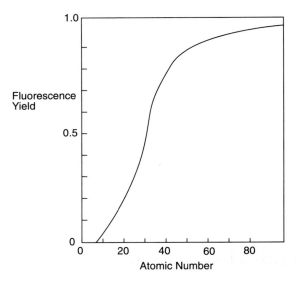

Figure 6-23 Relationship between the fluorescence yield and the atomic number of the specimen.

Fluorescence Yield.   The fluorescence yield (X-ray yield) is the fraction of ionizations of specimen atoms that actually results in the emission of an X-ray photon from the atom or, in different terms, the probability that a vacancy in an inner shell will produce an X-ray. When an electron transition fills a vacancy in an inner electron shell, there is a certain probability that the emitted X-ray photon will be absorbed within the atom. The absorbed energy of the X-ray photon results in the ejection of an electron from one of the outer shells as an Auger electron. The higher the atomic number of the specimen atoms, the greater the chance the X-ray photon will escape from the atom and the greater the fluorescence yield (Fig. 6-23). Conversely, the lower the atomic number of the specimen atoms is, the lower the fluorescence yield and the greater the Auger electron yield.

The preceding is a brief summary of X-ray production in the electron microscope. X-ray instrumentation is covered in Chapter 12.

Auger Electrons

Auger electrons were discovered by Auger in 1925 while examining the tracks produced by X-rays in a cloud chamber. The initial event in the formation of Auger electrons is the same as in the formation of X-rays. A primary electron (beam electron, backscattered electron, or high-energy secondary electron) strikes an inner shell electron in a specimen atom, ejecting the shell electron and creating a vacancy in the shell (Fig. 6-24). An electron from an outer shell of the ionized atom drops to the vacant electron shell. The atom undergoes de-excitation by emitting a characteristic X-ray photon with an energy approximately equal to the

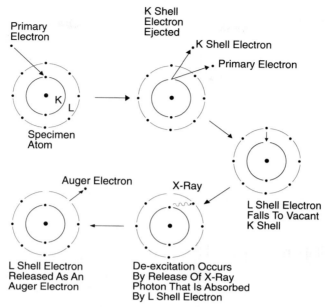

Figure 6-24   Process that results in the ejection of a KLL Auger electron from an oxygen atom.

energy difference between the two electron shells of the specimen atom. Up to this point, the sequence is the same as for the generation of characteristic X-rays. However, instead of the X-ray photon escaping the atom as a characteristic X-ray, the X-ray photon is absorbed by an electron in one of the outer shells of the specimen atom. Absorption of an X-ray photon is an all or nothing phenomenon, so all the energy of the X-ray photon is transferred to the outer orbital electron (this is different from the interaction of high-energy electrons with matter, where a portion of the energy of the high-energy electron can be transferred to the matter, with the high-energy electron traveling on to make other interactions). After absorption of the X-ray photon, the outer shell electron of the specimen atom is ejected from the specimen atom as an Auger electron, leaving a double-ionized specimen atom (with two vacancies in the shell[s]). The energy of the Auger electron is equal to the energy of the X-ray minus the binding energy of the ejected electron (ionization energy of the electron orbital of the ejected electron). Figure 6-25 shows a spectrum of an oxygen KLL Auger electron peak of 510 eV that resulted when an oxygen K_α X-ray of 531 eV was absorbed by an L orbital electron bound in the atom with an energy of 21 eV. The L orbital electron was ejected as an Auger electron with an energy of 510 eV. The energy of the Auger electron is characteristic of the atomic number of the specimen atom from which the Auger

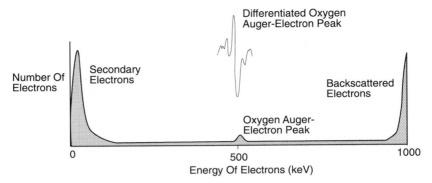

Figure 6-25   Emitted electron spectrum showing a small oxygen Auger electron peak at 510 eV. Also included is the same oxygen peak in the differentiated mode. Accelerating voltage of the electron beam is 1000 eV.

electron was ejected. Measurement of the energy of Auger electron peaks can characterize the type of elements in a specimen or, more specifically, the elements present in the specimen to a depth of a couple of nanometers.

Auger electrons are identified by a three-letter code. The Auger electron from Fig. 6-24 would have the code KLL, identifying the shell of the initial ionization (K) and the shell(s) containing the vacancies (L and L).

Auger electrons fall in the range from 0 to 2500 eV and appear as a small peak in an emitted electron spectrum (Fig. 6-25). Auger electron peaks are relatively small compared to the relatively large continuum of backscattered electrons and Auger electrons that have lost energy by interaction with specimen atoms before escaping from the surface of the specimen. The Auger electron peaks are frequently presented in the differentiated mode to enhance the size of the peaks against the background (Fig. 6-25).

Auger electrons with energies characteristic of the elements from which they arose have to escape the surface of the specimen without undergoing additional interaction with the specimen atoms that would reduce their energy and add to the continuum. Auger electrons in the energy range from 50 to 2500 eV have inelastic mean free paths between interactions with specimen atoms in the range of 0.1 to 2.0 nm. This means that only Auger electrons escaping from a depth of 0.1 to 2.0 nm (approximately five to ten atomic layers) will not have undergone additional inelastic interactions with specimen atoms after the Auger electrons have been generated.

Auger electron microanalysis and spectroscopy must be performed with very clean specimens under very high vacuum conditions, typically 10^{-8} mbar at the specimen. Poorer vacuum conditions can result in the deposition of a few monolayers of contaminants on the surface of the specimen. These contaminants absorb

Auger electrons and reduce the size of the already small Auger electron peaks. Few scanning electron microscopes operate at high enough vacuum conditions at the specimen, so Auger electron spectroscopy is not common in scanning electron microscopes. Auxiliary techniques, such as ion sputtering or cleavage of the specimen under high-vacuum conditions in the specimen chamber, can produce clean surfaces for Auger electron microanalysis.

Auger electron spectroscopy is most effective for low atomic number specimens. This is because the number of Auger electrons per incident beam electron increases as the atomic number of the specimen decreases. The number of Auger electrons produced is inversely proportional to the fluorescence yield (X-ray yield) from the specimens (Fig. 6-23), which increases as the atomic number of the specimen increases. The Auger electron yield can be expressed as

yield of Auger electrons = 1 − fluorescent yield of X-rays

Thus X-ray microanalysis is more effective for high atomic number specimens, while Auger electron microanalysis is more effective for low atomic number specimens.

Cathodoluminescence and Phonons

Cathodoluminescence is the production of photons of ultraviolet (10- to 400-nm wavelength; 1240- to 30-eV energy), visible (400- to 750-nm wavelength; 30- to 17-eV energy), or infrared (750-nm to 1000-μm wavelength; 17 to 1.24×10^{-2} eV energy) radiation on interaction of the specimen with the electron beam. The most common example of cathodoluminescence is encountered everyday in television sets and cathode ray tubes where the back of the viewing screen is covered with a phosphor that emits visible light photons when the phosphor grains are struck by the electron beam. Cathodoluminescence results when a high-energy electron interacts with the coulomb field of an electron in one of the outer shells of a specimen atom. The outer shell electron can be ejected out of the atom or it can be bumped to a higher shell. The subsequent return of an electron to the vacant shell releases the absorbed energy, commonly as ultraviolet, visible, or infrared light. The energy is given off as cathodoluminescence, instead of X-rays, since the energy difference between the outer shells is considerably less than the energy difference between the inner shells (for example, X-ray production). For example, a silicon atom giving up 1.1 eV of energy as an electron falls to a vacant outer shell would produce an infrared photon with a wavelength (λ) of 1130 nm according to Planck's equation:

$$E = \frac{hc}{\lambda}$$

where E is the energy of the radiation, h is Planck's constant and λ is the wavelength of the radiation. Rearrangement and substitution of values yields

$$\lambda = \frac{12.4}{E}$$

where the energy E is in kilo-electron volts (keV) and wavelength is measured in angstroms (1 Å = 0.1 nm). Therefore, to calculate the wavelength of a photon that is emitted when a silicon atom loses an energy of 1.1 eV,

$$\lambda = \frac{12.4}{1.1 \text{ eV} \times 10^{-3} \text{ eV/keV}}$$

$$= \frac{1.24 \times 10^{1}}{1.1 \times 10^{-3}}$$

$$= 1.13 \times 10^{4} \text{ Å}$$

$$= 11,300 \text{ Å} \quad \text{or} \quad 1130 \text{ nm}$$

Thus, an infrared photon with a wavelength of 1130 nm (1.13 μm) is emitted from the silicon atom.

Cathodoluminescence is common in insulators and semiconductors. These materials have a full valence band with all possible electron states occupied, a band gap of forbidden electron energies, and an empty conduction band (Fig. 6-26). Interaction of the electron beam with such a material results in inelastic events that cause electrons in the full valence band to be lifted, across the forbidden gap of energies, into the conduction band. This creates a positive hole in the valence band and results in electron–hole pairs. If a bias is applied to the specimen (one side of the specimen is made positive), the electron is swept away and is unable to recombine with the positive hole. However, without a bias, the negative electron and positive hole can recombine with the release of the excess energy (equal to the gap energy of the forbidden zone). The excess energy can be given off either in a radiative or nonradiative process. **Radiative** processes result in the release of energy as a photon of ultraviolet, visible, or infrared radiation (cathodoluminescence). Thus, in a silicon specimen, a photon of 1.1 eV and 1130-nm wavelength is emitted from the specimen.

Nonradiative processes result in the energy being absorbed by the specimen as phonons (heat). A **phonon** is a quantum of vibrational energy; it is the acoustical counterpart of the photon. In acoustics, a sound wave is a vibration or mechanical disturbance in an elastic medium such as a gas, liquid, or solid. The disturbance travels as a wave as one part of the material is forced to move, communicating motion to adjacent parts of the medium at a definite speed. The disturbance is a physical change (that is, in density, pressure, or particle displacement) in the medium. A phonon is the smallest unit of energy corresponding to a sound wave.

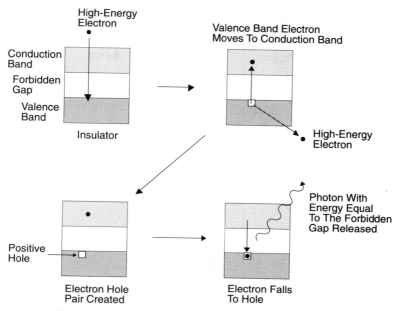

Figure 6-26 Cathodoluminescence production in an insulator.

Thus, though photons and phonons involve waves, there are basic differences between the two. Photons involve electrical and magnetic fields than can travel through space (no medium), while phonons involve a change in a medium. When a phonon is absorbed by a specimen, there is an increase in thermal motion (lattice vibration) of the specimen, which causes an increase in the temperature of the specimen. The atoms and molecules of the specimen are always moving and vibrating due to their internal energy. If the atoms and molecules move and vibrate slowly, the temperature is low and the specimen has a low level of internal energy. The opposite is true when the atoms and molecules move and vibrate violently. Heating of the specimen by phonon absorption can damage the specimen and cause it to drift while it is being examined in the electron microscope. This makes it difficult to view and photograph the specimen, especially at high magnification. Fortunately, phonon excitations are small, in the area of 0.02 eV, resulting in minimal heating of a well-grounded specimen.

ESCAPE DEPTHS OF SIGNALS FROM A SPECIMEN

The absolute depths from which the different types of signals escape from the specimen depend on the atomic number of the specimen and the energy of the electron beam. However, some generalizations can be made (Fig. 6-27). Auger

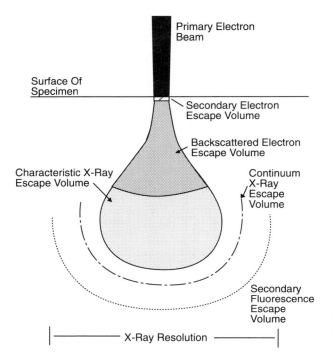

Figure 6-27 Relative escape depths of different types of signals from a specimen of low atomic number.

electrons escape from the shallowest depths. Only those Auger electrons that are generated within 0.1 to 2.0 nm beneath the specimen surface are able to escape without undergoing collision with specimen atoms. Secondary electrons escape from a depth of about 10 nm. In a low atomic number specimen, backscattered electrons are able to escape from a depth of up to 5 μm, while X-rays are able to escape from the whole of the beam penetration volume (about 10 μm for a low atomic number specimen).

7

Detectors

Detectors used in scanning electron microscopes accept radiation from the specimen and convert the radiation to an electrical signal that, after amplification, is used to modulate the gray-level intensities on a cathode ray tube, producing an image of the specimen.

The retina is the radiation detector in the human eye. The retina is only able to convert visible light radiation (from wavelengths of 400 to 770 nm) into electrical signals. These signals travel to the brain where an image is produced. Unlike the eye, scanning electron microscopes can use a large part of the electromagnetic spectrum (infrared, visible, ultraviolet, and X-rays), as well as electrons (Auger, secondary, backscattered, and conducted) to form an image, providing a detector system is available to convert the radiation into an electrical signal.

The types of detectors that are commonly used in scanning electron microscopes are (1) scintillator–photomultiplier systems, (2) solid-state detectors, (3) specimen current detectors, and (4) cathodoluminescence detectors. Solid-state detectors will be covered in Chapter 12. The remainder of the detector systems are discussed here.

SCINTILLATOR–PHOTOMULTIPLIER SYSTEMS (EVERHART–THORNLEY DETECTOR)

Scanning electron microscopes usually have an Everhart–Thornley detector system, which consists of four parts: (1) a Faraday cage, (2) a scintillator disc, (3) a light guide pipe, and (4) a photomultiplier (Fig. 7-1). The Faraday cage is usually biased to $+300$ V to attract the secondary electrons (less than -50-eV energy) emitted from the specimen. The scintillator disc, biased to $+10$ to $+12$ kV, is inside the Faraday cage. The secondary electrons collected by the cage and those backscattered electrons generated toward the scintillator disc are accelerated into the scintillator disc by the $+10$ to $+12$ kV. The high-energy electrons strike the scintillator and generate light photons by cathodoluminescence. The light photons pass down the light guide pipe to the quartz window of the photomultiplier. In the photomultiplier, the light photons generate electrons that are amplified to the order of 10^5 to 10^6, producing a signal that is used to construct an image of the specimen on the cathode ray tube.

Faraday Cage

The Faraday cage around the scintillator can be biased from -50 to $+300$ V and functions to shield the electron beam from the $+10$ to $+12$ kV on the scintillator, which would otherwise draw the electron beam toward the scintillator, causing astigmatism and imaging artifacts. The -50 to $+300$ V on the Faraday cage is such a low bias that the electron beam is minimally affected. The Faraday cage is a wire mesh with large openings that permits the passage of most of the backscattered and secondary electrons to the scintillator. During normal operation of the scanning electron microscope, the Faraday cage is biased to $+300$ V, which results in the collection of most of the secondary electrons (those with an energy less than -50 V). The backscattered electrons are of such high energy (most are near the

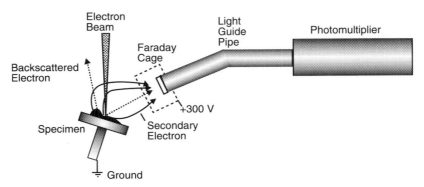

Figure 7-1 Everhart–Thornley detector setup with the Faraday cage biased to $+300$ V to collect secondary electrons ($2° e^-$).

accelerating voltage) that most are not significantly affected by the bias on the Faraday cage, and only those backscattered electrons generated toward the detector contribute to the image.

Scintillator

The scintillators are inorganic and characterized by the presence of heavy elements, which readily produce electron–hole pairs and emit light photons by cathodoluminescence (see the section on cathodoluminescence in Chapter 7). The most commonly used scintillator is P47 or P47P, consisting of Y_2SiO_5 doped with cerium, which is manufactured by Sylvania. This scintillator has a very long life and normally does not have to be replaced if the scanning electron microscope is properly operated. There may be a 10- to 70-nm coating of aluminum over the scintillator that assists in holding the positive potential. Some scintillator discs do not have the aluminum coating since about 2 kV of energy is necessary for an electron to penetrate the aluminum coating, reducing the number of electrons detected by the system by 20%. As the backscattered and secondary electrons from the specimen approach the detector, they are accelerated into the scintillator by the bias potential of +10 to +12 kV. About 2% (200 to 240 eV) of the bias energy of the secondary electrons is converted into light photons. Each light photon has an energy of about 3.1 eV (400-nm wavelength), which results in about 65 light photons emitted for each electron striking the detector. Most of the backscattered electrons have an energy around that of the accelerating voltage of the electron beam (usually 20 to 30 kV) (Fig. 6-10) and will produce light photons without the bias on the scintillator. However, most of the secondary electrons have an energy of only a few electron volts and do not have enough energy to generate light photons without the bias of +10 to +12 kV on the scintillator.

Light Guide Pipe

The scintillator is attached to a light guide pipe, which is a solid plastic or glass rod that has a high internal reflection of light photons. The light photons generated in the scintillator pass through the glass backing of the scintillator, into and through the light guide pipe and into the photomultiplier. The light guide pipe passes at least 40% of the photons generated in the scintillator through to the photomultiplier. Most of the light photons are lost at the junction of the scintillator disc and light guide pipe and at the junction of the light guide pipe and the photomultiplier; few are lost in any bends in the light guide pipe. The scintillator disc and inner portion of the light guide pipe are in the high vacuum of the specimen chamber, while the outside of the light guide pipe and the photomultiplier are at 1 atm, with an O-ring around the light guide pipe sealing the vacuum. A **light guide pipe** is a cylinder composed of clear glass or plastic and consists of two parts: (1) a **core** with a refractive index that is higher than the refractive index of the (2) sheath or **cladding** surrounding the core (Fig. 7-2). Light photons entering from the scintilla-

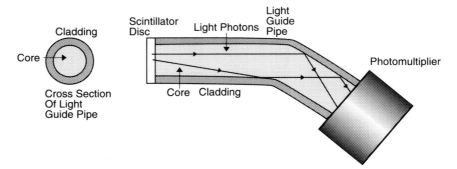

Figure 7-2 Construction of a light guide pipe. The core consists of glass or plastic with a high refractive index, which is surrounded by the cladding of a low refractive index. The scintillator disc generates light photons that pass through the light guide pipe by total internal reflection at the junction of the core and cladding.

tor travel the length of the light guide pipe after being reflected back into the core everytime they reach the core–cladding interface (Fig. 7-2). Total internal reflection occurs at the interface of the high-refractive index core and the low-refractive index cladding when the light is incident at an angle greater than the critical angle of incidence in the high-refractive index core. The light rays enter one end of the cylindrical light guide pipe and strike the cylindrical interface between the core and the cladding at an angle greater than the critical angle, advance along the cylindrical length by a series of internal reflections, and emerge at the other end. Even when the light guide pipe is bent around sharp corners, the light is transmitted only within the core.

Photomultiplier

A photomultiplier is a very sensitive detector of light containing a **photocathode**, a number of secondary-electron-emitting electrodes (**dynodes**), which amplify the electrons emitted from the photocathode, and an **output electrode** or anode that collects the electrons and provides the electrical output signal (Fig. 7-3). The front window of the photomultiplier tube is coated on the inside with a semitransparent photocathode material (usually an alkali antinomide such as Cs_3Sb or Na_2KSb; different photocathode materials absorb different wavelengths of light) that absorbs the light photons coming out of the light guide pipe and converts the light photons into electrons. The efficiency of the process (around 15%) is defined as the number of photoelectrons emitted from the photocathode per incident light photon from the light guide pipe. The photoelectrons produced from the photocathode are discharged into the vacuum of the photomultiplier tube and are accelerated toward the positive potential of the first dynode. Dynodes are plates composed of an alkali metal compound (such as cesium antimony) or a metal oxide layer (such as magnesium oxide on a silver–magnesium alloy). Each successive dynode is

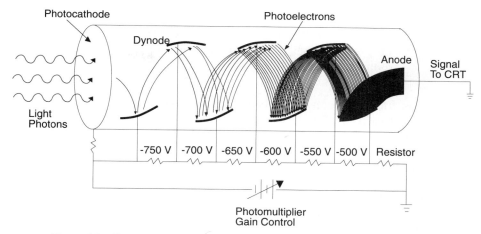

Figure 7-3 Construction of a photomultiplier tube. The light photons from the light guide pipe generate electrons on colliding with the photocathode. The electrons are cascaded through the photomultiplier tube to produce the amplified signal. The amount of amplification is determined by the voltage difference between the dynodes (gain).

kept at a more positive voltage than the previous dynode. The photoelectron from the photocathode strikes the first dynode with an energy equal to the accelerating voltage (potential difference between the photocathode and the first dynode) of the first dynode, liberating a number of electrons from the first dynode. These liberated electrons go on to strike the next dynode, liberating even more electrons, and so on. The voltage difference (**gain**) between the dynodes controls the amount of amplification of the electrons through the tube. The gain over the dynodes can be increased so that 1 million electrons leave the last dynode for each electron produced by the photocathode. However, increasing the gain across the photomultiplier also increases the electronic noise across the photomultiplier, resulting in a snowy image on the cathode ray tube. The noise is due to the continual field emission of a few electrons from the dynodes and from the photocathode, even in total darkness. These electrons are also amplified across the photomultiplier tube, resulting in background noise. However, the total amount of background noise is very low, resulting in an unacceptable signal to noise ratio only if the amplification is very great (above 10^6). A typical photomultiplier is capable of detecting a 1-candlepower lamp at a distance of 6 miles.

The efficiency of the detection process can be calculated as follows:

- One secondary or backscattered electron strikes the scintillator, producing 65 light photons per electron.
- The light guide pipe passes 40% of the light photons to the photomultiplier, resulting in 26 light photons reaching the photocathode of the photomultiplier.

- The photocathode of the photomultiplier is 15% efficient, resulting in the formation of four electrons from 26 light photons.
- The photomultiplier tube produces 10^5 to 10^6 electrons for each electron produced at the first electrode, resulting in the formation of 4×10^5 to 4×10^6 electrons.

Thus, for each secondary or backscattered electron accelerated into the scintillator of the Everhart–Thornley detector, 4×10^5 to 4×10^6 electrons are produced by the photomultiplier tube with a minimal amount of electronic noise.

Modification of the Everhart–Thornley Detector

During normal viewing of the specimen, the Faraday cage of the Everhart–Thornley detector is biased to $+200$ to $+300$ V. In this configuration, 90% to 95% of the electrons striking the scintillator are secondary electrons collected by the positive bias on the Faraday cage. The other 5% to 10% of the electrons striking the detector are backscattered electrons emitted from the specimen toward the detector. About 65% of the secondary electrons received by the detector are generated when the electron beam interacts with the specimen. The remaining 35% of the detected secondary electrons are generated when backscattered electrons strike the inside of the specimen chamber, creating secondary electrons that are collected by the positive bias on the Faraday cage.

There are two possible ways to generate an image from only backscattered electrons using an Everhart–Thornley detector (Fig. 7-4). The first way is to bias the Faraday cage to -50 V so that the secondary electrons (with an energy of less than -50 eV) are repelled by the Faraday cage. Only the higher-energy backscat-

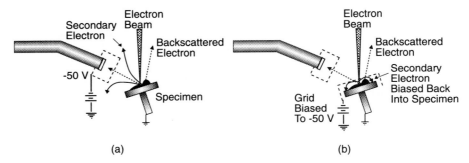

(a)   (b)

Figure 7-4 Two methods of obtaining a backscattered signal using an Everhart–Thornley detector. (a) The Faraday cage is biased to -50 V, resulting in deflection of the secondary electrons from the Faraday cage. Only electrons with more than 50 V (the backscattered electrons) can penetrate the field of the Faraday cage and be detected. (b) A biased grid of -50 V repels the secondary electrons back onto the specimen stub, while the higher-energy backscattered electrons pass through the biased grid, to be picked up by the Everhart–Thornley detector if they are generated in that direction.

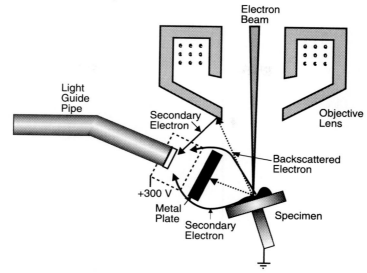

Figure 7-5 Obtaining a signal composed mostly of secondary electrons by placing a metal plate in the line of sight between the specimen and the Everhart–Thornley detector to absorb the backscattered electrons generated in that direction. Backscattered electrons generate secondary electrons by striking portions of the specimen chamber. Therefore, some backscattered electrons are still included in the signal.

tered electrons are then able to penetrate the Faraday cage to strike the scintillator. The second way is to place a biased grid with a potential of -50 V over the specimen, preventing the secondary electrons from escaping and forcing them back onto the specimen.

An image generated mostly from secondary electrons (most of the backscattered electron signal is removed) is obtained by placing a metal plate in the line of sight between the specimen and the detector to absorb the backscattered electrons passing toward the detector (Fig. 7-5). The Faraday cage is biased to $+300$ V to collect the secondary electrons emitted from the specimen, most of which are able to travel around the metal plate to the detector. However, the image is not constructed solely from secondary electrons, since backscattered electrons from the specimen strike the inside of the specimen chamber, creating secondary electrons that are picked up by the detector and contribute to the signal.

TAKE-OFF ANGLE AND SOLID ANGLE OF COLLECTION

The **take-off angle** is the angle that the detector makes with the surface of the specimen stub. Usually, Everhart–Thornley detectors are placed so that the take-off angle is about 30° (Fig. 7-6). The **solid angle of collection** is the three-dimensional

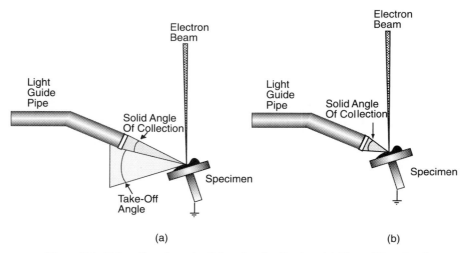

Figure 7-6 Take-off angle and solid angle of collection. (a) The solid angle of collection is small when the detector is far from the specimen. (b) Moving the detector close to the specimen results in a larger solid angle of collection.

angle that the detector face makes with the specimen and is measured in steradians (see Appendix I for the definition of steradian). The term solid angle of collection is valid for a collection of *backscattered electrons* since they are received by the detector more or less in a straight line from the specimen. The term solid angle of collection is not as easy to define for *secondary electrons* since the positive bias on the Faraday cage causes a large deviation in the trajectories of the secondary electrons toward the detector.

The closer the detector is to the specimen, the larger is the solid angle of collection (Fig. 7-6). A detector with a 1-cm diameter scintillator disc positioned 4 cm from the specimen will have a solid angle of collection of backscattered electrons of about 0.05 sr.

MAXIMIZING THE COLLECTION OF BACKSCATTERED ELECTRONS

It is often desirable to obtain an image composed mostly of backscattered electrons, particularly in specimens where atomic number contrast and topographic contrast need to be maximized. An Everhart–Thornley detector has a relatively small solid angle of collection of backscattered electrons, since only those electrons generated toward the detector from the specimen are picked up, resulting in a weak backscattered signal. There are basically three ways by which the backscattered electron signal can be increased: (1) increasing the solid angle of collection by using multiple detectors or using a detector with a large detecting face, (2) using a conversion detector, or (3) reversing the specimen current signal. If a signal composed exclusively of backscattered electrons is desired, then a biased grid at -50 V can be placed over the specimen to keep the secondary electrons on the specimen stub.

Multiple Detectors

The Jackman arrangement of multiple detectors has four detectors mounted on a ring above the specimen. Each detector consists of a scintillator disc mounted on a long light guide pipe that conveys the light photons to a photomultiplier (Fig. 7-7). The detectors can be moved in and out in the ring to bring the detector close or far away from the specimen. The signals from the detectors can be mixed by a switch to present different views of the specimen. The signal from a single detector gives an asymmetric image, while the signal from all four detectors gives a more symmetric image.

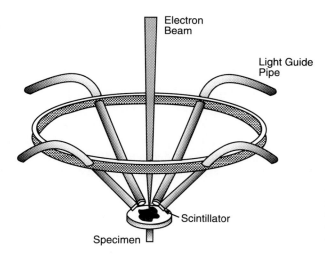

Electron Beam

Light Guide Pipe

Scintillator

Specimen

Figure 7-7 Jackman type of detector consisting of four detectors at the end of light guide pipes mounted on the ring above the specimen. The detectors can be moved farther away or closer to the specimen by manually changing their position in the ring.

Conversion Detectors

In a conversion detector, the backscattered electrons are converted into secondary electrons, and the secondary electrons are collected to produce the signal. In the normal operation of an Everhart–Thornley detector, approximately one-third of the secondary electrons received by the detector are generated by backscattered electrons striking the inside of the specimen chamber. A conversion detector increases the number of secondary electrons generated in this fashion by placing a material (target) that has a large secondary electron coefficient, such as MgO, in a position to intercept as many backscattered electrons as possible. Most backscattered electrons are generated in a direction parallel to the electron beam. There-

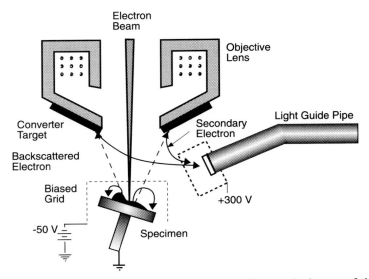

Figure 7-8 Detector setup with a converter target fixed to the bottom of the objective lens to convert backscattered electrons into secondary electrons. The secondary electrons from the specimen are forced back onto the specimen support by a biased grid of − 50 V.

fore, placing the converter target on the bottom of the objective lens pole piece results in a high conversion rate of backscattered electrons into secondary electrons (Fig. 7-8). These secondary electrons are collected by the positive bias on the Faraday cage of the Everhart–Thornley detector and produce a backscattered image of the specimen.

Reversing the Specimen Current

This method will be covered on the following section.

SPECIMEN CURRENT USED AS A DETECTOR

Illumination of the specimen with the electron beam results in the emission of backscattered and secondary electrons from the specimen. However, not all the electrons generated in the specimen are able to escape. Those backscattered and secondary electrons that are generated deep in the specimen–beam interaction zone and the low-energy secondary electrons generated closer to the surface with insufficient energy to overcome the barrier potential at the surface of the specimen do not escape from the specimen. If the specimen is a conductor of electrical current, the electrons are carried to the specimen support and to ground. This

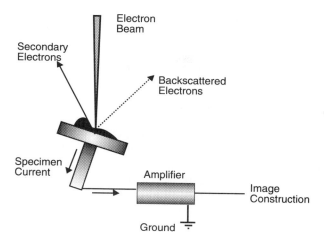

Figure 7-9 The specimen current is the difference between the beam current and the current of the secondary backscattered electrons that escape from the specimen. The specimen current can be amplified to construct an image of the specimen.

electrical current draining off the specimen is the **specimen current** and can be used to visualize the specimen (Fig. 7-9). If the specimen is not a good conductor of electrical current (an insulator), then at least some of the electrons accumulate on the specimen and make the specimen negative. This causes the specimen to charge, causing astigmatism of the beam and imaging artifacts. A specimen that is an insulator can be mounted with colloidal graphite or silver and coated with a conducting metal to prevent it from charging.

Thus the current flow to ground is the specimen current and, after amplification, can be used to form an image of the specimen. The specimen current varies inversely with the production of backscattered and secondary electrons. For example, an electron with a current of 10 picoamps (pA) striking a specimen of iron might produce secondary electrons that escape from the specimen with a current of 1 pA and backscattered electrons with a current of 3 pA. The specimen current flowing to ground is then 6 pA, or the difference between the beam current (10 pA) and the electrons exiting the specimen as backscattered and secondary electrons (4 pA). A current amplifier is placed in the line between the specimen and ground (Fig. 7-9) to produce a large enough current to form an image on the cathode ray tube, since the 6 pA is far too small a current to produce the image without unreasonably long dwell times at each pixel on the specimen.

The construction of an image in this manner is often difficult for the beginning student to comprehend. Remember that the image is built up on a pixel by pixel basis, each pixel having a particular gray level, depending on the current traveling to the cathode ray tube at that moment. Instead of the current coming from an Everhart–Thornley detector, as it would in a conventional image, the current is now coming from the specimen. A negative image of the backscattered and secondary electron signal is produced that can be easily made into a positive image by reversing the signal. The resolution obtained using the specimen current is the same as the diameter of the specimen–beam interaction zone in the specimen (around 1 μm at 30 kV), since this is the volume from which the electrons contributing to the specimen current are generated. This is poorer resolution than that obtained with the backscattered and secondary electron image from an Everhart–Thornley detector. However, the depth of focus using the specimen current as a detector is greater, since the beam spot diameter is generally much smaller than the large pixel size on the specimen.

   Specimen Current Used to Obtain an Image Based on 100% of the Emitted Backscattered Electrons.   Backscattered electrons can often provide more information about topographic and atomic number contrast than an image composed of *both* backscattered and secondary electrons. Unfortunately, it is usually difficult to collect a large number of the generated backscattered electrons because their high energy (most near the energy of the electron beam) prevents their collection by a biased Faraday cage. An indirect way of producing an image composed of 100% of the backscattered electrons is to place a biased grid at −50 V over the specimen (forcing the secondary electrons back onto the specimen) and to reverse the specimen current signal (Fig. 7-10). Such an image represents 100% of the backscattered electrons escaping from the specimen and is independent of detector position. For example, if the electron beam has a current of 10 pA and the backscattered electrons leaving the specimen through the −50-V biased grid have a current of 3 pA, the specimen current would equal 7 pA. The value of the specimen current is an inverse of the backscattered electron current. Therefore, reversing the specimen current results in an image constructed from 100% of the backscattered electrons emitted from the specimen.

AUGER ELECTRON DETECTORS

Auger electrons are usually collected with a cylindrical mirror analyzer that consists of an inner and outer cylinder directed toward the specimen (Fig. 7-11). The inner cylinder is held at ground, while the outer cylinder has a variable negative voltage applied to it. Some of the Auger, secondary, and backscattered electrons are generated toward an aperture (hole) covered with a grid in the inner cylinder. The electrons pass through this gridded aperture and into the electrostatic field generated by the negative voltage on the outer cylinder of the cylindrical mirror analyzer. The electrostatic field causes the Auger, secondary, and backscattered electrons to deviate from their original trajectories, the degree of the deviation depending on the energy of the electrons and the negative voltage applied to the outer cylinder (varying from 0 to −2000 V). The greater the energy of the electrons, the less the electrons are affected by the electrostatic field and the less they deviate from their original trajectories. The greater the negative voltage of the outer cylinder, the more the electrons deviate from their original trajectory. The trajectory of some of the electrons is such that they pass through a second gridded aperture and then through an annular (ring-shaped) exit pore (Fig. 7-11). This setup

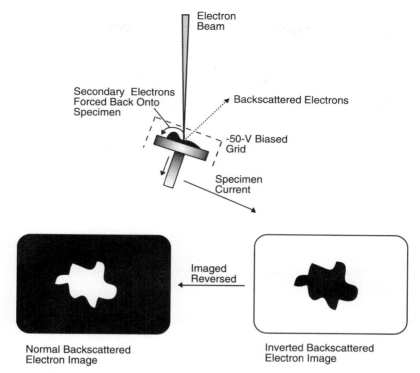

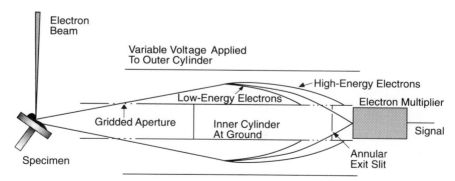

Figure 7-10 Configuration used to obtain an image representing 100% of back-scattered electrons emitted from the specimen. A -50-V biased grid is placed over the specimen to force secondary electrons back onto the specimen. The specimen current forms the image of the specimen. The image can be reversed to produce a normal image of the specimen.

Figure 7-11 Cylindrical mirror analyzer used to produce a spectrum of emitted electrons energies.

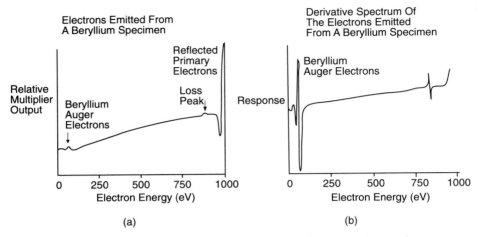

Figure 7-12   (a) Spectrum of electron energies emitted from a beryllium specimen, showing the small beryllium Auger electron peak superimposed on a large background. (b) The same peak after the background was removed and a first derivative of the spectrum was produced.

allows only electrons of specific energies to reach the first dynode of an electron multiplier positioned behind the annular exit slit. The signal from the electron multiplier passes through an amplifier to an oscilloscope that generates a spectrum of electron energies (Fig. 7-12). A metal blank in the direct trajectory between the specimen and the annular exit slit interrupts electrons generated from the specimen directly toward the electron multiplier.

When the cylindrical mirror analyzer is used as an Auger electron spectrometer, the negative voltage on the outer cylinder is varied slowly from 0 to 2000 V. This results in the electron multiplier collecting a spectrum of Auger, secondary, and backscattered electrons in the energy range of 0 to -2000 V. This occurs as the more energetic electrons are progressively repelled from the outer cylinder through the second gridded aperture in the outer cylinder and the annular exit slit into the electron multiplier.

The spectrum that is generated is complex and due to a number of different events (Fig. 7-12):

1. Backscattered electrons generated from the specimen form the greatest peak just below the accelerating voltage of the electron beam (usually 1000 to 2000 V). The backscattered electrons also form most of the background from -50 eV to the accelerating voltage of the electron beam (-2000 V).

2. Secondary electrons generated from the specimen form the background from 0 to -50 eV.

3. Auger electrons generated in the specimen that escape from the specimen after colliding with the specimen atoms will contribute to the background.

However, those Auger electrons that escape from the specimen without interacting with specimen atoms will produce characteristic Auger electron peaks. These peaks will be at the same electron energy regardless of the energy of the electron beam. Thus, in the spectrum in Fig. 7-12, the KLL peak for beryllium is always at 111 eV.

4. Loss peaks at a fixed energy from the accelerating voltage of the electron beam are from beam electrons that have lost a characteristic amount of energy after producing Auger electrons or X-rays through collisions with specimen atoms. With an accelerating voltage of 2000 V, the loss peak for beryllium (Fig. 7-12) is -2000 V minus -111 eV, or 1889 eV.

The Auger electron peaks generated in such a specimen are usually small and difficult to work with, so a first-derivative spectrum is usually generated. First, the background is removed, and then the first derivative of the spectrum is produced, resulting in peaks and valleys in the spectrum that are very accentuated and easier to work with (Fig. 7-12).

In addition to producing an Auger electron spectrum, Auger electrons can be used to image a specimen using scanning electron spectrometric microscopy (SESM). In this method a window of specific electron energies representing the Auger electron peak is selected, and this signal is used to modulate the pixel intensity in the video display. The difficulty with this type of a scan is the long dwell time on each pixel, resulting in long scan times (usually at least 2 min for a single scan of the specimen) and a large beam current at the specimen (at least 10^{-8} A). Specimen resolution is around 100 to 200 nm in such a system.

CATHODOLUMINESCENCE DETECTORS

Cathodoluminescence is the emission of electromagnetic radiation as ultraviolet, visible light, or infrared photons during bombardment of the specimen by beam electrons. A photomultiplier tube is used to collect the photons in cathodoluminescence detector setups (Fig. 7-13). The photocathode at the front of the photomultiplier converts the ultraviolet, visible, or infrared photons into electrons that are cascaded through the photomultiplier tube to produce a signal that is strong enough to construct an image of the specimen (see explanation of photomultiplier tubes in the section on the Everhart–Thornley detector). The most commonly used photocathode materials are able to collect light photons from the longer ultraviolet wavelengths to the infrared (up to 1200 nm). Variations in cathodoluminescence detectors consist of the methods used to direct the photons of cathodoluminescence onto the photomultiplier. These variations are (1) placing a large-diameter photomultiplier close to the specimen to directly collect the electrons, (2) using a quartz or plastic (Perspex) lens or light guide pipe to focus the light photons on the photomultiplier tube, and (3) using an ellipsoidal mirror to focus the photons onto the photomultiplier.

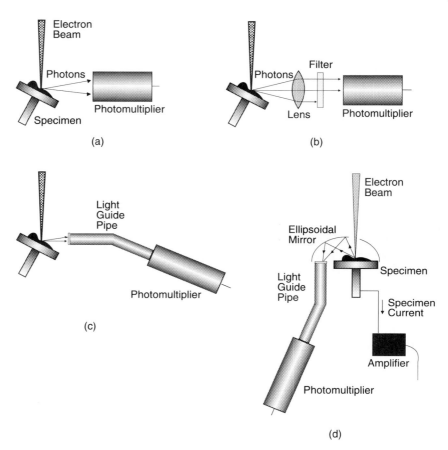

Figure 7-13 Cathodoluminescence detector setups. (a) A photomultiplier can be used alone to collect the photons of cathodoluminescence. (b) A lens can be used to concentrate the photons onto the photomultiplier. In the drawing a filter is inserted so that only selected wavelengths reach the photomultiplier. (c) A light guide pipe can be used to carry the photons to the photomultiplier. (d) An ellipsoidal mirror can be used to focus the photons onto a light guide pipe that carries the photons to the photomultiplier.

Directly Receiving the Photons with a Photomultiplier Tube.   In this setup, the specimen is angled toward the front of a photomultiplier tube so that the maximum number of light photons is received (Fig. 7-13a). To collect a sufficient fraction of the cathodoluminescence, a large-diameter photomultiplier tube (2- to 4-in. diameter) is placed about 6 in. from the specimen to produce a significant solid angle of collection (around 0.1 sr). A disadvantage of this method is the collection by the photomultiplier of other high-energy radiation (X-rays and secondary and backscattered electrons) produced from the specimen and from X-rays and backscattered electrons striking the inside of the specimen chamber. The photomultiplier is unable to distinguish between the photons of cathodoluminescence and the other types of radiation, with the result that the image that is produced is from all these types of radiation. However, this composite image can be compared to an Everhart–Thornley image (produced from backscattered and secondary electrons), and the difference between the two can be presumed to be due primarily to cathodoluminescence from the specimen.

Using a Lens or Light Guide Pipe to Focus the Cathodoluminescence onto a Photomultiplier Tube.   Some systems use a lens made of quartz or Perspex placed between the photomultiplier and the specimen to focus the photons of cathodoluminescence onto the photomultiplier tube (Fig. 7-13b). Other systems use a light guide pipe to carry the photons of cathodoluminescence to the photomultiplier tube (Fig. 7-13c). If a lens or light guide pipe is used, it must have transmission characteristics that pass all the wavelengths of cathodoluminescence that are generated in the specimen. Using a lens or light guide pipe makes it easier to shield the photomultiplier from other high-energy forms of radiation and to produce an image based primarily on cathodoluminescence.

Use of an Elliptical Mirror to Focus the Photons of Cathodoluminescence.   In this method, the specimen is placed at the focal point of an elliptical mirror. The electron beam bombards the specimen through a hole in the mirror, producing photons of cathodoluminescence that are reflected by the mirror to a second focal point, where the photomultiplier or light guide pipe is positioned (Fig. 7-13d). Collection of almost 100% of the emitted cathodoluminescence can be obtained with this system. The backscattered and secondary electron image of the specimen can be obtained for comparison by reversing the image from the specimen current (Fig. 7-13d).

An image generated from specific wavelengths of cathodoluminescence can be obtained in any of the preceding setups by placing filters in front of the photomultiplier tube (Fig. 7-13b).

The specimen resolution obtained from cathodoluminescence in a scanning electron microscope varies from about 100 to 500 nm. In comparison, the resolution obtained from a light optical system is about one-half the wavelength of the light used to image the specimen. Thus, if 400-nm light (blue-green) is used to view the specimen, the best resolution that can be obtained is one-half of the wavelength, or about 200 nm. In contrast, the resolution of a cathodoluminescence image in the scanning electron microscope is determined by the volume of the escape zone of the cathodoluminescence from the specimen at the moment the electron beam is striking the specimen at a particular pixel, since the image is constructed on a pixel by pixel basis.

The specimen and inside of the specimen chamber must be kept particularly clean when imaging with cathodoluminescence. This is because many vacuum oils are luminescent, as are many kinds of dust particles. In addition, light guide pipes made of Perspex or quartz are luminescent when bombarded by electrons, requiring that they be protected from these electrons.

8

Image Reconstruction

In the light microscope and the transmission electron microscope, a true image of the specimen is formed by the light or electron beam passing through the specimen. This is not the situation in the scanning electron microscope, where a true image is not formed. Instead, the scanning electron microscope reconstructs the image of the specimen on a point by point (pixel by pixel) basis in much the same manner as an image is reconstructed in a television set.

IMAGE CONSTRUCTION IN TELEVISION

In television, the image is formed on the phosphor viewing face of a vacuum tube that has an electron gun at the opposite end (Fig. 8-1). The electron gun consists of a heated tungsten cathode that is held at about $-10,000$ V to an anode. Electrons are emitted from the tungsten cathode by thermionic emission (see Chapter 2) and accelerated toward the anode by the 10,000-V difference between the two. The intensity of the electron beam is controlled by a grid between the tungsten cathode and the anode. The grid can be biased so that it is up to -150 V more negative than the tungsten cathode. At this bias, no electrons will penetrate through the electrostatic field of the grid to the anode, and there will be no electron beam passing down the vacuum tube. Reducing the bias on the grid allows more and more electrons to pass through the grid to the anode. Normally, the grid is varied

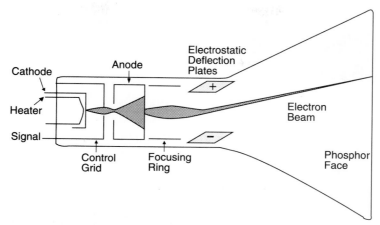

Figure 8-1 A cathode ray tube used to receive television transmissions.

in voltage by the signal input (ultimately from the television transmitter), and this controls the intensities of the electron beam at different positions on the phosphor viewing face of the vacuum tube. The phosphor converts the electron intensity to light intensity (by cathodoluminescence) to produce the image.

After acceleration through the anode, the electron beam passes through a focusing ring whose negative bias focuses the electron beam to a pinpoint on the face of the vacuum tube. The position of the pinpoint of electrons is determined by the deflection coils (magnetic deflection) or plates (electrostatic deflection) that move the electron beam in the X and Y direction. In a television receiver, the image is normally built up of 525 consecutive horizontal lines to a scan, with the scan occurring 60 times a second in two alternate scans of 30 lines each. Each line can be thought of as having 1000 **picture elements** or **pixels** per line, where the electron beam rests for a brief instant of time. The image on the face of the vacuum tube is formed from 525,000 pixels (1000 pixels times 525 lines), with the gray levels of the pixels (black to white) forming the image (Figs. 8-2 and 8-3). The gray

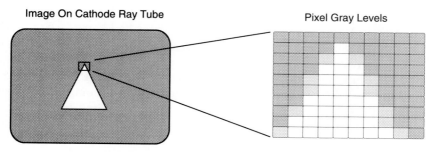

Figure 8-2 The image on a cathode ray tube is built up from a large number of pixels of different gray levels.

Figure 8-3 Gray levels. Normally, the eye can detect about 16 gray levels.

level of the pixels is determined by the amplitude of the signal from the television transmitter at that particular instant in time. This signal travels to the control grid, which modulates the intensity of the electron beam and therefore the gray level of the pixel on the phosphor face of the vacuum tube.

IMAGE CONSTRUCTION IN THE SCANNING ELECTRON MICROSCOPE

An image is constructed in a scanning electron microscope in essentially the same manner that it is constructed in a television set, except in a scanning electron microscope the signal is generated from the specimen instead of from the television transmitter. In the column of the scanning electron microscope, the electron beam is focused to a small pinpoint of electrons, 3 to 20 nm in diameter, on the specimen. This pinpoint of electrons is rastered over the specimen in a series of consecutive lines by the X and Y deflection coils in the objective lens (see Chapter 5). There is a scan generator that synchronizes the position of the electron beam on the specimen with the position of the electron beam in the cathode ray tube. There is a 1:1 position relationship between the position of the electron beam on the specimen and the position of the electron beam in the cathode ray tube. Thus, if the electron beam is finishing its scan on the specimen, then it is also finishing its scan on the cathode ray tube (Fig. 8-4).

The electron beam on the specimen and the electron beam in the cathode ray tube are moved in a raster of consecutive lines that can usually be varied by the operator from 250 to 4000 lines per scan. There are two different ways to drive the electron beam, (1) with an analog or (2) digital scanning system. Both systems produce identical images, but in different ways. In an analog scanning system the electron beam is swept continuously along each scan line. In a digital scanning system the whole scan pattern is divided into a large number of pixels, and the electron beam is directed to a specific pixel (Fig. 8-5). Normally, the number of pixels on a line is the same as the number of lines in the raster pattern. Therefore, there are 1000 pixels per line if the raster pattern is composed of 1000 lines. This results in an image on the cathode ray tube composed of 1 million pixels (1000 pixels times 1000 lines) every time the cathode ray tube is scanned once. The gray level of the pixel can be varied from black through white, and it is the gray level of the pixels that determines the actual image on the cathode ray tube. The gray

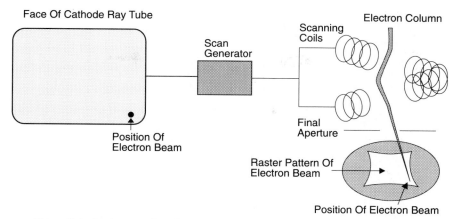

Figure 8-4 In the scanning electron microscope the scan generator drives both the electron beam in the cathode ray tube and the electron beam in the column. There is a 1:1 relationship between the position of the beam on the specimen and the position of the beam on the face of the cathode ray tube. Thus, when the beam is at a particular location on the specimen, it will be at the same pixel location in the cathode ray tube.

level *within a pixel* is uniform and is obtained by integrating the signal over the time the electron beam is on this area of the specimen (Fig. 8-6). Each pixel can be given a consecutive number of 1 to 1 million (if a scan of 1000 lines is used). When the electron beam in the cathode ray tube is at pixel 022 (Fig. 8-5), the electron beam on the specimen will also be at pixel 022. A large amount of radiation from the specimen striking the detector at pixel position 022 produces a large pulse of electrical current that is amplified to produce a bright pixel 022 on the cathode ray tube. However, if the opposite happens and there is only a small amount of radiation leaving the specimen at pixel 022, then a weak electrical pulse is produced from the detection system and a dark pixel is produced at pixel 022 on the cathode

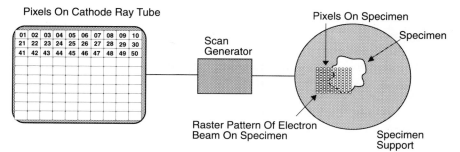

Figure 8-5 The image on the cathode ray tube is composed of a large number of pixels. For each pixel position on the cathode ray tube, there is a corresponding pixel position on the specimen.

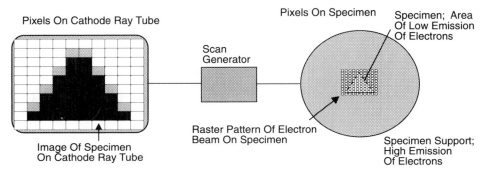

Figure 8-6   Image reconstruction in a scanning electron microscope. The specimen is rastered by the electron beam on a pixel by pixel basis. The secondary and backscattered electrons emitted from a single pixel on the specimen produce the signal for the corresponding pixel on the cathode ray tube. The gray level within a pixel on the cathode ray tube is uniform and represents the signal received from the corresponding pixel on the specimen. The number of pixels represented in the drawing is far less than the number of pixels in an ordinary scan of a specimen. The low number of pixels is used for illustration purposes.

ray tube. After producing a gray level at pixel 022 on the cathode ray tube, the electron beam moves to pixel 023 on the specimen and on the cathode ray tube, and the process is repeated. The image is constructed from 1 million of these pixels (if there are 1000 scan lines). Those areas of the specimen that emit more radiation appear bright, while those that emit little radiation appear dark (Fig. 8-6).

The variation in the brightness of each pixel on the cathode ray tube is called Z modulation (the electron beam is scanned in the X and Y directions; the brightness is amplified in the Z direction toward the operator). The image constructed in the scanning electron microscope is remarkably easy for the novice operator to interpret since it is usually very similar to the image that would be obtained using a light optical system (of course, the image generated in the scanning electron microscope has much better resolution).

Line Scan

The image construction method just described is called an **image** or **area scan** and utilizes scanning in the X (horizontal) and Y (vertical) directions. In a **line scan**, the electron beam is rastered only in one direction over the specimen, usually the X direction, and the intensity of the radiation generated from the specimen is modulated as peaks in the Y direction on the cathode ray tube. Thus, *in the cathode ray tube*, as the electron beam is rastered continuously over a single line by the X deflection coil, the Y deflection coil is driven by the intensity of the signal from the detector, producing a Y-modulated line scan. In Fig. 8-7, a scan line is produced

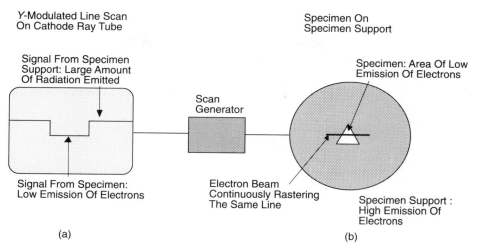

Y-Modulated Line Scan
On Cathode Ray Tube

Specimen On
Specimen Support

Signal From Specimen
Support: Large Amount
Of Radiation Emitted

Specimen: Area Of Low
Emission Of Electrons

Scan
Generator

Signal From Specimen:
Low Emission Of Electrons

Electron Beam
Continuously Rastering
The Same Line

Specimen Support :
High Emission Of
Electrons

(a)   (b)

Figure 8-7 (a) In a line scan the same raster line is continuously rastered over the specimen. (b) Those areas of the specimen that emit a large amount of radiation when bombarded by the electron beam produce areas in the Y-modulated line scan of the CRT, which are higher than those areas of the specimen that emit small amounts of radiation.

by the X deflection coil. The peaks on the line scan represent areas where large amounts of radiation are being emitted from the specimen, while the valleys in the line scan represent areas where small amounts of radiation are emitted from the specimen. Thus the position on the horizontal line on the *cathode ray tube* is proportional to the position on the scan line on the *specimen*, and the vertical position of the *cathode ray tube* is proportional to the signal strength emanating from the *specimen*. Line scans accurately display small changes in radiation emitted from a specimen, changes that would be difficult to notice in a Z-modulated image scan. Line scans are commonly used during normal operation of a scanning electron microscope to perform two functions: (1) to saturate the filament current, and (2) to obtain proper contrast and intensity on the photographic cathode ray tube for taking a photograph.

Use of a *Y*-modulated line scan to saturate the filament heating current. The image on the cathode ray tube in a scanning electron microscope is composed of two components: (1) the base-level brightness, and (2) the signal from the detector system.

1. The *base-level brightness* of the cathode ray tube represents the lowest intensity of the electron beam in a Z-modulated area scan. The base-level brightness is the bias on the control grid between the cathode and anode in the cathode ray tube when there is no signal input from the detector into the control grid (Figs.

Y-Modulated Line Scans

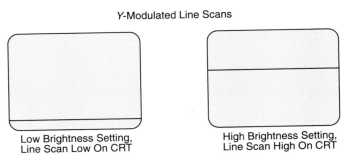

Low Brightness Setting,
Line Scan Low On CRT

High Brightness Setting,
Line Scan High On CRT

Figure 8-8 *Y*-modulated line scans on a cathode ray tube. (a) At a low brightness (black level) setting, the *Y*-modulated line scan will be low on the cathode ray tube since the height of the *Y*-modulated line scan represents the brightness. (b) At a high brightness setting, the *Y*-modulated line scan will be high on the cathode ray tube.

8-1 and 8-8). The more negative the control grid is in comparison to the tungsten cathode, the less electrons pass through the control grid, the weaker the electron beam in the cathode ray tube, and the lower the base-level brightness on the cathode ray tube. The bias of the control grid, and therefore the base-level brightness of the cathode ray tube, is controlled by the operator with a knob that is usually labeled ''brightness'' or ''black level'' on the scanning electron microscope. The brightness of the cathode ray tube is represented by the height of the line in a *Y*-modulated line scan (Fig. 8-8). The greater the brightness of the cathode ray tube is, the higher the position of the line scan.

    2. The *signal from the detector system* modulates the bias on the control grid of the cathode ray tube and is superimposed on the base-level brightness. If there is no signal input from the detector system in an intensity or *Z*-modulated area scan, the gray level of the pixels is the same as the base-line brightness. If there is a large amount of signal input, the bias on the control grid is reduced, increasing the intensity of the electron beam in the cathode ray tube and resulting in pixel gray levels that are much whiter. In a *Y*-modulated line scan, the superposition of the signal from the detector is manifested as peaks over the base-line brightness level (Fig. 8-9). The *Y*-modulated line scan provides an accurate means of determining the saturation point of the filament current. Increasing the current through the filament results in an increase in electron emission, an increase in electrons impinging on the specimen, and an increase in radiation emanating from the specimen. This increase continues until the saturation point is reached, at which point an increase in current through the filament results in no increase in electron emission (see Chapter 2). The saturation point is usually determined by means of a *Y*-modulated line scan. As the filament current increases, there is an increase in the height of the peaks in the *Y*-modulated line scan as more signal

Y-Modulated Line Scans On Cathode Ray Tube Of An SEM

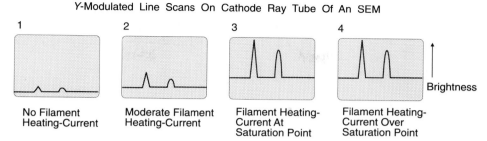

Figure 8-9 The appearance of Y-modulated line scans during the saturation of the filament heating current. The Y-modulated line scans increase in height as the filament heating current and electron emission from the filament increase. At the saturation point, there is no longer any increase in electron emission as the filament heating current increases. Oversaturation of the filament heating current does not result in an increase in electron emission or height of the Y-modulated line scan.

from the detector reaches the Y deflection coil in the cathode ray tube. Once the filament current is at the saturation point, the peaks in the Y-modulated line scan no longer rise, since there is no increase in signal from the detector setup to the Y deflection coil to drive the peaks higher. It is at this filament heating current (or slightly less) that the scanning electron microscope should be operated to ensure maximum life of the filament.

Use of a Y-modulated line scan to determine the correct contrast and brightness for taking a photograph.   Normally, in a scanning electron microscope the contrast and brightness setting on the photographic cathode ray tube is kept the same for a particular type of film. The correct contrast and brightness for the exposure are adjusted by varying the signal from the specimen using a Y-modulated line scan (Figs. 8-10 and 8-11). The bottom of the valleys in the Y-modulated line scan is used to determine the base-line brightness of the photograph. The distance from the top of the peaks to the bottom of the valleys of the Y-modulated line scan is used to determine the contrast (Fig. 8-11). The tip of the peaks represents the lightest pixel gray levels, while the bottom of the valleys represents the darkest pixel values. *The range of gray levels is the definition of contrast; the greater the contrast is, the greater the range of gray levels.*

While these are the two common uses of line scans, they can also be used in any situation where an accurate indication of pixel values along a line is needed. Area scans of a specimen can also be obtained by modulating each line in the scan in the Y direction instead of the normally used Z modulation (Fig. 8-12). In this method the specimen is scanned in a conventional raster pattern with the series of consecutive line scans in the cathode ray tube modulated in the Y direction. Y-modulated area scans are useful in enhancing small changes in the surface structure of the specimen that are difficult to see in Z-modulated area scans.

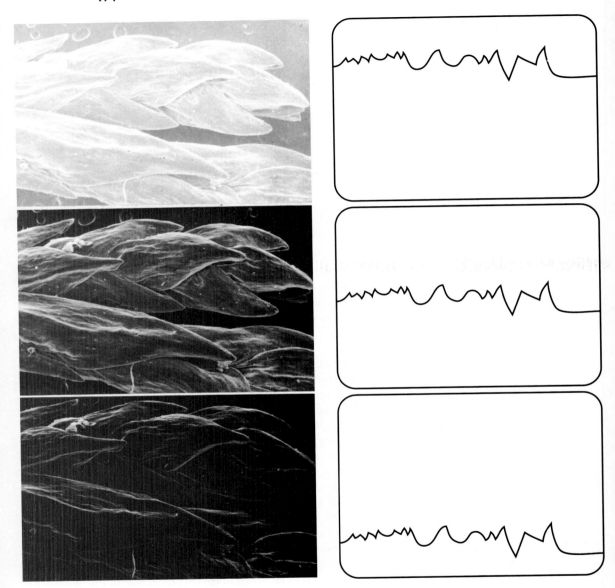

Figure 8-10 Micrographs of the same specimen at three different brightness (black level) settings. The micrograph at the top is too bright and has a Y-modulated line scan high on the cathode ray tube. The middle micrograph is at optimal brightness with a Y-modulated line scan near the center of the cathode ray tube. The bottom micrograph is too dark and has a Y-modulated line scan near the bottom of the cathode ray tube.

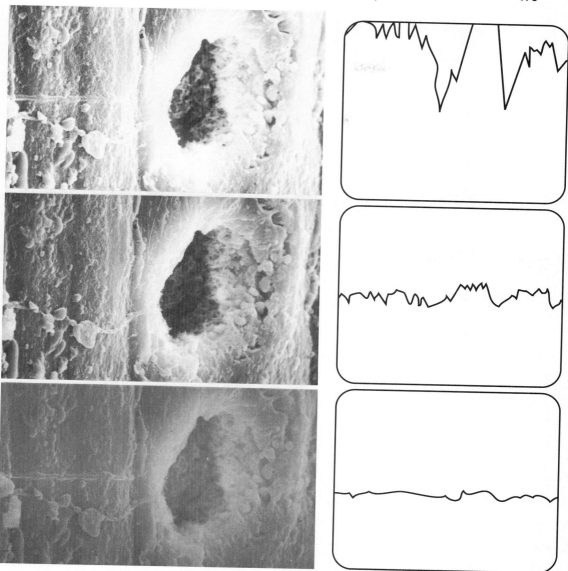

Figure 8-11   Micrographs of the same specimen at different contrast (gain) settings. The bottom specimen has too little contrast. A *Y*-modulated line scan of a line from this micrograph shows little difference between the top of the peaks (whitest gray level) and the bottom of the valleys (blackest gray level). The middle micrograph is of optimal contrast. A *Y*-modulated line scan from this image shows a good range between the top of the peaks and the bottom of the valleys. The top micrograph has too much contrast. The *Y*-modulated line scan shows peaks that are being clipped off, resulting in loss of information from pixels saturated at the white end of the gray level range.

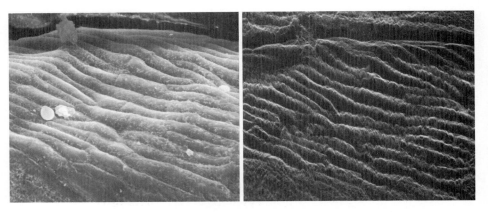

Figure 8-12 (a) An intensity or *Z*-modulated image of a specimen. (b) The same specimen imaged with an area scan of a large number of *Y*-modulated line scans.

MAGNIFICATION

Magnification is varied in the scanning electron microscope by varying the area rastered on the specimen (Fig. 8-13 and Table 8-1). The larger the area rastered on the specimen, the lower is the magnification of the image on the cathode ray

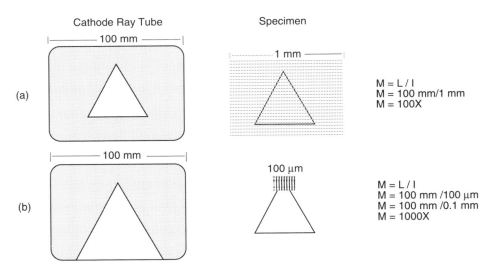

Figure 8-13 An example of how magnification changes are affected in the scanning electron microscope. (a) At 100×, a raster line 1 mm long on the specimen is reconstructed on a 100-mm-wide cathode ray tube. (b) At 1000× a raster line 100 μm long on the specimen is reconstructed on the 100-mm-wide cathode ray tube.

TABLE 8-1 AREA RASTERED ON SPECIMEN AT DIFFERENT MAGNIFICATIONS (ASSUMING A CATHODE RAY TUBE 100 MM BY 100 MM)

Magnification	Length of raster line on specimen	Area rastered on specimen
10×	10 mm	1 cm²
100×	1 mm	1 mm²
1,000×	100 μm	10,000 μm² (0.1 mm²)
10,000×	10 μm	100 μm²
100,000×	1 μm	1 μm²

tube. The smaller the area rastered on the specimen, the greater is the magnification of the image on the cathode ray tube. Remember that the size of the raster on the cathode ray tube is fixed, with the length of the line usually around 120 mm, with each line having around 1000 pixels of information. The gray level of each pixel in the cathode ray tube is determined by the amount of radiation emitted from the corresponding pixel on the specimen. Shorter raster lines of the electron beam on the specimen mean that the information for the 1000 pixels per line is obtained from short raster lines on the specimen, resulting in greater magnification of the specimen on the cathode ray tube.

Magnification in the scanning electron microscope can be expressed by the following formula:

$$\text{magnification on the CRT} = \frac{\text{length of raster line on CRT}}{\text{length of raster line on the specimen}}$$

The method that the scanning electron microscope uses to magnify an image has two important effects: (1) an image that is in focus at high magnifications stays in focus as the magnification is decreased, and (2) the image does not rotate as the magnification changes (Fig. 8-14). Focus is determined in the scanning electron microscope by the electrical current passing through the objective lens. The image is in focus when the electron beam is at crossover on the specimen. During *focusing*, the image will rotate on the cathode ray tube. This occurs because the electrons spiral down the column of the scanning electron microscope (Fig. 8-15), resulting in different focal positions at different rotation positions. However, once focus is obtained on a particular area of the specimen, there is no rotation or change in focus of that area as the magnification is changed. This is because *changes in magnification occur by changing the excitation of the scanning coils in the column*, resulting in the raster of larger or smaller areas of the specimen by the electron beam. These changes do not involve any changes in the position of crossover of the electron beam on the specimen (except at very low magnifications), so it is not necessary to change the excitation of any of the lenses during changes in magnification. This, in turn, means that there is no change in the rotation or focus of the specimen as the magnification is changed. A focused

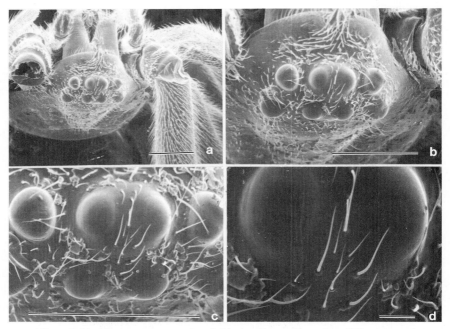

Figure 8-14 Electron micrographs of a black widow spider at four different magnifications. The image does not rotate as the magnification is changed. Bars on a, b, and c equal 1 mm. Bar on d equals 0.1 mm.

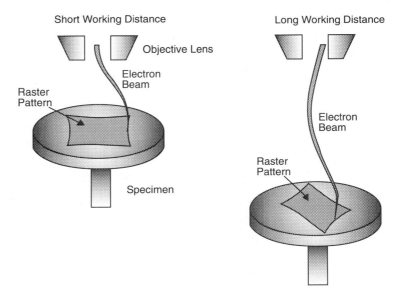

Figure 8-15 Changing the working distance results in a change in the crossover position of the electron beam on the specimen (focus) and a rotation of the raster pattern on the specimen. The rotation is due to the spiraling of the electrons down the column.

Figure 8-16 Micrographs of the same specimen at two different working distances showing rotation of the image.

specimen is routinely obtained in the scanning electron microscope by obtaining focus at a higher magnification and then bringing the magnification down to the desired magnification.

Changing the *working distance* results in a change in the focal plane, which requires a change in focus. This results in a rotation of the image, since a change in focus requires a change in the current to the objective lens, causing the electrons to spiral to a different position on the specimen (Figs. 8-15 and 8-16).

RESOLUTION AND IMAGE QUALITY IN THE SCANNING ELECTRON MICROSCOPE

The best resolution that can be obtained in the scanning electron microscope is determined by the size of the escape volume of the secondary and backscattered electrons from the specimen. The electron beam spreads as it enters into the specimen, producing the specimen–beam interaction volume (see Chapter 6). The secondary electrons generated on interaction of the electron beam with the specimen escape from a volume whose surface area is slightly larger than the area of the beam striking the specimen (Fig. 8-17). The backscattered electrons and the secondary electrons generated by the backscattered electrons on interaction with the specimen atoms arise from a larger surface area. This is because the backscattered electrons are able to escape from a greater depth and are able to travel farther sideways from the point of entry of the electron beam. A cross section of the escape volume shows the bulk of the emitted electrons coming from a central spot slightly larger than the beam spot size, with the remainder of the electrons producing a sloping tail at greater distances from the entry of the electron beam (Fig. 8-17c).

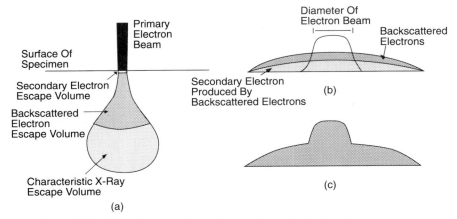

Figure 8-17 (a) In a low atomic number specimen, the electron beam penetrates into a light bulb-shaped volume of the specimen. (b, c) Secondary electrons, produced by the electron beam, escape from an area slightly larger than the diameter of the electron beam. The backscattered electrons and the secondary electrons produced by the backscattered electrons escape from a larger area.

The diameter of the electron beam striking the specimen can be used to estimate the escape volume of backscattered and secondary electrons from the specimen. If the diameter of the electron beam is 6 nm, then the diameter of the area of the escape zone on the specimen (and therefore the resolution) would be about 10 nm. Thus the approximate resolution that can be obtained is the diameter of the beam spot multiplied by $1\frac{2}{3}$.

The pixel diameter on the specimen determines the best possible resolution of the scanning electron microscope. As the magnification increases, the pixel diameter on the specimen decreases, as can be seen in Table 8-2. Smaller areas on the specimen are scanned by the electron beam as the magnification increases. However, the same number of pixels is included within the area of the raster pattern. Therefore, the size of the pixels must decrease in order to fit into the smaller area (Fig. 8-13). From Table 8-2, it would appear that the best possible

TABLE 8-2 PIXEL DIAMETER ON THE SPECIMEN
AT DIFFERENT MAGNIFICATIONS

Pixel diameter on specimen μm (nm)		Magnification
10	(10,000)	$10\times$
1	(1,000)	$100\times$
0.1	(100)	$1,000\times$
0.01	(10)	$10,000\times$
0.001	(1)	$100,000\times$

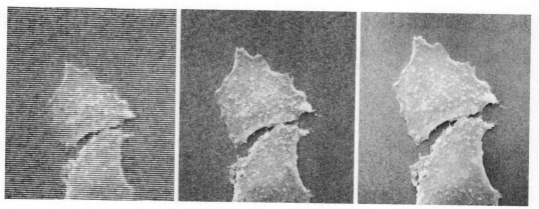

Figure 8-18 The same specimen imaged with rasters of (a) 1 line/mm, (b) 2.5 lines/mm, and (c) 4.0/mm.

magnification with a 6-nm beam spot diameter (about the smallest beam spot diameter that can be obtained with thermionic emission from a tungsten cathode) would be about 10,000× (assuming a 6-nm beam spot diameter produces a 10-nm escape volume of backscattered and secondary electrons). However, another factor that must be considered is the resolving ability of the human eye. The human eye can only resolve about four to five lines per millimeter on a cathode ray tube (Fig. 8-18). If the cathode ray tube actually has 10 lines/mm (1000 lines/raster on a 100-mm cathode ray tube), then the magnification can be increased by two to three times before the human eye is able to detect the hollow magnification. This means that photographs at magnifications of 20,000 to 30,000 times normal can be routinely obtained, providing the image is properly focused and stigmated. Higher magnifications can be obtained if smaller beam spot diameters can be obtained (with lanthanum hexaboride or field emission electron guns).

The relationship between pixel size on the specimen, beam spot diameter, and magnification is an important concept in operating the scanning electron microscope as can be illustrated with the following three examples.

Example 1 (Fig. 8-19)

> Magnification: 10,000×
> Pixel diameter on the specimen: 10 nm
> Beam spot diameter: 6 nm
> Diameter of escape area of backscattered and secondary electrons: 10 nm

In this example, the *escape volume of secondary and backscattered electrons from the specimen has the same surface diameter, 10 nm, as does the pixel diameter on the specimen.* Therefore, as the focused electron beam travels over the specimen, the amount of information from one pixel on the specimen is transferred to

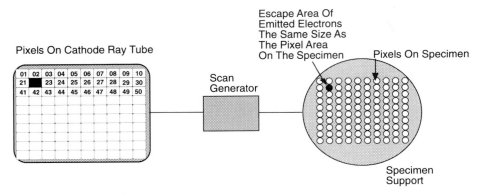

Figure 8-19 A specimen is in focus on the cathode ray tube when the escape area of backscattered and secondary electrons is the same size as the pixel area on the specimen. The gray level of pixel 22 on the cathode ray tube is being derived only from the emitted electrons of pixel 22 on the specimen.

one pixel on the cathode ray tube, the specimen is in focus, and there is a maximum amount of signal to produce the gray level of the pixel. In Fig. 8-19, the information from pixel 22 on the specimen is transferred to pixel 22 on the cathode ray tube. The instrument is operating at maximum resolution, although there is no depth of focus (if the working distance of the specimen is changed, it is necessary to bring the electron beam back to a minimum diameter, that is, true focus).

Example 2 (Fig. 8-20)

Magnification: 10,000×
Pixel diameter on specimen: 10 nm
Beam spot size: 12 nm
Diameter of escape area of backscattered and secondary electrons: 20 nm

In this example, *the diameter of the escape area of secondary and backscattered electrons at a particular pixel location is 20 nm, twice the diameter of the 10-nm pixel diameter on the specimen.* Figure 8-20 illustrates this situation when the electron beam is resting on pixel 22 on the specimen and on pixel 22 in the cathode ray tube. The electron beam striking the specimen creates an escape zone of backscattered and secondary electrons that, in addition to including the targeted pixel 22 on the specimen, also includes part of the surrounding pixel locations: 01, 02, 03, 21, 23, 41, 42, and 43. The information from all nine of these pixels is transferred to pixel 22 on the CRT. The aggregation of information from adjacent pixels on the specimen into a single pixel on the cathode ray tube continues as the electron beam scans the specimen. This results in a loss of resolution, because the information from three scan lines on the specimen contributes to one line on the cathode ray tube. In actual fact, the operator would not be able to resolve this loss of detail, since the image is usually built up of 10 lines/mm and the operator can

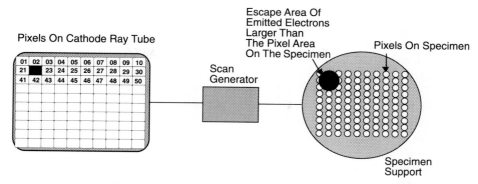

Figure 8-20 The specimen is out of focus on the cathode ray tube when the escape area of the secondary and backscattered electrons is greater than the pixel area on the specimen. The gray level of pixel 22 is being derived from information derived from pixels 01, 02, 03, 21, 22, 23, 41, 42, and 43 on the specimen.

only see 4 to 5 lines/mm. However, if a negative was made and enlarged or if the diameter of the beam spot was increased further, blurring of the image and loss of detail could be seen.

Example 3 (Fig. 8-21)

Magnification: $10,000 \times$

Pixel diameter on the specimen: 10 nm

Diameter of the beam spot: 3 nm

Diameter of the escape zone of backscattered and secondary electrons: 5 nm

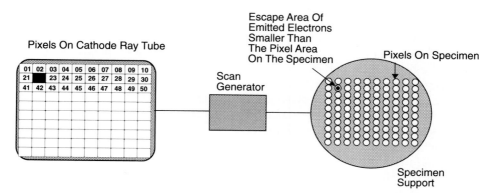

Figure 8-21 The escape area of backscattered and secondary electrons on the specimen is smaller than the pixel area on the specimen. There is a great depth of focus, since the topography of the specimen will remain in focus until the point is reached when the escape area of the emitted electrons is the same size as the pixel size on the specimen. A weaker signal is produced because a smaller escape area of emitted electrons results in less electrons to produce the signal.

Here, *the diameter of the escape area of the backscattered and secondary electrons from the specimen is smaller than the pixel diameter on the specimen.* Thus there is no overlap of adjacent pixels on the specimen, and the resolution is optimal since the information on one pixel on the specimen is included in one pixel on the cathode ray tube. However, the number of electrons in the electron beam is relatively low, since the number of electrons in the electron beam falls off approximately as the square of the beam spot diameter. This results in a relatively small number of backscattered and secondary electrons that reach the detector during the brief interval that the electron beam rests on this pixel. A large degree of amplification of the signal is required to obtain an adequate image on the cathode ray tube (unless a lanthanum hexaboride or field emission source is used to provide a large number of electrons in the electron beam). The image on the cathode ray tube is snowy due to the electronic noise that accompanies amplification of the signal. It therefore would probably be advantageous to increase the beam spot diameter slightly to increase the signal, since it could be done without a loss of resolution. A large depth of focus (field) is the only advantage to having the escape area of the backscattered and secondary electrons smaller than the pixel area on the specimen. If the scanning electron microscope is operated so that there is a great depth of focus and the specimen has a good deal of topography, an in-focus image would still be obtained as long as the features of the specimen had an escape area of backscattered and secondary electrons that was smaller than the pixel diameter on the specimen at this point (Fig. 8-26).

DEPTH OF FOCUS (FIELD)

The depth of focus in a scanning electron microscope is typically 500 times greater than in a light microscope at the same magnification (Figs. 8-22 and 8-23). The scanning electron microscope is able to use a longer working distance and a thinner electron beam (smaller semiangle), which accounts for the greater depth of focus. The depth of focus in a scanning electron microscope is determined by the size of the beam spot, the area of the escape zone of secondary and backscattered electrons, and the pixel size on the specimen. *The specimen will be in focus as long as the electron beam produces an escape zone of secondary and backscattered electrons that is the same size or smaller than the pixel area on the specimen* (Fig. 8-24). The smaller the escape zone of secondary and backscattered electrons is when compared to the pixel area on the specimen, the greater is the depth of focus and the more vertical topography on the specimen that is in focus. To determine the depth of focus, it is necessary to find the vertical distance along the electron beam where the electron beam diameter is smaller than the escape area of secondary and backscattered electrons on the specimen. It is this vertical distance that represents the depth of focus. Thus a single pixel area on the specimen would

(a)   (b)

Figure 8-22 Photographs of the same specimen taken on (a) a light microscope and (b) an electron microscope, showing the greater resolution and depth of focus in the scanning electron microscope.

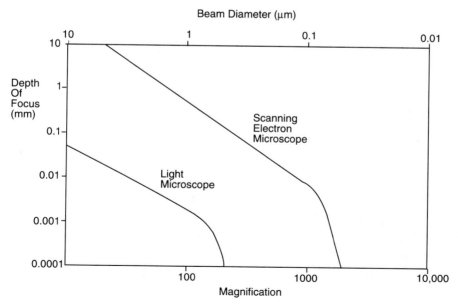

Figure 8-23 Comparison of the depth of focus in a light microscope and a scanning electron microscope at different magnifications.

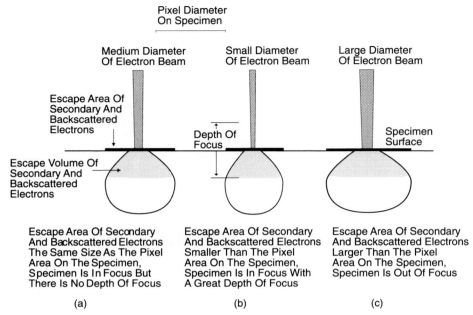

Figure 8-24 The size of the electron beam striking the specimen surface affects the depth of focus. The smaller the diameter of the electron beam, the smaller is the area of the secondary and backscattered electrons escaping from the surface of the specimen. (a) The specimen is in focus if the escape area of the emitted electrons is the same size as the pixel diameter on the specimen, but there is no depth of focus. (b) The specimen is in focus and there is depth of focus if the escape area of the emitted electrons is smaller than the pixel diameter on the specimen. (c) The specimen is out of focus if the escape area of the emitted electrons is larger than the pixel diameter on the specimen.

remain in focus if the working distance of the specimen were changed throughout the depth of focus. For example, in Fig. 8-24b the following apply:

Magnification: $10,000 \times$

Beam spot diameter at focus: 3 nm

Escape area of backscattered and secondary electrons: 5 nm

Pixel diameter on the specimen: 10 nm

Depth of field: 1 μm

In this situation, the depth of field is 1 μm, so the depth of focus extends 0.5 μm on each side of true focus (when the electron beam is at its minimal diameter on the specimen). Therefore, a portion of the specimen 0.5 μm above or below the portion of the specimen at true focus would still be in focus. While 1 μm is a common depth of focus at high magnifications, at lower magnifications the depth of focus can extend to 4 mm (Table 8-3).

The depth of focus is affected by three factors: (1) the beam spot area at focus, which affects the size of the escape zone of secondary and backscattered

TABLE 8-3   DEPTH OF FIELD AT A 10-MM WORKING DISTANCE IN A SCANNING
ELECTRON MICROSCOPE

Magnification	100-μm aperture ($\alpha = 5 \times 10^{-3}$ rad)	200-μm aperture ($\alpha = 10^{-2}$ rad)	600 μm aperture ($\alpha = 3 \times 10^{-2}$ rad)
10×	4 mm	2 mm	670 μm
50×	800 μm	400 μm	133 μm
100×	400 μm	200 μm	67 μm
500×	80 μm	40 μm	13 μm
1,000×	40 μm	20 μm	6.7 μm
10,000×	4 μm	2 μm	0.67 μm
100,000×	0.4 μm	0.2 μm	0.067 μm

electrons, (2) the working distance, and (3) the diameter of the final (probe-forming or objective) aperture.

Area of the beam spot at focus: As explained previously, the depth of field increases as the beam spot size decreases (Fig. 8-24).

Working distance: The depth of field increases as the working distance increases. This is because the semiangle (α) decreases as the working distance increases, leading to a narrower beam that has less variation in beam area along the length of the beam (Fig. 8-25).

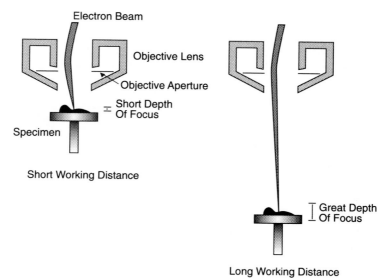

Figure 8-25   The shorter the working distance, the greater is the semiangle of the electron beam and the shorter the depth of focus. The depth of focus ceases on each side when the diameter of the escape area of the secondary and backscattered electrons from the specimen becomes the same size as the pixel area on the specimen.

Diameter of the final (probe-forming or objective) aperture: The smaller the
final aperture is, the narrower is the electron beam and the greater the depth
of focus (Figs. 8-26 and 8-27).

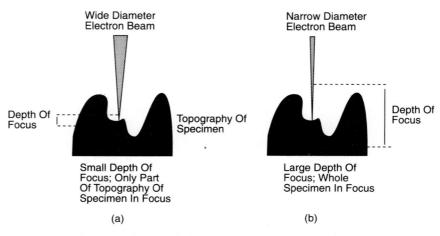

Figure 8-26   (a) An electron beam with a large semiangle will have a small depth
of focus. Only a part of the specimen will be in focus if the specimen has a considerable amount of topography. (b) Decreasing the semiangle of the electron beam
results in an increase in the depth of focus, resulting in all the topography of the
specimen being in focus.

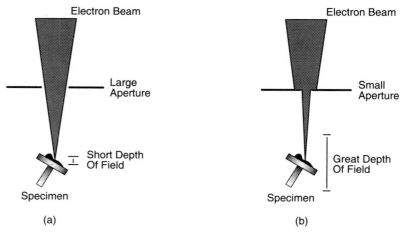

Figure 8-27   (a) A large objective aperture produces an electron beam with a large
semiangle and a small depth of focus. (b) A small objective aperture results in an
electron beam with a small semiangle and a larger depth of focus.

CONTRAST

Adjacent pixels on the *cathode ray tube* will be of the same gray level if the same pixels on the *specimen* have approximately the same number of backscattered and secondary electrons escaping from the specimen during bombardment by the electron beam. Even though the scanning electron microscope is operating at optimal resolution, the operator will see no detail, since the pixels or group of pixels has the same gray level. Generally, it is possible for an individual to discern about 16 gray levels from black to white. This means that contrast levels below about 6% are not seen by humans. In general, the maximum information is obtained about the specimen when the gray levels of the pixels on the cathode ray tube are throughout the whole gray-level range from white, through the grays, to black.

The contrast of the pixels in the image on the cathode ray tube can be varied by processing the signal after it has left the detector. In this process the signal from the detector is expanded or contracted to produce the maximum number of gray levels between the whitest and blackest pixels. This processing can also increase the amount of noise in the image if it is performed with the photomultiplier. Ideally, it is preferable to increase the contrast at the specimen level by manipulating the radiation leaving the specimen during electron bombardment.

Contrast from the specimen results when different parts of the specimen emit different amounts of radiation (such as secondary and backscattered electrons) under bombardment by the electron beam. The emission of different amounts of radiation at different places on the specimen is due to (1) a difference in the topography of the specimen (**topographic contrast**) or (2) a difference in atomic number in different parts of the specimen (**atomic number contrast**).

Topographic Contrast

Topographic contrast is the contrast mechanism usually used to image the specimen in the scanning electron microscope. Topographic contrast from a detector (such as an Everhart–Thornley detector) is based on two different principles: (1) more backscattered and secondary electrons escape from a rough surface than a smooth surface, and (2) more backscattered and secondary electrons are received by portions of the specimen angled toward the detector.

Surface of the Specimen. Smooth specimen surfaces have the escape zones for backscattered and secondary electrons uniformly deep in the specimen, resulting in minimal escape of electrons from the specimen per unit area. Irregular surfaces have the escape zones of backscattered and secondary electrons thrown off the horizontal, producing more volume of escape zone per unit area and resulting in the escape of more electrons during bombardment by the electron beam (Fig. 8-28). A larger number of backscattered and secondary electrons escape from a rough surface because more of the generated electrons are able to reach and escape from the surface.

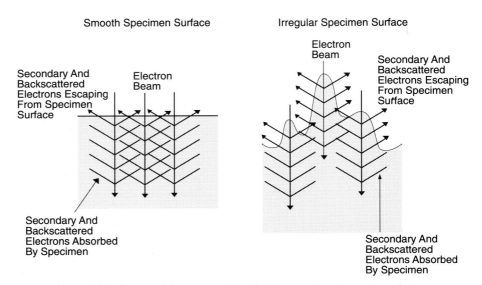

Figure 8-28 Topographic contrast. More backscattered and secondary electrons escape from an irregular surface than from a smooth surface. This results in areas of the specimen with irregular topography appearing brighter in the viewing cathode ray tube.

Angle of the Specimen.   The detector will collect a greater number of backscattered electrons, and to a lesser extent secondary electrons, from facets of the specimen that are directed toward the detector (Fig. 8-29). The facets of the specimen directed away from the detector will have lower collection efficiencies. The effect is more pronounced for backscattered electrons, which are not collected by a positive bias on a Faraday cage or focusing ring and are only collected if they are generated toward the detector. The effect is less pronounced for secondary electrons because they are collected with a high efficiency by the bias on the Faraday cage. Therefore, a backscattered image (usually generated by placing a -50-V bias on the Faraday cage) produces a high-contrast (hard) image of blacks (facets of the specimen directed away from the detector) and whites (facets of the specimen directed toward the detector). An image produced by topographic contrast from backscattered and secondary electrons produces a softer (less contrast) image because of the collection of secondary electrons from almost all the surfaces.

While viewing the specimen, the operator will often want to know the orientation of the specimen in relation to the detector. Topographic contrast allows an easy interpretation of this orientation (Fig. 8-30). Using a cone-shaped specimen as an example, those areas of the cone directed toward the detector are bright on the cathode ray tube. In the light microscope analogy, the eye is in the position of the electron beam, and the light illuminating the specimen is in the position of the detector.

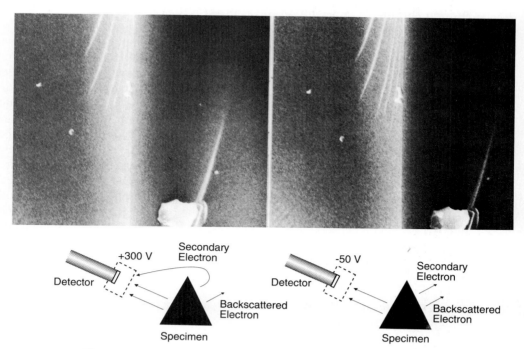

Figure 8-29 Topographic contrast. Facets of the specimen directed toward the detector appear brighter in the viewing cathode ray tube. (a) Almost all the secondary electrons are collected if the Faraday cage of an Everhart–Thornley detector is biased to +300 V. However, almost all the backscattered electron signal consists of those backscattered electrons generated toward the detector. (b) No secondary electrons are collected by the detector when the Faraday cage is biased to −50 V. Those facets of the specimen that are directed away from the detector now appear darker in the viewing cathode ray tube.

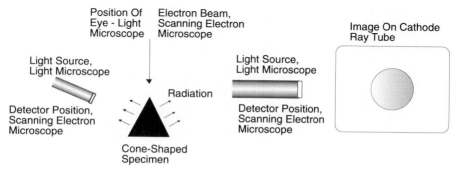

Figure 8-30 Analogy between the image generated in a light microscope and that generated in a scanning electron microscope. The detector in the scanning electron microscope is in an analogous position to the light source in a light microscope.

Atomic Number (Compositional) Contrast

A *smooth* specimen with no *topographic relief* composed of a single element will produce pixels on the viewing cathode ray tube of approximately the same gray level, resulting in a homogeneous image. This is because each pixel on the specimen is emitting approximately the same number of backscattered and secondary electrons, which are collected by the Everhart–Thornley detector. However, different amounts of secondary and particularly backscattered electrons are emitted from areas of the specimen that contain different elements (Fig. 8-31). The number of backscattered electrons is strongly dependent on atomic number (see Chapter 6). Because the number of backscattered electrons emitted from the specimen (backscattered coefficient) increases with atomic number, portions of the specimen containing higher atomic number elements are brighter in the viewing cathode ray tube. The greater the difference in atomic number, the greater is the difference in contrast. Elements separated by only one atomic number have a minimal contrast difference. Aluminum ($Z = 13$) has a contrast difference of only 6.7% from silicon ($Z = 14$). Elements with greater differences in atomic number have greater contrast differences. Aluminum ($Z = 13$) has a 69% contrast difference from gold ($Z = 79$).

From the preceding discussion, it is seen that a usable image may not be obtained if the specimen does not have a certain degree of inherent topographical and/or atomic number contrast, even when the scanning electron microscope is operating at maximum resolution.

Figure 8-31 Atomic number contrast. This specimen contains aluminum on the left and copper on the right. The surface topography is approximately the same in both portions. The copper (right) portion of the specimen is ligher than the aluminum portion because copper is of a higher atomic number and emits more backscattered electrons per beam electron bombarding the specimen.

Dwell Time.   The amount of time (dwell time) that the electron beam rests on a particular pixel on the specimen will determine the number of secondary and backscattered electrons that are available to construct the gray level of the pixel on the cathode ray tube. The longer the dwell time is, the more secondary and backscattered electrons that are emitted from the specimen, the less the signal for that pixel has to be amplified by the photomultiplier, and the less electronic noise that is introduced.

There are two opposing factors in producing an image on the cathode ray tube: (1) a slow raster is necessary to produce sufficient signal to construct the image, and (2) a relatively fast raster is necessary to view the specimen on the face of the cathode ray tube. At low magnifications, the pixel size on the specimen is large, so the beam spot size can be large, resulting in the production of a large number of secondary and backscattered electrons. This produces a large amount of signal that can be used to construct the gray level of each pixel. Relatively fast scan rates can be used here, since it is not necessary to have long dwell times on each pixel. At high magnifications, the beam spot size is small and contains few electrons. This requires slower scan rates to create sufficient secondary and back-scattered electrons at each pixel location to produce the required amount of signal for the construction of the gray level of each pixel without large amplification of the signal by the photomultiplier, which leads to noise in the signal.

DISTORTIONS OF THE IMAGE

The image on the cathode ray tube in a scanning electron microscope is a two-dimensional reconstruction of the three-dimensional specimen. The image is made up of a number of raster lines and picture elements (pixels). As such, there are three types of image distortions that can appear during reconstruction of the image: (1) projection distortions, (2) tilt distortions, and (3) moiré effects. However, these image distortions are only seen if the specimen is a grating or some similar object with regular spacings. Most specimens examined in the scanning electron micro-scope are irregular, and these distortions are not seen.

Projection Distortions

The distance from the objective aperture to the *center* of the raster pattern on the specimen is less than the distance from the objective aperture to the *edge* of the raster pattern on the specimen. Therefore, the electron beam travels a greater distance to the edge of the raster pattern than to the center (Fig. 8-32). Instead of being a symmetrical rectangle, the raster pattern is drawn out at the corners, so it covers a distorted rectangle. This produces a change in magnification across the raster pattern as the image is constructed on the cathode ray tube. The amount of magnification of a portion of the specimen is inversely proportional to the close-ness of the raster lines on the specimen. The closer the raster lines are on the

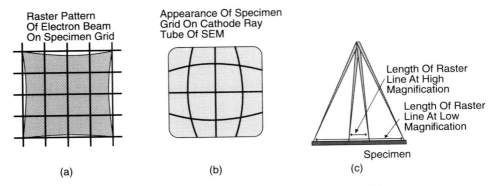

Raster Pattern Of Electron Beam On Specimen Grid

Appearance Of Specimen Grid On Cathode Ray Tube Of SEM

Length Of Raster Line At High Magnification

Length Of Raster Line At Low Magnification

Specimen

(a)   (b)   (c)

Figure 8-32 Projection distortion. (a) The farther the position of the specimen from the optic axis, the longer is the raster line on the specimen. This leads to an asymmetrical raster pattern. (b) An image of the specimen in (a) on the cathode ray tube in the scanning electron microscope. The closer the raster lines of the electron beam on the specimen, the farther apart are those features of the specimen on the cathode ray tube. (c) At low magnifications a large area is rastered on the specimen and the electron beam is far from the optic axis during the beginning and end of the raster. At high magnifications the electron beam rasters a small area on the specimen, and the electron beam is relatively close to the optic axis during the whole raster, resulting in little projection distortion of the specimen.

specimen, the greater is the magnification of that area of the specimen on the cathode ray tube. In projection distortion, the raster lines on the specimen are closest at the optic axis, resulting in greater magnification of these specimen features on the cathode ray tube. The raster lines are farther apart at the edge of the raster pattern on the specimen, resulting in the least magnification of these areas on the cathode ray tube. At a magnification of $10\times$ and a working distance of 10 mm, there is a 20% distortion in magnification at the edge of the raster pattern relative to the center of the raster pattern. At magnifications above $100\times$, the projection distortion becomes negligible because of the small scan angle at these magnifications.

Tilt Distortions

Tilting a specimen results in portions of the specimen being closer to the objective lens, while other portions of the specimen are farther from the objective lens. The closer an area of the specimen is to the objective lens, the closer together and shorter the raster lines are on the specimen and the greater the magnification of that portion of the image on the cathode ray tube (Figs. 8-33 and 8-34). The magnification reading on the scanning electron microscope is only accurate for that area of the tilted specimen at optimal focus (at crossover of the electron beam). Most modern electron microscopes have **tilt correction**, which corrects the tilt distortion of the image by reducing the length of the scan perpendicular to the

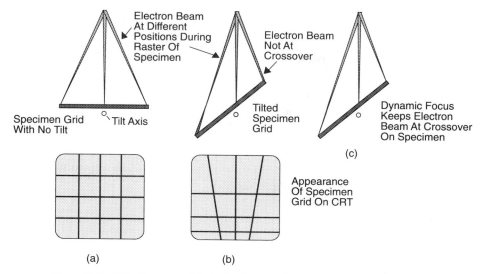

Figure 8-33 Tilt distortion. (a) At 0° tilt a specimen grid is accurately reconstructed on the cathode ray tube in the scanning electron microscope (ignoring
projection distortion). (b) When the specimen grid is tilted, those areas of the
specimen closer to the objective lens will be rastered by lines that are closer
together, while the opposite happens with areas of the specimen that are far from
the objective lens. This results in features of the specimen close the objective lens
being reconstructed farther apart on the cathode ray tube than areas that are far
from the objective lens. Tilting the specimen will probably result in those areas of
the specimen farthest from the tilt axis being out of focus because the electron
beam is no longer at crossover on these areas. (c) Dynamic focusing. In dynamic
focusing the strength of the objective lens is varied as the electron beam is rastered
over the specimen. This results in the electron beam staying at crossover on the
specimen (providing the specimen is flat).

tilt axis by a factor cos θ (the magnification perpendicular to the tilt axis is less
than the magnification parallel to the tilt axis by a factor equal to cos θ, where θ
is the tilt angle).

Tilting the specimen results in a change in the working distance of various
parts of the specimen relative to the objective lens. This results in various parts
of the specimen being at different focal lengths. If the specimen tilt is great enough,
then the depth of focus of the electron beam will be exceeded at the edges of the
raster pattern perpendicular to the tilt axis, and these parts of the specimen will
be out of focus on the cathode ray tube. Some scanning electron microscopes have
dynamic focusing, which compensates for differences in the focal length of parts
of the specimen. In dynamic focusing, the strength of the objective lens is varied
linearly as the raster pattern progresses over the specimen, keeping the specimen
at the crossover point of the electron beam (and therefore at focus). The lens has
great strength at the top of the raster pattern (on the portion of the specimen

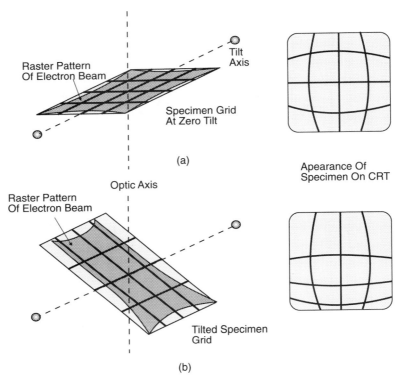

Figure 8-34 Projection and tilt distortion together. (a) At 0° tilt, only projection distortion is seen with a specimen grid. (b) Both projection and tilt distortion are illustrated with a specimen grid.

closest to the objective lens), with the lens strength decreasing as the raster pattern proceeds over the specimen. Dynamic focusing works well on a flat, planar object, but not very well on a rough, irregular specimen.

Moiré Effects

The image on the cathode ray tube of the scanning electron microscope has a regular periodicity determined by the number of lines and picture elements (pixels) of the image. Moiré fringes can appear on the image if the specimen has a periodicity that approaches that of the image on the cathode ray tube (Fig. 8-35). The moiré fringes appear wherever there is superposition of the grating of the specimen on the grating of the image (for example, the raster lines and pixels). Three methods can be used to determine if the features seen on the cathode ray tube are moiré fringes or real features of the specimen: (1) Rotating the specimen will result in a change of position of the moiré fringes, but not of real features on the specimen.

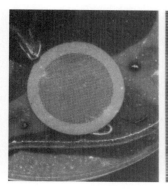

Figure 8-35 (a) Micrograph of a 500-mesh grid at 10 lines/cm. (b) Micrograph of the same grid at 2.5 lines/cm, showing moiré effects.

(2) Changing the magnification should eliminate or change the appearance of the moiré fringes. (3) Changing the number of scan lines per frame should result in elimination of the moiré fringes (Fig. 8-35).

THE HARD COPY

The operator frequently needs a hard copy after obtaining an image of the specimen on the viewing cathode ray tube of the scanning electron microscope. This hard copy usually consists of a photograph or a print from a videoprinter.

Photographing the Image

Scanning electron microscopes have two cathode ray tubes. The viewing CRT has a long-retention phosphor that holds the image for a comparatively long time for viewing by the operator. In comparison, the photographic CRT has a short-retention phosphor. The short-retention phosphor produces a higher-quality image on film since there is no long-term emission of photons to excite adjacent pixels and produce a more diffuse image. A camera box with a film holder, lens, and aperture fits over the photographic CRT (Fig. 8-36). The lens focuses the light rays from the photographic CRT onto the film in the film holder. The aperture controls the amount of light reaching the film from the photographic CRT. Usually, the aperture is left at a particular setting and is not changed by the operator. Instead, the amount of light reaching the film is controlled by the brightness and contrast setting on the photographic CRT and the output from the photomultiplier tube.

The film in the film holder can be either conventional photographic film or Polaroid film. Conventional film is economical, but it requires processing and printing. Polaroid film is more expensive, but the print is obtained within a minute. Despite its cost, Polaroid film is used by most operators of scanning electron microscopes.

Film
Holder

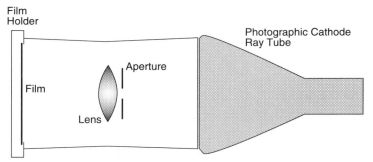

Photographic Cathode
Ray Tube

Aperture

Film

Lens

Figure 8-36 Photographic apparatus on a scanning electron microscope. The image is carried to the film in the film holder by a lens. The aperture controls the amount of light reaching the film.

Conventional film.   Conventional film consists of a plastic or cellulose acetate base that supports an emulsion containing silver halides (Fig. 8-37). The silver halides are compound salts of silver, usually silver bromide, silver iodide, and silver chloride. On the other side of the base is an antihalation dye that helps to prevent light from reflecting back from the rear of the film to form halos. The antihalation dye gives the back of the film its characteristic dark appearance before processing. Processing removes this dye, leaving the film colorless.

Each silver halide crystal in the emulsion contains silver and bromide (iodide or chloride) ions in a cubical arrangement (Fig. 8-38). The crystals are not perfect structures; if they were, they would not react to light. Some of the silver ions are not held in the cubical structure and are able to move about. There are also impurities, such as silver sulfide, in the crystal that are referred to as sensitivity specks. The film is exposed to a single slow raster of the photographic cathode ray tube. The slower the raster is, the more time that is available to produce the gray level of each individual pixel, resulting in less noise and better quality of the photograph. A photon of light from the photographic cathode ray tube strikes a silver bromide crystal. The energy of the light photon is transferred to an electron of a negative bromide ion, knocking the electron out of the ion. This electron roams the crystal

Gelatin Coat

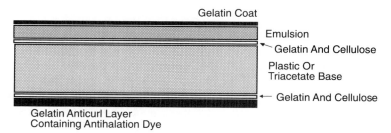

Emulsion
Gelatin And Cellulose

Plastic Or
Triacetate Base

Gelatin And Cellulose

Gelatin Anticurl Layer
Containing Antihalation Dye

Figure 8-37 Structure of photographic film.

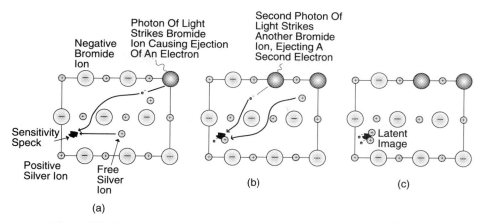

Figure 8-38 Formation of a latent image. (a) A photon of light strikes a bromide ion in a silver bromide (halide) crystal, ejecting an electron from the bromide ion. The electron travels to an impurity (sensitivity speck), making it negative. A positive free silver ion is attracted to the negative sensitivity speck. (b) The process is repeated as a second photon of light strikes the crystal. (c) The silver bromide crystal now has a latent image consisting of the negatively charged sensitivity speck.

until it stops at a sensitivity speck. The sensitivity speck is now negatively charged and pulls a free positive silver ion to the sensitivity speck. The process is repeated as other photons of light strike the crystal, bringing more silver ions into the sensitivity speck. The sensitivity speck with the attached silver ions is the **latent image**, an invisible chemical site that initiates the conversion of the whole crystal to metallic silver during the next process.

The film is taken to the darkroom, where it is developed. In the developing process the film is first placed in a developer (metol, phenidone, or hydroquinone) that donates electrons to the development centers formed by the latent images. The donated electrons cause the formation of metallic silver from the silver halides in these areas. There is typically an amplification of 10 million times in the amount of metallic silver in the areas of the film with latent images. Those silver halides without a latent image are not transformed into metallic silver by the developer. At the end of the developing, the film is placed in a weak acetic acid stop bath that halts the developing process. The undeveloped silver halides are still in the emulsion and will slowly form metallic silver if left there. In the next step, the silver halides are washed out of the emulsion with a fixer, usually sodium thiosulfate, although the more rapid acting ammonium thiosulfate is used in rapid fixer solutions. The resulting negative is washed with water to remove any fixer associated with the negative. Finally, the negative is placed in a weak detergent solution (such as Kodak Photoflo) that causes the water to evenly run off the negative instead of forming water droplets during the subsequent drying.

The resulting dry negative contains a negative image of the specimen. There are black areas on the negative (areas with metallic silver) wherever the light from the photographic cathode ray tube has struck the film. The negative is printed in the darkroom to produce the positive print of the specimen.

A number of types of film can be used for recording images off scanning electron microscopes. The films vary in (1) sensitivity to different wavelengths of light, (2) sensitivity to the total number of light photons striking the film, and (3) the size of the metallic silver grains in the final negative.

Sensitivity to Different Wavelengths of Light. Black and white films can be either panchromatic or orthochromatic. **Panchromatic films** are sensitive to all colors of light and have to be handled and processed in total darkness. **Orthochromatic films** are sensitive to only certain wavelengths of light and can be handled and processed under a safelight that emits wavelengths of light that do not sensitize the film. Orthochromatic black and white film is primarily sensitive to invisible ultraviolet, violet, and blue wavelengths, and the film is processed under a yellow safelight. Orthochromatic film containing only silver bromide in the emulsion layer is sensitive to wavelengths of light up to 520 nm. The addition of 5% silver iodide sensitizes the film up to 560 nm.

Sensitivity to the Total Amount of Light. Different types of film have different sensitivities to the total number of light photons striking the film. Currently, three different methods are used for rating the sensitivity of a film: (1) the **ASA number** from the American National Standards Institute, where the doubling of the speed of the film is indicated by the doubling of the ASA number, (2) the **DIN number** from the Deutsche Industrie Norm, where the doubling of the number indicates a rating increase of 3, and (3) the **ISO number** from the International Standards Organization. In an ISO number as 25/15°, the first number (25) refers to the equivalent ASA number, while the second number, containing the ° (15°), is equivalent to the DIN number (Table 8-4). With either system, the higher the number

TABLE 8-4 EQUIVALENT SPEED RATINGS OF FILM IN ASA AND DIN

ASA	DIN
16	13
32	16
64	19
125	22
200	24
400	27
800	30
1250	32
2000	34

is, the more sensitive the film is to light and the less light that is necessary to obtain the proper exposure.

Size of Metallic Silver Grains in the Final Negative. The size of the silver halide crystals in the emulsion will determine how large the metallic silver grains are in the developed negative. Films that have large silver halide crystals produce negatives with large metallic silver grains. The positive print produced from such a negative is relatively grainy. Negatives with large metallic silver grains cannot be enlarged very much before the size of the silver grains limits the resolution of the features in the print. *There is an inverse relationship between the size of the metallic silver grains in the final negative and the speed of the film.* The larger the silver grains are, the less light that is needed to produce an image. Conversely, the smaller the silver grains are, the more light that is needed to produce the image. Usually, relatively large negatives (4 by 5 in.) are used in scanning electron microscopes, so little enlargement of the negative is required to produce the final print. As such, relatively fast films (ASA 320) can be used. However, if 35-mm film is used (requiring a large magnification to produce a usable print), slower, finer-grained film is required.

Polaroid Film. Polaroid film consists of three basic parts (Fig. 8-39): (1) a *negative* consisting of light-sensitive silver halides spread on a transparent film base (Polaroid type 55 positive negative film) or on a paper base (Polaroid type 52 or 53 film), (2) a *positive print* that consists of paper coated with gelatin in which there is a trace of silver or silver sulfide atoms (the amount of silver is so small that the paper appears white), and (3) a *pod of chemicals* that contains a silver halide solvent plus a powerful alkali in a sticky, thick solution that spreads easily.

To take the photograph (Figs. 8-39 and 8-40), the film is inserted into the Polaroid film holder until the metal strip at the far end of the Polaroid film is engaged by a spring lock. The cardboard covers of the film are partially withdrawn. The cardboard covers carry the positive with them as they are withdrawn, leaving the negative exposed to the darkened photographic cathode ray tube through the lens and aperture of the camera box. A single, slow raster of the cathode ray tube is made, with the image being carried to the Polaroid negative by the lens. The raster on the photographic cathode ray tube usually consists of only 1000 lines if only a positive print of the image is required. This produces about 10 lines/mm, which is greater than the 4 to 5 lines/mm that the human eye can perceive, so the lines forming the image cannot be seen. An image of 2000 lines per raster should be generated if a photographic negative is needed that can be later enlarged on a print. The photographs are taken at a slow raster patter of around 32 ms/per line in order to ensure adequate signal to produce the gray levels of the pixels.

After the single, slow raster of the photographic cathode ray tube is complete, the Polaroid negative contains the latent image of the specimen (Fig. 8-41c). Next, the cardboard backing and the positive are pushed back into the camera back (Fig. 8-41d). A lever on the film holder is pushed down from the load position to the

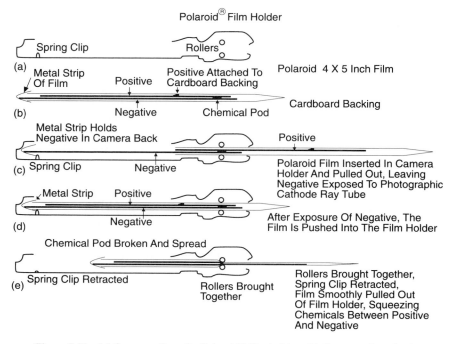

Polaroid® Film Holder

(a)

Spring Clip   Rollers

Metal Strip Of Film   Positive   Positive Attached To Cardboard Backing   Polaroid 4 X 5 Inch Film

(b)

Negative   Chemical Pod   Cardboard Backing

Metal Strip Holds Negative In Camera Back   Positive

(c)

Spring Clip   Negative   Polaroid Film Inserted In Camera Holder And Pulled Out, Leaving Negative Exposed To Photographic Cathode Ray Tube

Metal Strip   Positive

(d)

Negative   After Exposure Of Negative, The Film Is Pushed Into The Film Holder

Chemical Pod Broken And Spread

(e)

Spring Clip Retracted   Rollers Brought Together   Rollers Brought Together, Spring Clip Retracted, Film Smoothly Pulled Out Of Film Holder, Squeezing Chemicals Between Positive And Negative

Figure 8-39   (a) Cross section of a Polaroid® film holder. (b) Cross section of a 4-by 5-in. Polaroid film. (c) Polaroid film ready for the image to be carried from the photographic cathode ray tube of the scanning electron microscope. The Polaroid film has been inserted into the film holder and the cardboard enclosure retracted. The metal strip of the film and also the negative are held by the spring clip. The cardboard enclosure and the attached positive are retracted from the film holder, leaving the negative open to the lens. (d) After the latent image has been formed in the negative, the cardboard enclosure is pushed into the film holder. (e) The spring clip is retracted, releasing the metal strip, and the negative and the rollers are brought together. The film is removed from the film holder and the rollers squeeze the pod of chemicals and spread the chemicals between the positive and negative, beginning the development process.

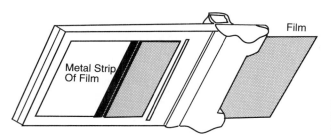

Film

Metal Strip Of Film

Figure 8-40   A 4- by 5-in. Polaroid® film holder with a piece of Polaroid film partially inserted.

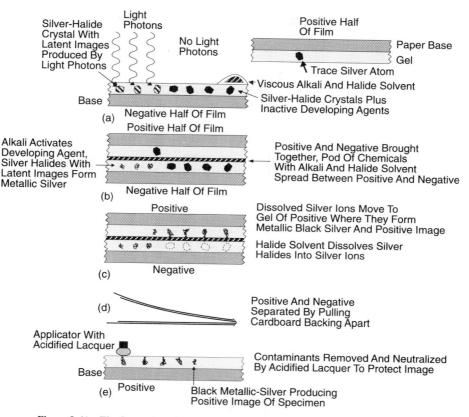

Figure 8-41 The formation of an image in Polaroid 4- by 5-in. film. (a) The positive is retracted from the film back before taking a photograph, leaving the negative material exposed to the light from the photographic cathode ray tube. The light photons strike portions of the negative, producing latent images in the silver halide crystals in those areas. (b) The positive is pushed back into the film holder, and the whole film is removed from the film back. During this process the pod of chemicals is broken and the alkali and halide solvent are spread between the positive and negative. The alkali activates the developing agent in the negative, producing metallic black silver grains in those silver halide crystals with a latent image. (c) The silver halide solvent dissolves those remaining silver halides (those not formed into metallic black silver grains) into silver ions, which move into the positive. The activated developing agents produce black metallic silver grains in the positive in these areas. (d) The positive and negative are pulled apart. (e) An acidified lacquer is applied is some types of Polaroid film (types 52 and 55) to protect the image in the positive.

process position. Two things happen when this level is pushed down: (1) The lever pushes a trigger to the left, which retracts the spring clip that holds the metal strip of the Polaroid film, allowing the film to be retracted from the film holder. (2) A pair of rollers come together behind the pod of chemicals.

The whole film is now retracted from the film holder by pulling the film smoothly out to the right (Fig. 8-41e) (jerking the film out unevenly results in uneven spreading of the chemicals and uneven development). During this process, the pod of chemicals is broken by the rollers and spread between the positive and negative. The negative contains a developing agent in addition to silver halide crystals. Up to this point the developer has been inert because it needs alkaline conditions to be activated. The pod of chemicals contains alkali that, after spreading, activates the developer in the negative (Fig. 8-41b). The developer causes those silver halide crystals with latent images to form metallic black silver, producing a negative image of the specimen. This process occurs rapidly, within seconds of mixing. The pod of chemicals also contains a solvent that relatively slowly dissolves the silver halide crystals that do not have latent images into silver ions. The developing agents, followed by the dissolved silver ions, diffuse out of the negative and into the positive. The developing agents first donate electrons to the relatively small number of silver atoms that were always in the positive, making them negative. A few seconds later, the dissolved silver ions diffuse over from the negative emulsion and are attracted to the charged silver atoms, combining to form metallic black silver and the image of the specimen. A positive print of the specimen is produced in the paper.

It takes 15 to 60 s (depending on the type of film) from the time the chemicals are spread between the positive and negative until the image is complete on the film. The positive print will turn eventually brown with certain types of Polaroid film (for example, types 52 and 55). This is due to the presence of the silver solvent and the alkali-activated developing agents, which soon oxidize. Also, the positive print can easily be scratched. To prevent these processes, the positive print is swabbed with an acidified fast-drying lacquer. This neutralizes the alkali developing agents, removes contaminants, and protects the print. Most of the images photographed on the scanning electron microscope now use type 53 Polaroid black and white film, which does not require swabbing with the lacquer. In type 53 film, the positive is formed on white plastic, which has an underlying layer into which alkali and other chemicals sink and become neutralized, leaving the metallic silver image in the layer above. The developing agents that are used in type 53 film oxidize to a colorless product.

Videoprinters

Hard copies from videoprinters offer a fast, economical alternative to photographic film. Prints can be obtained within a minute from a videoprinter for 5% to 10% of the cost of photographic film. Recent advances in videoprinter technology

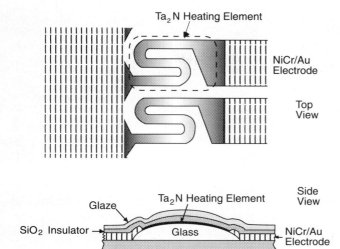

Figure 8-42 A top and side view of resistive heating elements used to produce the image in videoprinters using thermal paper.

have resulted in hard copies of sufficient quality for some applications, resulting in significant cost savings.

Videoprinters use thermal printing technology to produce an image. Thermal printers use heat to produce a chemical change that gives rise to visible dyes on paper. Videoprinters have printheads with 300 heating elements per inch that use the conversion of electrical energy into thermal energy by joule heating within an electrically resistive material. A common design for a printhead is shown in Fig. 8-42. The main component of the printhead is the NiCr/Au electrode that supplies electrical current to the resistor heating element, which in this case is Ta_2N. The resistor heating element is a thin film that has been given a meandering shape by photolithography. The heating element has a relatively large resistance to electrical current. The NiCr/Au electrode supplies electrical current to the resistor heating element, causing the latter to heat (typically to about 200°C) because of its relatively large resistance. Individual heating elements are energized by providing current through each NiCr/Au electrode. The thin-film resistor heating element has a layer of electrically insulating glass on the bottom and SiO_2 on the top. Glaze on the top of the printhead acts as a protective coating that minimizes wear of the printhead. This is important because the printhead is brought into close mechanical contact with the thermal paper during printing.

In printing, the thermal paper is moved over the printhead (Fig. 8-43). Pulses of current are applied selectively to the heating elements. The electrical current to each of the heating elements is directed by a print controller that contains a digital bit map of the image of the specimen (see Chapter 9). The pulse of current

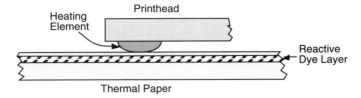

Figure 8-43 In a videoprinter, the thermal paper is drawn tightly over the heating elements. Black dyes are formed if the heating elements are hot enough to cause the binder in the reactive dye layer to melt.

results in rapid heating of the resistor heating elements. This heat is conducted to the thermal paper, where a significant temperature rise occurs in the paper. The thermal paper contains a layer of dye-forming chemicals within binder material. At room temperature the dyes are in a leuco (colorless) form, and the paper has the appearance of white smooth paper (the paper is smooth to reduce friction with the printhead). There are usually two components to the dye, a **dye precursor** and a **color former**, which are trapped in separate binders. When they are heated by the heating elements to the "blush" temperature, the binders melt, combining the reactants to a colored dye. The reaction between a fluoran dye precursor and bisphenol A color former to yield a black dye is shown in Fig. 8-44. The intensity

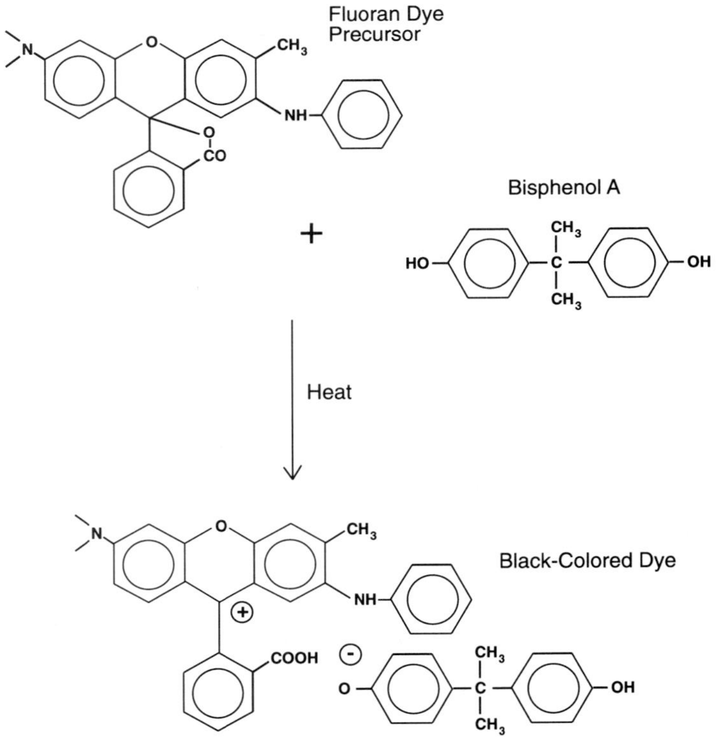

Figure 8-44 Reaction between a leucodye (colorless dye precursor) and bisphenol A (color former) to produce a colored dye.

of the dye on the paper is a function of the temperature of the heating element and the length of time that the heating element is hot. Following the heating pulse, the heating elements have to cool before the next set of pixels (pels) is printed on the paper. Most of the energy of the heating elements moves to the printhead support. The thermal absorbing ability of the printhead determines the printing rate, which is usually on the order of 1 ms (a couple of pages a minute).

The formation of a gray scale in the print can be accomplished in one of two ways. In the first, a 4 × 4 array of heating elements (16 heating elements) in the printhead is treated as a matrix to form a 16-unit gray scale. The heating of progressively more of the elements in the 16-pixel matrix results in a darker tone. For example, if a gray scale of $\frac{5}{16}$ is required, 5 of the 16 heating elements in the matrix are heated. This method results in an image of relatively poor resolution, because the information of the 16 pixels of the image is average to one tone. The second method uses a printhead that contains heating elements that change shape as they are heated. The greater the temperature of the heating element is, the greater the surface area of the heating element that presses against the surface of the thermal paper and the darker the gray scale. This method results in images of higher resolution.

STEREO PHOTOGRAPHS

Stereo photographs are taken by moving the camera about 5 in. (the distance between the left and right eye) between photographs and viewing the photographs in a stereo viewer. Stereo photographs can also be taken in the scanning electron microscope using one of two methods: (1) moving the specimen between photographs or (2) tilting the specimen between photographs.

Moving the Specimen between Photographs. This method works only at low magnifications (below 20×). After taking the first photograph, the specimen is moved approximately half the distance of the cathode ray tube and a second photograph is taken.

Tilting the Specimen between Photographs. This method works at all magnifications. The first photograph is taken, and a portion of the specimen is outlined on the cathode ray tube using a grease pencil. The specimen is tilted 4° to 10°. The specimen will move laterally during the tilting, unless the photographed area of the specimen is directly on the tilt axis (Fig. 8-45). The X and Y translation knobs are used to bring the specimen back into the areas of the specimen outlined on the cathode ray tube. However, the working distance has changed during the tilting (unless the photographed area of the specimen is on the tilt axis). Changing the working distance results in a change in the focus. It is now important that the focus *not* be adjusted by changing the current to the objective lens; otherwise, the image will rotate and the specimen will be at a different magnification on the cathode ray tube. Instead, the image is brought back to the same working distance

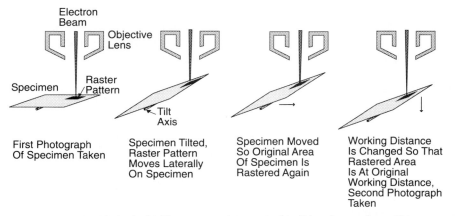

Figure 8-45   Method of taking stereo micrographs by tilting the specimen 4° to 10° between photographs. Tilting the specimen results in lateral movement and a change in the working distance (unless the specimen is on the tilt axis). The specimen has to be moved laterally back to the original working distance and the original rastered area that was in the first photograph before the second photograph is taken.

and focus by bringing the specimen back to the working distance of the first photograph. This is done by varying the working distance with the Z (working distance) knob on the door of the specimen chamber. Once the image is in focus using this method, the second photograph is taken and the two photographs are viewed with a stereo viewer (Fig. 8-46). A stereo viewer is an arrangement of lenses or mirrors that presents one photograph to each eye. The combined image from the two photographs results in parallax error that the brain interprets as depth. During viewing, it is important that the photographs be oriented properly. If the photographs are reversed, then the projections in the photographs will appear as depressions.

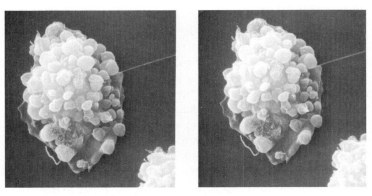

Figure 8-46   Stereo photographs of a cultured Chinese hamster ovary cell.

9

Image Processing

The image obtained from a scanning electron microscope can be manipulated to optimize the information available to the investigator. There are three basic ways that the image can be processed: (1) using **optical photographic** methods; (2) using **electrical analog** methods, or (3) using **digital** methods.

OPTICAL PHOTOGRAPHIC METHODS

This method basically involves printing a negative of the image using lenses available in the photographic enlarger. The contrast and brightness of the image are controlled by such factors as the darkness of the negative, the type of photographic paper, and the length of developing of the paper. The darkness of parts of the image can be changed by "dodging" portions of the negative during printing. The main advantage of processing the image using optical photography is its economy. However, only relatively crude image processing can be carried out.

ELECTRICAL ANALOG METHODS

In this method the electrical signal supplied from the detection system of the scanning electron microscope is manipulated to process the image. Most scanning electron microscopes have the ability to process the analog signal. These manipulations involve the whole image. More complex image-processing operations, such as manipulating a portion of the image, changing gray levels of the individual pixels, or colorization of an image, are not easily performed using the analog signal. Commonly used manipulations of the analog signal include (1) linear amplification, (2) nonlinear amplification (gamma), (3) contrast reversal, (4) area scan using Y-modulated raster lines, and (5) derivative signal used to produce an area scan.

In the following discussions, the contrast leaving the detector will be referred to as **natural contrast**, while the contrast produced after signal processing on the viewing or photographic cathode ray tube will be referred to as **image contrast**.

Linear Amplification

The photomultiplier takes the natural contrast leaving the detector and linearly amplifies it. This method produces satisfactory image contrast on the viewing cathode ray tube if the natural contrast leaving the detector has a satisfactory contrast range. For example, two pixels with a natural contrast difference of 10% leaving the detector will still have a contrast difference of 10% after undergoing linear amplification. A Y-modulated line scan (intensity modulated in the vertical direction) will have the same contours before and after linear amplification, although the heights of the peaks will be higher after linear amplification. However, the gray-level range between the two peaks is much greater after linear amplification, allowing the operator to more easily discern the two areas in an area scan.

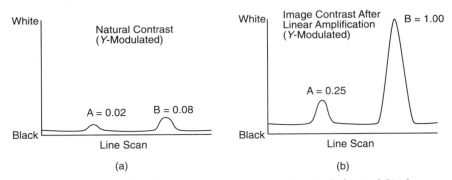

Figure 9-1   *Y*-modulated line scans of the same specimen (a) before and (b) after linear amplification of the signal. Linear amplification amplifies both peaks to the same extent.

In Fig. 9-1 the gray-level value of peak A is 0.02, while the gray-level value of peak B is 0.08. A difference in gray levels of 0.06 is just barely noticeable by the human eye, presuming that the eye can detect 16 gray levels and that 0.00 to 1.00 represents the range of gray levels. The contrast difference between the peaks is as follows:

$$\begin{aligned}
\text{contrast} &= \frac{\text{maximum value} - \text{minimum value}}{\text{maximum value}} \\
&= \frac{0.08 - 0.02}{0.08} \\
&= \frac{0.06}{0.08} \\
&= 75\%
\end{aligned}$$

The natural contrast as the signal leaves the detector is 75% between peaks A and B. The proper amount of linear amplification of this line scan would bring peak B to a value of 1.0, resulting in a value of 0.25 for peak B, still a contrast difference of 75%. However, the gray-level difference of 0.75 between peak A and B is now considerable, representing 12 detectable gray levels. This difference in gray-scale levels is readily seen by the eye. This linear amplification results in the maximum number of gray-scale levels without any clipping or blooming of the signal. Peak B has been pushed to a gray-scale level 1.0 (white). Pushing the gray-scale level of peak B further results in the peak flattening at the gray-scale level of 1.0 and a loss of contrast information from these pixels. This is called *clipping* since the top of the peaks in the *Y*-modulated line scan are removed. *Blooming* also results when the peaks are pushed over into the pure white. In blooming, the electron beam inside the cathode ray tube is so intense that it drives the phosphor of the adjacent pixels into the purest white, even though the adjacent pixels should have lesser gray-scale levels. Also, electronic noise can appear on the viewing cathode ray tube as a speckled snowy image if the signal is greatly amplified. A large amount of electronic noise during the brief period that the gray-scale level of a pixel is being determined can produce a large enough electrical pulse to drive the gray-scale level of the pixel enough so that it is more than 7% of the gray-scale level without the noise component. This would result in the pixel(s) being visible to the eye as lighter speckles or snow against the darker, true image of the specimen. This noise can only be reduced by increasing the dwell time on each pixel to average out the noise over a greater time period. This is normally evident to the operator, who finds that photographs (taken at a slower raster time and therefore longer dwell time) have less snow than the image on the viewing cathode ray tube (with faster raster times and shorter dwell times). Natural contrast levels as low as 0.1% can normally be linearly amplified before noise levels produce an image that has unacceptably high levels of noise.

Linear amplification with an Everhart–Thornley detector is accomplished by adjusting the gain across the photomultiplier (see Chapter 7). The scanning electron microscope will normally have a knob labeled **contrast** or **gain** with which the operator can linearly amplify the signal by boosting the gain across the dynodes of the photomultiplier. Linear amplification also results in an increase in the overall brightness, as well as contrast, of the signal. The brightness has to be adjusted after linear amplification by means of a knob usually labeled **brightness** or **black level** to reduce the brightness. The signal in the cathode ray tube has two compo-

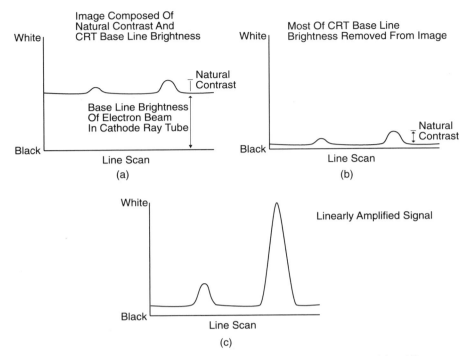

Figure 9-2   Procedure used in linearly amplifying a signal. (a) A *Y*-modulated line scan shows the image to be composed of natural contrast superimposed on the base line brightness of the cathode ray tube. (b) In linear amplification, the base line brightness is lowered so that the peaks will not clip after linear amplification. (c) The natural contrast is linearly amplified, resulting in a greater range of gray levels in the signal.

nents: (1) the natural contrast, which is superimposed on (2) the base line brightness of the cathode ray tube (Fig. 9-2). In linear amplification, the base line brightness is first largely removed; otherwise, an image that is too bright on the cathode ray tube will result after linear amplification.

Nonlinear Amplification (Gamma)

In nonlinear or gamma amplification, the gray-scale levels are amplified to different degrees. In Fig. 9-3 the straight line between signal in and signal out shows the direct relationship that occurs in linear amplification, where the contrast difference of two pixels going into the amplifier is the same as the contrast difference of the pixels coming out of the amplifier. In nonlinear amplification, the contrast difference of the pixels going into the amplifier is not the same as the contrast difference of the pixels leaving the amplifier. The degree of amplification is along a specified curve, the degree of curvature being under the control of the operator.

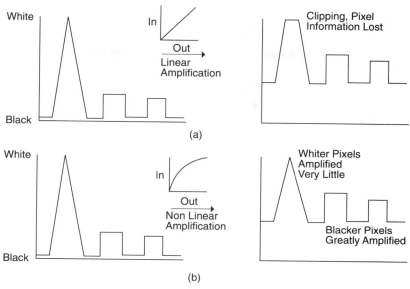

Figure 9-3 *Y*-modulated line scans of a specimen that has one white feature in an otherwise dark background. The features at the black end of the gray scale have insufficient contrast difference to be recognizable. (b) Linear amplification helps to separate the dark features, but the white feature is clipped, leading to loss of information. (b) Nonlinear (gamma) amplification of the same line scan amplifies the whiter pixels only a small amount, while greatly amplifying the blacker pixels.

Nonlinear amplification is usually used when there is a dark feature in an otherwise bright image, or vice versa. In Fig. 9-4 there is a "cave" in the specimen from which only a few secondary and backscattered electrons escape to form the image. This results in a dark area in an otherwise bright image. There is detail in the cave, but it cannot be seen. If linear amplification is used to bring out the detail in the cave, the rest of the image would be clipping at the white end of the gray-scale range, resulting in loss of information from the remainder of the image. Nonlinear amplification is used to increase the contrast of details in the cave without producing clipping at the white end of the gray scale (Fig. 9-4).

Contrast Reversal

In contrast reversal, the contrast level of the pixels is subtracted from 1 to give a negative image of the specimen. Almost all scanning electron microscopes have a switch that allows the operator to obtain contrast reversal. Contrast reversal is usually used to produce negative images that can be directly photographed with a negative film to produce positives that can be used for projection slides. Contrast reversal is also used when a scanning electron microscope utilizes specimen current to produce an image of the specimen. The specimen current forms a negative

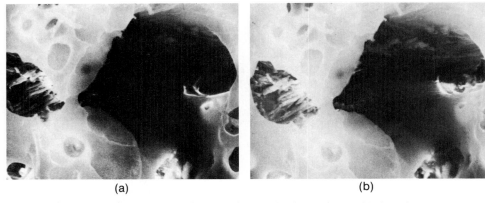

(a) (b)

Figure 9-4 (a) A scanning electron micrograph of a specimen with deep depressions that are black in a normal Everhart–Thornley image. (b) Nonlinear (gamma) amplification results in very little amplification of the whiter pixels, with a large amount of amplification of the blacker pixels, thereby bringing out detail in the depressions.

image of the backscattered and secondary electron image (see Chapter 7). Contrast reversal of the negative specimen current image produces a positive image representing the backscattered and secondary electrons escaping from the specimen.

Area Scan Using *Y*-modulated Raster Lines

Normally, an area scan is made with the intensity modulated in the *z*-direction (toward the operator) to produce the image of the specimen on the cathode ray tube. However, it is also possible to produce an area scan using a number of *Y*-modulated raster lines. The height of the peaks in each *Y*-modulated line represents the intensity of the radiation received by the detector (Fig. 8-12).

Derivative Signals

Taking a first derivative of the signal allows the operator to emphasize areas of the specimen that produce relatively rapid changes in signal as the specimen is scanned by the electron beam. Figure 9-5 shows a *Y*-modulated line scan and the first derivative of the same line scan. The first-derivative signal produces large changes in gray-scale level when there is a change in the contrast as the beam moves over the specimen. Flat areas occur in the first-derivative signal when there are no changes in contrast as the electron beam moves over the specimen. A first-derivative area scan produces a signal that results in an image that has the changes in signal accentuated. Unfortunately, the features of the specimen (such as sharp edges) must be perpendicular to the direction of the movement of the electron beam in order to be accentuated (for example, with the electron beam moving from left to right, the specimen feature needs to be oriented top to bottom) (Fig. 9-6). If the specimen feature is oriented parallel to the scan lines, the feature will

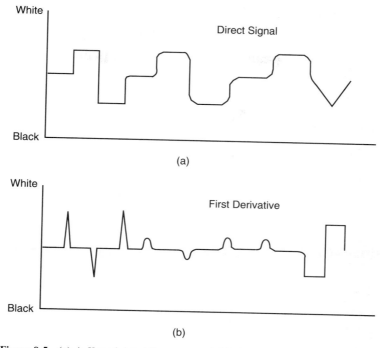

(a)

(b)

Figure 9-5 (a) A *Y*-modulated line scan and (b) the first derivative of the same line scan.

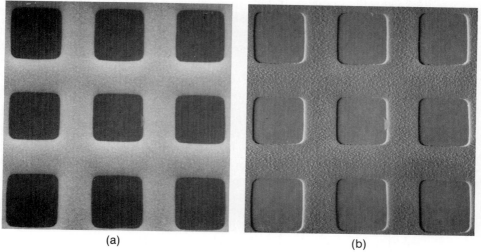

(a)

(b)

Figure 9-6 (a) A secondary and backscattered electron image of a copper grid. (b) A first-derivative image of the same copper grid.

not produce a significant first-derivative signal and will be difficult to see. This problem can be eliminated by the use of **orthogonal scanning** in which the specimen is scanned twice at right angles. First, the specimen is scanned from left to right and then from top to bottom, with the signals from the two scans summed to produce the final image.

DIGITAL IMAGE PROCESSING

Digital image processing has three advantages. First, there is much better definition of gray levels because the gray levels of the pixels are delineated as numbers that define distinctions too fine to be seen by the naked eye. Second, a digital image does not deteriorate, and a copy of the digital image stored as a magnetic or optical disc is as good as the original. Third, it is possible to manipulate the image through a large number of operations with a precision unattainable with optical or analog systems. Digital image processing, however, does have two disadvantages. The image-processing operations can take some time and the necessary equipment is more expensive.

There are four basic parts of a digital image-processing system: (1) the image processor, (2) the digital computer, (3) the display and recording devices, and (4) the storage device (Fig. 9-7).

Image Processor

The signal from the detection system of the scanning electron microscope is fed into the image processor. The digitizer in the image processor divides the signal into a specified number of time intervals (Fig. 9-8). The intensity of the signal at each time interval is used to assign a number to that time interval. The number assigned to each time interval is called an **image element** or **pixel element** (**pixel** or **pel**). The pixel numbers are arranged in a two-dimensional map, with pixel positions synchronized with the electron beam scanning the specimen. In the example in Fig. 9-8, the signal at each pixel is digitized into one of eight (0 to 7) different numbers, depending on the intensity of the signal from the detection system.

Bits and Bytes.   In digital terms, the eight numbers associated with the signal in Fig. 9-8 are a three-bit binary representation of the range of input signals. A bit can have a value of 0 or 1. Therefore, three bits (2^3) combined together can have eight different combinations representing eight different numbers (Fig. 9-8).

The digital number of each pixel represents the gray-scale level of each pixel. A gray-scale with only eight values, however, produces a poor quality image. Fortunately, digital computers usually store pixel gray-scale values as bytes composed of eight bits (2^8) or 256 values. This provides a gray scale ranging from 0 (black) to 255 (white), with the numbers in between corresponding to the gray scale between black and white. This results in a sufficient number of gray-scale values to provide a digital image with more than enough gray-scale tones.

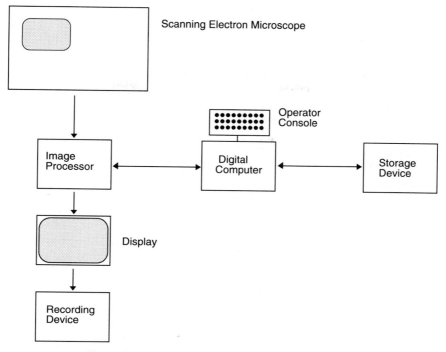

Figure 9-7 Components of an image-processing system.

Binary Code: 000 001 010 011 100 101 110 111
Numbers: 0   1   2   3   4   5   6   7

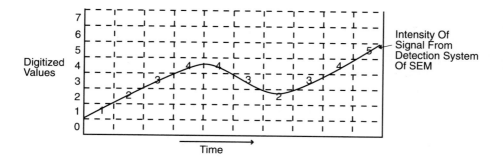

Pixel Values From Above Graph: 1   2   3   4   4   3   2   3   4   5

Pixel Values In Binary Code: 001 010 011 100 100 011 010 011 100 101

Figure 9-8 Process of digitization of an analog signal. The analog signal is analyzed at discrete time intervals. The strength of the analog pulse determines the digital number assigned to the time interval. The digital numbers are in the form of a three-bit (2^3) binary code in this example.

Bit Maps.   The image processor contains a **framer grabber** that digitizes a single frame (scan) of the specimen. The digitizer sends the image to the digital computer as a bit map, where it is stored in the storage module. The bit map contains the pixels used to make up the image. The most commonly used pixel array in image processors is 512 by 512 or, in other words, 512 lines per image with 512 pixels per line. Some systems use 256 by 256 arrays, resulting in individual pixels that are too large to produce good resolution images on monitors. Excellent resolution can be attained by 1024 by 1024 arrays. However, increasing the number of pixels in an array to higher numbers has two disadvantages. The first is the increased time necessary to process the image and the second is the large amount of storage space that is required. An image composed of 256 by 256 eight-bit bytes requires 65,536 bytes of storage, an image of 512 by 512 bytes requires 262,144 bytes (0.26 megabyte; a megabyte is 10^6 or 1 million bytes) of storage, and a 1024 by 1024 byte image requires 1,048,576 bytes (about 1 megabyte) of storage.

Each pixel in a digitized image has an x (horizontal) and y (vertical) address in addition to a number that indicates the gray-scale value of the pixel. Thus every pixel in Fig. 9-9 has an address and gray-scale value, which can be written as $f(x,$

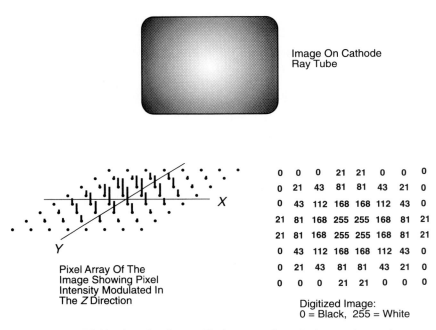

Figure 9-9   Digitization of an image. The image on the cathode ray tube consists of a bright spot in the center with progressive darkening toward the edges. Modulation of this image in the Z-direction (normally toward the viewer) shows the brightest pixels to be in the center of the image. The image is digitized into an 8 by 8 pixel array with a gray scale from 0 (black) to 255 (white), an eight-bit (2^8) gray scale.

y). For example, the pixel position at an *x* of 4 and a *y* of 3 has a gray-scale value (*f*) of 168, or in other words $f(x, y) = 168$ (4, 3).

Up to now it has been assumed that the scanning electron microscope is producing the same number of lines as the digitizer. In actual fact, this is rarely so. Therefore, the image processor has to contain algorithms (mathematical calculations) that convert the image of the specimen in the scanning electron microscope to the number of lines used by the image-processing unit.

Digital Computer

An image processor can be equipped with internal processing abilities or the image processor can be interfaced to a general-purpose computer to provide versatility as well as ease of programming. A well-equipped microcomputer (basically a personal computer) or minicomputer (a small mainframe) is usually dedicated to an image-processing system and provides sufficient computing capability to process images from the scanning electron microscope at a reasonable cost.

Display And Recording Devices

Color television monitors are the principal devices used to display the processed image. The monitors are driven by the output of the image display module in the image processor. The hard copy from the display monitor is either done photographically or from a videoprinter, as described in Chapter 8.

Storage Devices

Providing adequate bulk storage facilities is an important aspect of an image-processing system. A digital image of 512 by 512 pixels quantitized into eight bits requires 0.26 megabyte of storage. The two most commonly used storage media are magnetic discs and optical discs. Magnetic discs that store from 10 to 700 megabytes are commonly used. A 10-megabyte disc would store about forty 512 by 512 images composed of eight-bit pixels, while a 700-megabyte disc could store about 2800 images. Optical discs, based on laser writing and reading technology, can approach a storage of four gigabytes (giga = 10^9 or 1 billion), or 16,000 images.

PROCESSING THE IMAGE

Five basic image-processing operations are used with digital images from scanning electron microscopes (Fig. 9-10): (1) **point-processing operations** that vary the value of *each pixel* according to some arithmetic or logical scheme; (2) **spatial-processing operations** in which the value of each pixel is considered as part of a *family of pixels*; (3) **geometric-processing operations** that produce new images whose *dimensions, orientation, or shape* is different from the original image; (4) **multiple-image**

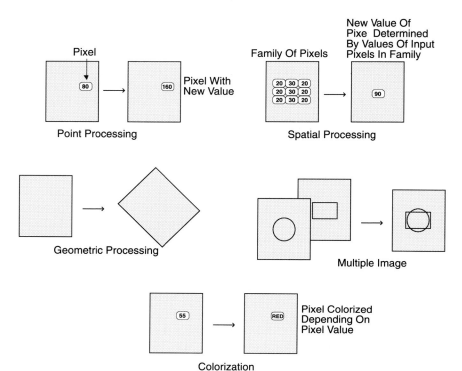

Figure 9-10   The five basic image-processing operations performed on digital images produced by scanning electron microscopes.

operations that *combine two or more digital images*; and (5) **colorization operations** by which different colors are assigned to specific gray-scale levels of the image.

Point-processing Operations

Point-processing operations process the value of each pixel according to an arithmetic or logical scheme. Adding a constant value to the gray-scale level of each pixel to make the image uniformly brighter is an example of a point-processing operation.

Look-up Tables.   Image-processing operations, such as point processing, are almost always implemented as table look-up operations applied to the data as they are being displayed (Fig. 9-11). The actual pixel values of the image are not changed, but are translated by the look-up table as they are sent to the display. The digital values from the refresh memory for the display are sent to a look-up table, which requires only one entry for each input pixel value, instead of a value for each pixel in the image array. Thus, if a 256-value gray scale (0 to 255) is used, only 256 entries are needed in the look-up table, compared to 262,144 if each pixel

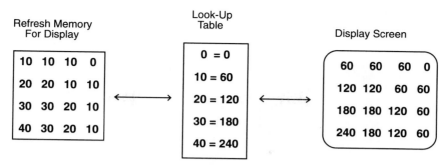

Figure 9-11 Digital image-processing operations usually use a look-up table to produce a processed image of the specimen on the display.

of a 512 by 512 array were treated individually. Another advantage of using a look-up table is that no data are lost, since the process is reversible.

The way a look-up table is used in processing data can be seen in Fig. 9-11. In this example the image is simplified to 16 pixels. The unprocessed image has pixel gray-scale values ranging from 0 to 40 out of a possible 256. Such a small range of gray-scale values means that the unprocessed image would be displayed as an image with low contrast. A look-up table that expands the range of contrast values is loaded into the system. The resulting output image is of higher contrast, with a pixel gray-scale range of 0 to 240.

Examples of Point-processing Operations

The following commonly used point-processing operations will be used as illustrations of the technique.

Brightness (bias, black level) modification. The brightness of the image can be increased or decreased by uniformly adding or subtracting a number to the gray-scale value of each pixel in the image (Figs. 9-12 and 9-13). A disadvantage

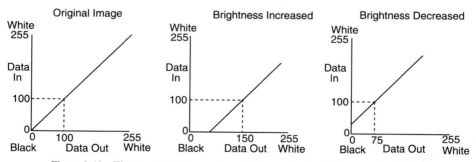

Figure 9-12 The process of increasing and decreasing the brightness of a digitized image. The digital gray-scale value of an input pixel is followed horizontally to the diagonal line. This point on the diagonal line represents the gray-scale value of the output pixel.

221

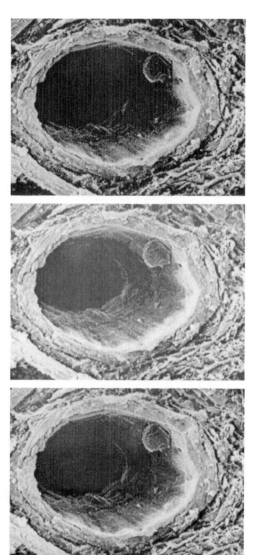

Figure 9-13 Digitized micrograph of a blood vessel, illustrating how the brightness can be increased or decreased by uniformly manipulating the gray-scale values of the pixels. The gray scales of the three micrographs are represented by the graphs in Fig 9-12.

of increasing or decreasing the brightness in this manner is that it is possible to clip pixel values. Decreasing the pixel brightness in Fig. 9-14 takes all the pixels with a value less than 70 and makes them all zero. This results in a loss of information from these pixels.

Contrast (gain) modification.   In this processing mode, numbers are added or subtracted from the gray-scale pixel values to increase or decrease con-

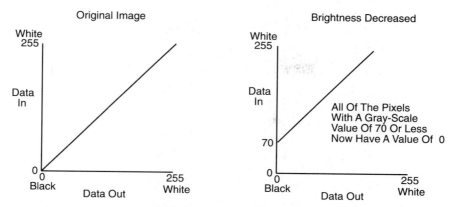

Figure 9-14 Some pixel values can be clipped when the brightness of an image is decreased (or increased). In this case, decreasing the brightness has taken all the input pixels with a gray-scale value from 0 to 69 and assigned them a number of 0.

trast. Figure 9-15 is a histogram showing the number of pixels in an input image in each of the gray-scale levels. The gray-scale levels are grouped between 150 and 200, meaning that the image encompasses only 50 gray-scale levels out of a possible 256. The means that the histogram is derived from a relatively low contrast image (Fig. 9-16). To obtain the maximum contrast, the distribution of pixels in the gray-scale levels is changed so that the pixels with a value of 150 have a new value of 0, and the pixels with a gray-scale value of 200 have a new gray-scale value of 255. The pixels with gray-scale values between 150 and 200 are scaled appropriately to yield an image with maximum contrast (Fig. 9-16).

Gamma (nonlinear) modification. Gamma (γ) refers to the slope of the curve or line that relates the input gray-scale values to the output gray-scale values. If the gray-scale values of input pixels are the same as the gray-scale values of the

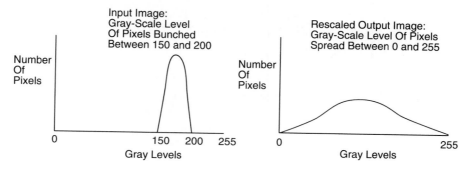

Figure 9-15 Example of rescaling of the gray-scale values to produce an image with the maximum number of gray-scale values.

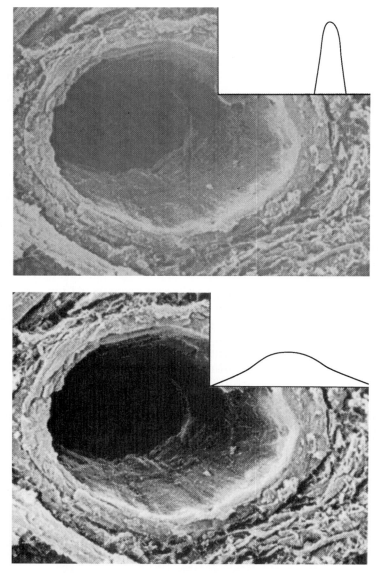

Figure 9-16   Image rescaled according to the diagrams in Fig. 9-15. The original image is composed of only 50 gray-scale levels. After rescaling, the image is composed of 256 gray-scale levels and the contrast is maximized.

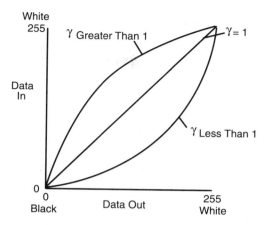

White
255

γ Greater Than 1

γ = 1

Data
In

γ Less Than 1

0
0
Black

Data Out

255
White

Figure 9-17 Gamma is the slope of the line that represents the relationship between the gray-scale levels of the input and output pixels.

output pixels, then the relationship is linear and gamma is equal to 1 (Fig. 9-17). Gamma modifications change this linear relationship to a curved relationship. Common gamma modifications include logarithmic, exponential, square, square root, inverse, and variable modifications of the input data.

Logarithmic. The logarithmic transform is of the general form

$$\text{output data} = \log (\text{input data})$$

This transformation (Figs. 9-18 and 9-19) spreads out the gray-scale values of the pixels with low numbers and compresses the gray-scale values of the pixels with high numbers. The output image thus has details in the dark areas that are more visible at the expense of detail in bright areas.

Exponential. This is the inverse of the logarithmic transformation and is written as

$$\text{output data} = \text{exponential (input data)}$$

This transformation (Figs. 9-18 and 9-19) enhances detail in light areas, but decreases detail in dark areas.

Inverse. In an inverse transformation, a negative image of the specimen is produced by subtracting the gray-scale value of each pixel from the highest possible value in the gray scale (Figs. 9-18 and 9-19). Thus, if an eight-bit gray-scale of 0 to 255 is used, the equation would be

$$\text{output data} = 255 - \text{input data}$$

Square Root. In a square root transformation (Figs. 9-20 and 9-21), the output data are equal to the square root of the input data:

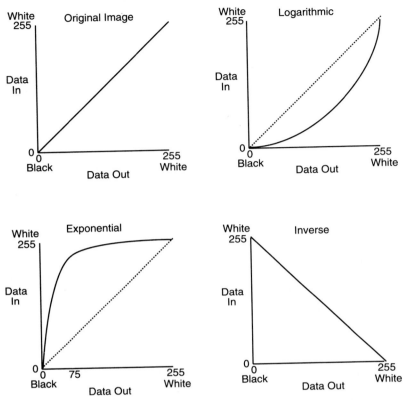

Figure 9-18 Graphs representing logarithmic, exponential, and inverse transforms of data.

$$\text{output data} = \sqrt{\text{input data}}$$

The transformation brings out detail in the light areas of the image, but not to the extent of the exponential transformation.

Square.   In this transformation (Figs. 9-20 and 9-21) the gray-scale values of the pixels are squared to yield the gray-scale values of the output pixels:

$$\text{output data} = (\text{input data})^2$$

Similarly to the logarithmic transformation, squaring the input data brings out detail in the dark areas of the image.

Variable.   Any gray-scale value (within the range being used) can be as-

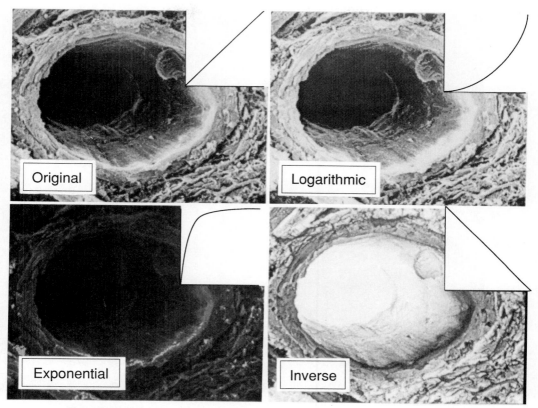

Figure 9-19   Micrographs of logarithmic, exponential, and inverse transforms of an image.

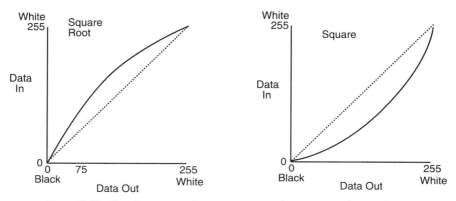

Figure 9-20   Graphs representing square root and square transforms of data.

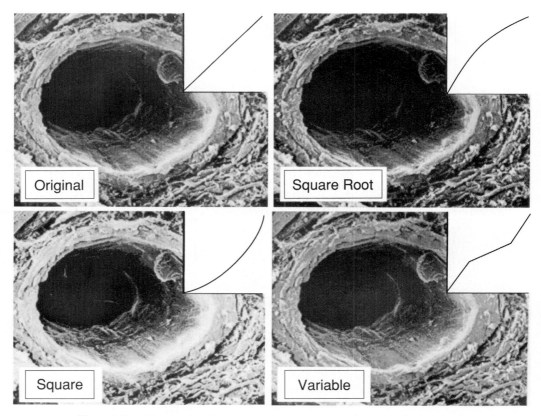

Figure 9-21   Micrographs of square root, square, and variable transforms of an image.

signed to the pixels of the input data. This process is done on a look-up table and is commonly controlled by the operator with a mouse. The mouse is used to draw the relationship between the gray-scale values of the input and output data on a graph that is displayed on the monitor screen. Thus a gray scale of 0 to 255 in the look-up table can have any value assigned to it. In Figs. 9-21 and 9-22, the input to the look-up table is manipulated so that the gray-scale values of the pixels between 0 and 75 and 180 and 255 are reduced, while the pixel gray-scale values between 75 and 180 are expanded.

Spatial-processing Operations

In this technique, the gray-scale level of each pixel is manipulated according to the gray-scale level of the pixel *and its neighbors*. For every pixel in the input image, a value for the output image is calculated by taking a weighted average of the pixel and its neighbors. The square array of the pixel and its neighbors is called

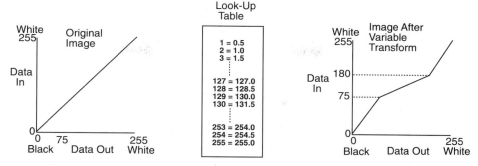

Figure 9-22 A variable transform does not modify the image according to a single mathematical formula. Instead, the output gray-scale values are fed into the look-up table by the operator, who usually uses a mouse to determine the slope of the relationship.

a **kernel**. The number of pixels in the kernel can vary, although a 3 by 3 pixel array is usually used. Increasing the size of the kernel beyond a 3 by 3 pixel kernel greatly increases the computation time. Even a 3 by 3 pixel kernel requires nine multiplications and nine additions for each pixel in the image. A 515 by 512 pixel image using an eight-bit gray scale requires 2,359,296 multiplications and 2,359,296 additions for a kernel of 3 by 3 pixels. Such a manipulation can take some time. The amount of time required to process an image is usually divided into three different classes of time (the actual time interval will depend on the capacity of the computer):

1. *Real time*: Here the operator initiates a function and the displayed image changes smoothly as the operator changes the parameters. Image processing in real time fully exploits hand to eye coordination for the rapid optimization of the image. Some point-processing operations occur in real time.

2. *Near-real time*: In near-real time the image changes within 3 s after operator input. This is fast enough to avoid operator frustration or boredom but forces conscious comparison of the image, which is less effective than the subconscious human feedback of real-time responses. Kernel operations usually occur in near-real time.

3. *Nonreal time*: These operations take more than 3 s to implement.

The mechanics of spatial processing are fairly straightforward. For a 3 by 3 pixel kernel, an array of nine weighted coefficients, called a **mask**, **template**, or **filter**, is defined. The mask is placed over the image with the pixel to be modified in the center of the mask. The gray-scale levels of each of the nine pixels in the mask is multiplied by the factors in the mask, and the results are added to yield the output gray-scale value of the pixel in the center of the mask. The operation is repeated for each pixel in the input image.

A large number of different types of masks can be used to spatially process images; only a few will be discussed here.

High-pass filter. A high-pass filter *accentuates* areas of the image that have large changes in gray-scale levels from one pixel to the next. These areas are called **high-frequency spatial components** in the jargon of image processing. Those areas that have little or no change in gray-scale values from one pixel to the next are called **low-frequency spatial components**.

A common high-pass mask is composed of a 9 in the center location with -1 in the surrounding pixel locations (Fig. 9-23). The large positive number in the center of the mask says that the center pixel is carrying a high weight, while the surrounding pixels carry less weight in opposing the center pixel. If the center pixel has a gray-scale level much greater than its neighbors, the surrounding pixel effect becomes negligible, and the output value of the center pixel becomes a brightened version of the input center pixel. Thus, in Fig. 9-23, by placing the high-pass filter over a pixel, the gray-scale level of the pixel is raised from its input

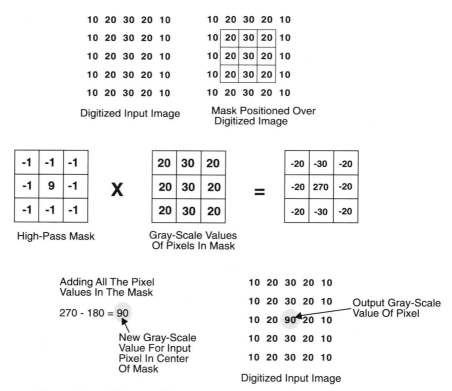

Figure 9-23 A high-pass filter accentuates areas of the image that have a change in gray-scale values.

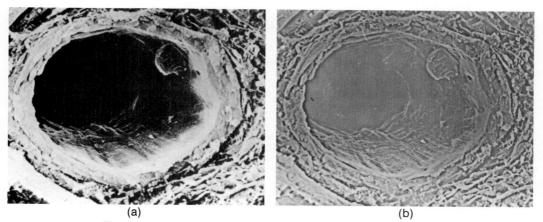

(a)   (b)

Figure 9-24 (a) A scanning electron micrograph of a vessel and (b) the same micrograph after subjection to a high-pass filter.

value of 30 to an output value of 90 (on a gray scale of 0 to 255). Repeating the process for each pixel in the original input image results in an image in which the high-frequency components have more contrast and stand out in the image (Fig. 9-24).

The high-pass filter does not accentuate low-frequency spatial components. In Fig. 9-25, all the pixels covered by the 3 by 3 pixel high-pass mask have the same gray-scale value. After manipulating the pixel values according to the mask formula, the output gray-scale value for the center pixel is the same as the input value. Thus a high-pass filter emphasizes high-frequency components while leaving low-frequency components unchanged.

High-pass filters are used in edge detection systems, an important part of enabling machines to recognize objects by their shapes. There are a large number of high-pass filters that can be used in edge detection systems.

Low-pass filter. A low-pass filter has an inverse effect to that of a high-pass filter. Low-pass filters *attentuate* the high-frequency components (areas of the image that show large contrast changes). A common low-pass filter is shown in Fig. 9-26. As the low-pass mask moves over high-frequency areas, the gray-scale value of each pixel is replaced in the output by the average of the pixel gray-scale values in the mask, producing an image with the high-frequency detail removed (Fig. 9-27). The visual effect is that of blurring to produce a mellowed image. Like the high-pass filter, a low-pass filter has less effect on low-frequency details in the image.

Low-pass filters are used when an investigator is interested in analyzing the low-frequency components of an image. A common undesirable high-frequency component of scanning electron micrographs is noise or snow. Noise is represented as white dots on micrographs and is commonly the dimension of a single pixel. Since they are white (usually equal to a gray-scale level of 255 on a 0 to 255

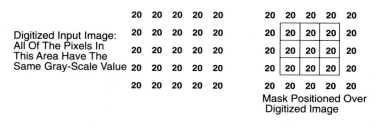

Digitized Input Image:
All Of The Pixels In
This Area Have The
Same Gray-Scale Value

Mask Positioned Over
Digitized Image

-1	-1	-1
-1	9	-1
-1	-1	-1

High-Pass Mask

X

20	20	20
20	20	20
20	20	20

Gray-Scale Values
Of Pixels In Mask

=

-20	-20	-20
-20	-180	-20
-20	-20	-20

Adding All The Pixel
Values In The Mask

180 - 160 = 20

New Gray-Scale
Value For Input
Pixel In Center
Of Mask

Output Gray-Scale
Value Of Pixel Same
As Input Gray-Scale
Value

Digitized Input Image

Figure 9-25   A high-pass filter leaves unchanged areas of the image that have no change in gray-scale levels.

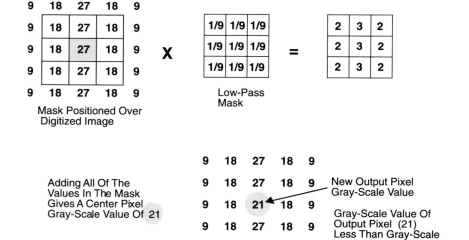

1/9	1/9	1/9
1/9	1/9	1/9
1/9	1/9	1/9

Low-Pass
Mask

X

=

2	3	2
2	3	2
2	3	2

Mask Positioned Over
Digitized Image

Adding All Of The
Values In The Mask
Gives A Center Pixel
Gray-Scale Value Of 21

New Output Pixel
Gray-Scale Value

Gray-Scale Value Of
Output Pixel (21)
Less Than Gray-Scale
Value (27) Of Input Pixel

Figure 9-26   A low-pass filter reduces the contrast in areas of the image with changes in gray-scale values.

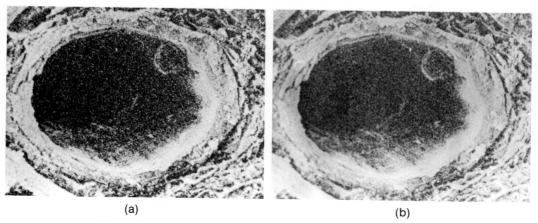

(a)   (b)

Figure 9-27   (a) A scanning electron micrograph of a vessel that contains a large amount of noise, and (b) the same micrograph after one scan with a low-pass filter. The low-pass filter blurs the details and makes the noise less noticeable.

gray scale), they usually have a gray-scale value very different from the gray-scale value of neighboring pixels. Such high-frequency components are attenuated by low-pass filters.

Geometric-processing Operations

Geometric-processing operations result in new output images of dimensions, shapes, or orientation different from the input images. Pixel data from an input image is transformed into new spatial locations according to a geometric algorithm, producing an output image of altered characteristics.

Image scaling.   Scaling deals with enlarging and shrinking an image or portion of an image. In scaling an image, a **scaling factor** is multiplied by the x and y address of the input pixel location to yield the x and y address of the output pixel. A scaling factor of 2 yields a magnification of 2, while a scaling factor of $\frac{1}{2}$ produces an image one-half the size of the original. In Fig. 9-28, an 8 by 8 pixel array is subjected to a scaling factor of 2. A pixel is retrieved from the input image at (x, y) coordinates of $(4, 4)$. The scaling factor multiplies x and y by 2 to yield new pixel locations of $(8, 8)$. This process is repeated for all the pixels to produce the new image. Theoretically, the new image contains information only in every other line and every other pixel since the new 8 by 8 pixel array is derived from only a 4 by 4 pixel array in the original image. To alleviate this problem during processing, the pixel to the immediate left of the scaled pixel is replicated into the empty pixel location. Similarly, the line of pixels above an empty line is replicated into the empty line. A micrograph composed of a 512 by 512 pixel array has had a scaling factor of 2 applied to it in Fig. 9-29.

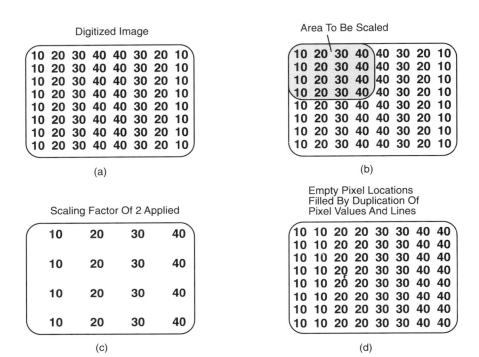

Figure 9-28   Process of scaling an image.

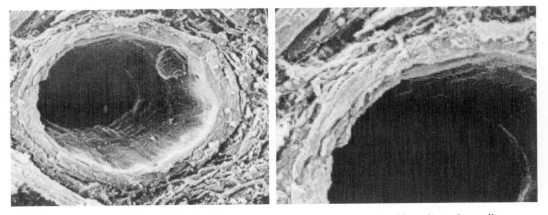

Figure 9-29   Scanning electron micrograph of a blood vessel subjected to a 2 × scaling.

Shrinking an image follows the same principles as magnifying an image. If the scaling factor is ½, then a 512 by 512 pixel array is shrunk to a 256 by 256 pixel array.

Image translation.   Image translation results in the up and down and side to side movements of an image (Figs. 9-30 and 9-31). In an image translation a number is added or subtracted to the x coordinates of all the pixels in the image to move the image horizontally. A number is added or subtracted from all the y coordinates of all the pixels in the image to move the image vertically. For example, suppose the image is to be moved 100 pixel positions to the right and 200 pixel positions up (Figs. 9-30 and 9-31). Each x pixel coordinate has 100 added to its address, while each y coordinate has -200 added to its pixel address. Figure 9-30 has a 512 by 512 pixel array making up the image. A pixel at an (x, y) address of $(137, 377)$ is moved to the location $(137 + 100, 377 - 200)$ or $(237, 177)$.

Image rotation.   This function allows the user to rotate the image about a central point (Fig. 9-32). Rotations of 90° (90°, 180°, 270°) are relatively easy since the algorithm maps an input pixel to an output pixel. There is a problem when the rotation angle is not a multiple of 90° because the input pixel is placed into four output pixel locations. This can be handled in one of two ways. In the first, the input pixel is assigned to the output pixel that is closest to the true output location. However, this results in straight lines in the input image becoming stepped lines, and the zigzag effect results in an undesirable image. The second method divides the total value of the gray scale of the input pixel and places the value into the four valid output pixel locations. Since other pixels are also mapped to fractional pixel locations in the output image, there can be as many as four input pixels contributing gray-scale values to one output pixel location.

Rubber sheet transformations.   This type of manipulation of image data is used to correct for rubber sheet distortions of the image. Rubber sheet distortions can be likened to having an image of the specimen on a sheet of rubber and

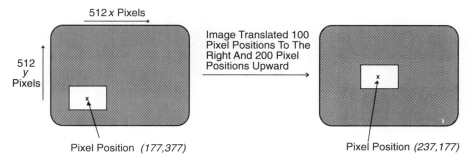

Figure 9-30 Process of translating an image.

then stretching the rubber sheet by pinning it down at selected points so that the image is geometrically contorted. Projection and tilt distortions of the image in the scanning electron microscope (see Chapter 8) are examples of rubber sheet transformations.

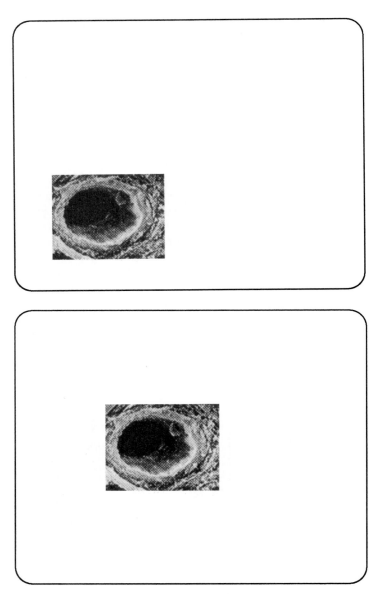

Figure 9-31   Micrograph translated according to the scheme in Fig. 9-30.

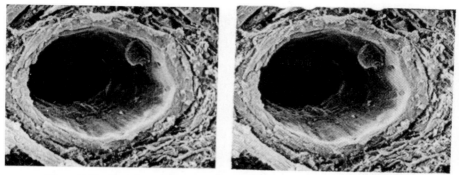

Figure 9-32   Rotation of a scanning electron micrograph of a blood vessel.

Rubber sheet transformations are usually not easy to produce or to subsequently correct. It is difficult to accurately define the mapping of the distorted input pixel locations to produce output pixel locations that accurately reflect the shape of the original image. Rubber sheet transformations can be carried out by using one of two methods. In the first, a massive look-up table is created that maps input pixels to new output locations. This is a difficult and time consuming procedure. In the second method, the rubber sheet transformation is reduced to a straightforward mathematical model by envisioning what areas of the image are to be stretched and to what degree. The stretching is often thought of as being carried out by pinning points of the rubber sheet down. The points are called control points. An input to output pixel transformation is calculated by defining mathematically where the control points are located and what degree of stretching has occurred between them. Such rubber sheet transformations are not easy to produce and often must be tailored to specific problems at hand.

Multiple-image Operations

These operations involve the combination of two or more digital images. The gray-scale level from two or more digital images is entered into the same (x, y) pixel coordinates to produce the pixel value. A number of **combination functions** are used to produce combined images, including the simple mathematical operators + (add), − (subtract), × (multiply), and / (divide). A common multiple-image addition/division operation is used in **frame averaging** to reduce random picture noise. Multiple images can be used to add an enhanced image of a specimen to the original to produce an image that is easier for the operator to visualize (Fig. 9-33).

Colorization of the Black and White Image

The black and white image from the scanning electron microscope can be colorized much the same way that old black and white movies are colorized. Each pixel in

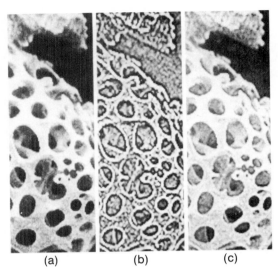

(a)   (b)   (c)

Figure 9-33 A micrograph of a
radiolarian (a) that was subjected to a
high-pass filter to accentuate the edges
of the specimen (b). (c) A multiple
image of the first two micrographs
showing enhanced edge detail on the
original micrograph. (*Micrographs
courtesy of Kevex Instruments, Inc.*)

the digitized image has a number assigned to it depending on the gray level of the
pixel in the analog image (Fig. 9-34). Once the image is digitized, a color is assigned
to a number or group of numbers to colorize the parts of the image with the
appropriate digital numbers. The process is repeated with other colors and shades
for other pixel numbers.

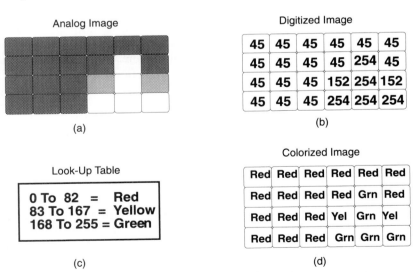

Figure 9-34 Colorization of an analog image. First the analog image is digitized
by assigning numbers from 0 (for black pixels) to 255 (for white pixels) to the gray-
scale levels of the analog images. The image is colorized by assigning colors to
groups of numbers in the digitized image.

10

Vacuum

The operation of the electron microscope requires that the electron beam have unimpeded passage down the column, through the lenses and apertures, to collide with the specimen. It is not possible to maintain a coherent electron beam down the column if a significant number of gas molecules are present; the collisions of the electrons with the gas molecules result in a diffuse electron beam that is unusable. In addition, a poor vacuum results in a large number of gas molecules in the area of the electron-emitting cathode. If thermionic emission from a tungsten cathode is used, the oxygen molecules quickly combine with the hot tungsten atoms, producing tungsten oxides that evaporate at the operating temperature of the tungsten cathode, causing failure of the cathode. Poor vacuum in electron microscopes using lanthanum hexaboride or field emission cathodes results in layers of contaminating molecules coating the cathode, restricting the emission of electrons. Electron microscopes using thermionic emission from tungsten cathodes require vacuums of 10^{-6} mbar (10^{-4} Pa). Electron emission from a lanthanum hexaboride cathode requires 10^{-7} mbar (10^{-5} Pa), while field emission requires vacuums of 10^{-9} mbar (10^{-7} Pa) (see Chapter 2). A number of different types of pumps, gauges, seals, and valves are present in electron microscopes to maintain the required vacuum. These will be discussed in the following. However, before discussing the components of the vacuum system, some of the principles of vacuum need to be considered.

GASES AND VAPORS

Gases consist of molecules traveling in straight lines at a high rate of speed within a containing space. The speed of gas molecules is quite high. At room temperature (20°C or 68°F), a hydrogen molecule has an average speed of about 1800 m/s, whereas a nitrogen molecule has a speed of about 500 m/s. The gas molecules collide frequently with each other and with the walls of the container. The force exerted per unit area on the walls of the container by the colliding molecules is the gas pressure. Pressure of fluids (gases and liquids) is defined as the quotient of the perpendicular force on a surface and the area of this surface (unit of force per unit surface area). According to the metric system (Système Internationale d'Unités), the only recognized unit of gas pressure is the **pascal (Pa)**, which is equal to one newton per square meter (N/m²). A newton is equal to a body whose mass is one kilogram experiencing an acceleration of one meter per second per second. The derived unit **millibar (mbar)** (1/1000 bar) is also acceptable, with 100 mbar = 1 Pa. The units millibar and pascal are the only units used in this text. Other units (such as the commonly used torr) are given with their equivalents in Appendix II. Even though the pressure unit torr is no longer valid, a brief description is warranted because of its frequent use. One torr is that gas pressure able at 0°C to maintain a mercury column 1 mm in height. Normal atmospheric pressure is 760 torr or 760 mm of mercury (Hg). The difference between mbar and torr is often insignificant (1.333 torr = 1 mbar), especially at low gas pressures.

    Gases are defined as materials in the gaseous state at room temperature. Gases cannot be compressed into the liquid state or solid state at room temperature. **Vapors** are also material in the gaseous state. However, vapors are condensable into liquid at room temperature. Dry air is an example of a gas, while water vapor is an example of a vapor. To produce liquid air, it is necessary not only to use a high degree of compression but also to do so at a low temperature.

Effect of Pressure and Temperature on a Gas

Gas molecules are separated from each other by large distances. Within a gas, the space actually occupied by gas molecules is small; most of the space is free space. The large amount of free space within a gas makes compression of the gas fairly easy. The compression reduces the large free space and brings the gas molecules close together. Atmospheric gas can be compressed to a liquid equal to $\frac{1}{1000}$ of the volume of the gas.

    As the pressure on a gas increases at a constant temperature, the volume of the gas decreases. In 1662, Robert Boyle discovered that at a constant temperature the volume of a gas decreases inversely as the pressure is increased. This generalization became known as Boyle's law:

$$\text{pressure} \times \text{volume} = \text{constant}$$

As the temperature of a gas increases, the volume of the gas increases if it is maintained at constant pressure. In 1882, Gay Lussac quantified the effect of temperature on gas volume. He found that for all gases the increase in volume for each degree centigrade rise in temperature was equal to approximately $\frac{1}{273}$ of the volume of the gas at 0°C. This was quantified as follows:

$$\text{volume of a gas} = \text{volume at 0°C} + \frac{\text{temperature}}{273.16} \times \text{volume at 0°C}$$

Total Pressure and Partial Pressure

The random motion of gas molecules results in the gas molecules completely filling a containment vessel. Two different gases, such as nitrogen and oxygen or any number of nonreactive gases, will by their motion mix with one another quickly. This mixture of gases will in many respects behave like a single gas, and the molecules of the various gases will collide with each other regardless of the similarity or dissimilarity of the gases. Furthermore, the total pressure exerted by the mixture will be determined by the total number of collisions between the molecules of all kinds and the walls of the container. The **partial pressure** of a given gas or vapor is that pressure that a gas or vapor would have if present alone in a container. The **total pressure** in a container is equal to the sum of the partial pressures of all the gases and vapors within it. In the electron microscope, it is atmospheric air that is removed by the vacuum pumps. The composition of atmospheric air and the partial pressure of each component are presented in Table 10-1.

TABLE 10-1 COMPOSITION OF ATMOSPHERIC AIR

	% by weight	% by volume	Partial pressure, mbar
N_2	75.51	78.1	792
O_2	23.01	20.93	212
Ar	1.29	0.93	9.47
CO_2	0.04	0.03	0.31
Ne	1.2×10^{-3}	1.8×10^{-3}	1.9×10^{-2}
He	7×10^{-5}	7×10^{-5}	5.3×10^{-3}
CH_4	2×10^{-4}	2×10^{-4}	2×10^{-3}
Kr	3×10^{-4}	1.1×10^{-4}	1.1×10^{-3}
N_2O	6×10^{-5}	5×10^{-5}	5×10^{-4}
H_2	5×10^{-6}	5×10^{-5}	5×10^{-4}
Xe	4×10^{-5}	8.7×10^{-6}	9×10^{-5}
O_3	9×10^{-6}	7×10^{-6}	7×10^{-5}
	Σ 100%	Σ 100%	Σ 1,013

Diffusion

In a container with a mixture of two or more kinds of gases, the gas molecules move randomly by the process of diffusion to produce a homogeneous molecular distribution of the gases. If gas A is introduced to one end of a container and gas B is introduced to the other end, more molecules of gas A will travel to regions of lower gas A concentration than travel in the opposite direction because of random molecular motion. The same occurs for gas B until the two gases are evenly mixed within the chamber.

Flow of Gases

There are two basic types of gas flow in a vacuum system: (1) **viscous flow**, where gas flows from a region of high pressure to low pressure, and (2) **molecular flow**, where gas flow depends only on the random movement of gas molecules out of a container. A transition region also exists at pressures where viscous flow merges into molecular flow.

Viscous Flow. Viscous flow occurs at gas pressures from 1 atm to about 10^{-3} mbar (10^{-1} Pa), the lower end depending on the size of the container. Viscous flow is easy to visualize; it is similar to the flow of water through a hose from an area of high pressure to low pressure. The flow occurs because the gas molecules are in constant and rapid motion, colliding with one another and moving in the downstream direction where fewer gas molecules are present. When moving toward an area of lower gas pressure, the colliding gas molecules have fewer collisions and move farther between collisions. The net effect is a rapid flow of molecules in the direction of lower gas pressure.

Molecular Flow. As the gas pressure is reduced, a point occurs (around 10^{-3} mbar [10^{-1} Pa] depending on the size of the vacuum system) at which the great majority of collisions of the gas molecules are with the walls of the container and no longer with other gas molecules. At this point, the **mean free path** of the molecules (the distance a gas molecule travels before colliding with another gas molecule) is greater than the dimensions of the vacuum container. When this occurs, the gas molecules no longer flow toward the area of low pressure because they are not primarily colliding with other gas molecules and no longer have a way of "sensing" the region of lower gas pressure. Only those gas molecules that happen by chance to move into the outlet of the vacuum system are captured by the pump and removed from the system, reducing the total gas pressure. This condition is known as **molecular flow**, and the flow of gas depends only on the dimensions of the evacuated area, the kinds of molecules present, and the temperature of the gas.

Outgassing

Gases are readily adsorbed to the surface of metals and are not easily removed, even in high vacuums. The **outgassing rate** is the rate at which a gas is desorbed from the walls of the inside of the vacuum system by the vacuum pumps. Electron microscopes are continually maintained at a high vacuum in order to prevent significant amounts of gas and vapors adsorbing to the interior surface and causing slow pumpdown rates during pumping of the vacuum system.

VACUUM PUMPS

A number of different types of vacuum pumps are used in electron microscopes. Rotary oil pumps are used to "rough" electron microscopes from 1 atm to about 10^{-3} mbar (10^{-1} Pa). At this gas pressure, either a turbomolecular pump or a diffusion pump, sometimes in conjunction with an ion getter pump, is used to bring the electron microscope to operating vacuum. Rotary oil pumps and diffusion pumps use fluids. The best possible vacuum that can be obtained with these pumps is the vapor pressure of the fluid used. Turbomolecular pumps and ion getter pumps do not use fluids and are able to attain better vacuums.

Mechanical Rotary Oil Pump

The rotary oil pump was designed by W. Gaede around 1905. In electron microscopes, the rotary oil pump is used as a **fore pump** to **rough** the vacuum system to 10^{-1} to 10^{-2} mbar (10 to 1 Pa) and to **back** a diffusion pump at the same pressures. The pump (Fig. 10-1) consists of a steel rotor that is placed eccentrically inside a steel casing (**stator**). The rotor has two opposed vanes that are held tightly against the inner wall of the casing by springs. The eccentrically placed rotor forms a third point of contact with the casing, dividing the pump into three enclosed spaces. A motor drives the rotor in the casing, sweeping the gas molecules from the vacuum intake to the exhaust portal. As the rotor turns, the gases trapped in the casing become compressed as the available space decreases. The compression ratio is very high, typically around 10^5. The exhaust valve opens when atmospheric pressure is exceeded (actually slightly above atmospheric pressure, since the exhaust valve is held by tension or by a spring) by compression of the gases by the rotor, and the gases are expelled to the outside of the pump. The exhaust valve is usually a feather or poppet valve. A **feather valve** consists of a strip of metal over the exhaust opening. A **poppet valve** consists of a housing containing a sealing metal disc with a spring to hold the disc in place.

The capacity of rotary oil pumps is expressed in terms of the volume of gas displaced per unit time (l^3/min or ft^3/min). A rotary pump used in a scanning

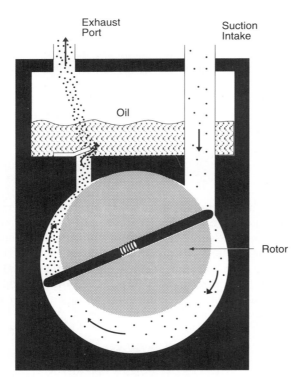

Exhaust
Port

Suction
Intake

Oil

Rotor

Figure 10-1 Diagram of a rotary oil
pump with spring-loaded vanes.

electron microscope has a capacity of about 100 l³/min. The rated pumping speed
is true only at inlet pressures near 1 atm. As the intake pressure is reduced, the
density of the gas being pumped is lower, so the amount of gas moved per sweep
is less (Fig. 10-2).

The oil in the pump is the lubricant and helps to prevent air from leaking past
the casing and the rotor and vanes. An oil reservoir is usually above the pump
from which oil flows by gravity through various oil feeds to the pump. The oil
circulates through the system continuously, being swept out of the exhaust and
into the reservoir. A good quality light petroleum oil (SAE of 15 to 20) with the
high pressure fractions removed is usually used. *The best possible vacuum that
can be obtained by a rotary oil pump is determined by the vapor pressure of the
oil, around 10^{-3} to 10^{-4} mbar (10^{-1} to 10^{-2} Pa).* However, because of leakage
of gases past the vanes and bearings, the actual pressure of the rotary oil pump
never reaches this good a vacuum. The rotary oil pump, used in either **roughing**
or **backing** the vacuum system in scanning electron microscopes, is usually only
required to obtain a pressure of 10^{-1} mbar (10 Pa). A relatively large amount of
oil is actually included between the casing and the rotor when the air is taken from
the intake of the pump. As the oil and air are compressed by the rotation of the
rotor, the amount of air volume is reduced very nearly to zero before it exits
through the exhaust valve, since the oil itself is noncompressible.

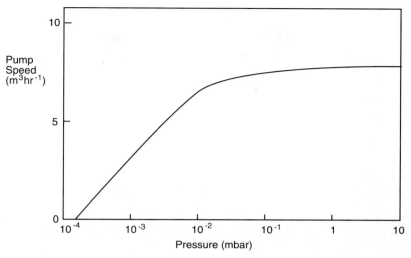

Figure 10-2   Pump speed compared to pressure for a typical rotary oil pump.

Oil contamination is the principle cause of unsatisfactory operation and sometimes total failure of rotary oil pumps. Oil contamination can usually be attributed to two factors: (1) oxidation of the oil and pump parts, and (2) water contamination of the oil.

1. Strong oxidizing agents such as sulfur dioxide, chlorine, fluorine, and acids cause oxidation of the oil and corrosion of the steel shafts, valves, and springs and bronze bearings. Oxidation of the oil results in the deposition of a gum residue on the pump surfaces, which may ultimately cause the rotor to stick tightly. The pump must then be taken apart and each part carefully cleaned.

2. Water vapor liquefies in the oil under high pressure because water vapor is relatively insoluble in oil and forms an emulsion. Vacuum pumps spend most of their time pumping water vapor, since the common gases such as nitrogen and oxygen are quickly pumped out of vacuum chambers. However, vapors from water and hydrocarbons are absorbed onto the surfaces of the vacuum chamber and are relatively difficult to remove. During pumping, the water vapor is pulled through the suction intake, swept around the compression side by the rotating vanes, and compressed into the space under the exhaust valve. The exhaust valve opens when atmospheric pressure is exceeded, and compression of the gases must proceed to this degree. On pumping the intake gases, a vapor, such as water vapor, liquefies on compression well before atmospheric pressure is reached. Water vapor liquifies when it is compressed to a volume approximately one-fifteenth of the original volume. The liquefied water passes through the exhaust valve and mixes with the oil in the main reservoir. The oil in the reservoir becomes more and more contaminated

Gas Taken In
Through Intake

Gas Enters Through
Gas Ballast During
Compression Phase

Exhaust Valve Opens,
Releasing Gas

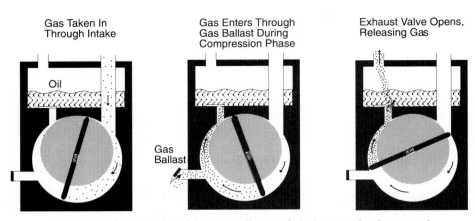

Figure 10-3   Gas ballasting with a rotary oil pump. Only the gas molecules pumped
in one compression phase are shown.

with water and reenters the pump through the oil ways. The best possible
vacuum that can now be obtained is poorer, since it is determined by the
vapor pressure of the water vapor (about 10 mbar, depending on the temperature), rather than the oil. A gas ballast system is often added to rotary oil
pumps to minimize water contamination of the oil. In gas ballasting (Fig. 10-
3), atmospheric air is admitted to pump during the compression stage to
increase the proportion of noncondensable gas (for example, nitrogen or
oxygen) in the pump. The total pressure in the compression chamber is the
sum of the partial pressures of the water vapor and introduced air. Most of
the total pressure is contributed by atmospheric air, with a smaller amount
by the water vapor. By the time atmospheric pressure is reached (mostly due
to the partial pressure of the introduced air from the gas ballast) during
compression, and the exhaust valve opens, the partial pressure of the water
vapor is not high enough to cause liquefication of the water vapor. The
exhaust valve opens sooner in the compression phase because of the increased amount of gas in the chamber, releasing all the gases. Gas ballasting
also requires extra work to compress the greater amounts of gases involved.
This extra work causes a temperature rise in the pump, which assists in
preventing water liquefication in the oil. The ultimate pressure obtained in
the compression phase of the pump is considerably poorer than that would
be obtained if there were no gas ballasting. This results in a poorer vacuum
obtained in the intake area, since some of the gas molecules from the
compression phase leak through the contact points between the vanes, rotor,
and casing and are included in the intake phase. Depending on the amount
of gas ballasting, this can reduce the pressure at the intake vacuum side to
10^{-1} mbar (10 Pa) or even 1 mbar (100 Pa).

Fortunately, oil contamination is usually not a problem with a well-maintained scanning electron microscope since the rotary oil pump spends most of the time backing the oil diffusion pump. Even when roughing the specimen chamber, little contamination is introduced by a responsible operator who controls the type of specimen viewed in the microscope.

When a rotary oil pump is turned off and the scanning electron microscope remains under vacuum, there is the danger that oil will continue to enter the pump from the oil reservoir through the oil ways and perhaps even through an imperfectly sealed exhaust valve. The oil is driven by the atmospheric pressure on the oil reservoir and can flood the pump and back into the vacuum of the electron microscope. This process can take only a minute or two and is usually prevented by inserting a solenoid valve, open to the air, near the intake of the pump. The solenoid remains closed as long as electrical current is supplied. When the scanning electron microscope is turned off or if there is power failure, air is admitted to the pump and prevents suck-back of the oil into the pump.

Oil vapor is carried out of the pump during operation, particularly during roughing of a system. This oil vapor is well atomized and floats everywhere in the air, constituting a health hazard. For this reason, the oil vapor should be fed to an outside vent. The pipe to the outside should not be vertical to the pump exhaust because of oil condensation in the pipe. The temperature of the pump exhaust is about 130°F and the temperature of the pipe is room temperature (70°F), resulting in condensation of the oil vapor in the pipe.

Diffusion Pumps

Diffusion pumps were developed between 1913 and 1916 by W. Gaede and I. Langmuir. The pumps have a high-velocity, high-density jet of mercury or oil that pushes gas molecules away from the pump intake toward the exhaust, where the gas molecules are removed from the system.

A diffusion pump is usually placed near the base of the column of the scanning electron microscope, and the two are separated by a valve that can be opened or closed. The diffusion pump (Fig. 10-4) consists of a smoothly finished stainless steel body with a backing (fore) pipe extending from the base of the casing to the backing rotary oil pump. The upper two-thirds of the casing, as well as the backing pipe, are surrounded by water cooling coils soldered or brazed in place. The flow of water in the coils is from the top of the casing to the backing pipe. Inside the casing is a three-stage chimney jet assembly of stainless steel containing three annular (ring-shaped) jets. A heater under the bottom of the casing heats the mercury or oil fluid in the bottom of the casing.

The operation of the diffusion pump commences with a rotary oil pump backing the oil diffusion pump through the backing pipe to a pressure of 10^{-1} mbar (10 Pa). At this point, the heater under the diffusion pump turns on, heating the

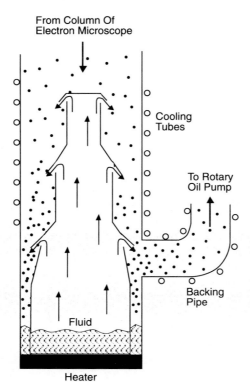

From Column Of
Electron Microscope

Cooling
Tubes

To Rotary
Oil Pump

Backing
Pipe

Fluid

Heater

Figure 10-4 Diagram of a diffusion pump.

mercury or oil in the bottom of the casing. The heated fluid evaporates, forcing vapor up the chimney, where it streams at a high velocity out of the annular jets. The jets direct the vapor stream down toward the area of higher pressure (poorer vacuum) at the base of the pump. When the vapor stream passes from the pressure of 2 mbar (200 Pa) inside the chimney to the pressure of 10^{-2} mbar (1 Pa) outside the jets, the vapor stream expands and converts its pressure energy into velocity energy. The vapor passing out of the jets is at a supersonic velocity (around Mach 2) in relation to the velocity of sound in the vapor. Gas molecules from the column of the scanning electron microscope diffuse into the upper throat of the diffusion pump, where they collide with the rapidly moving mercury or oil vapors and are carried downward with the mercury or oil vapors into the vapor stream of the second annular jet and then the third. Each succeeding jet compresses the gas molecules from an area of low pressure (good vacuum) to an area of high pressure (poor vacuum). A properly operated diffusion pump can produce compression ratios of more than 1,000,000 : 1 as the gas is pumped. Compression of the gases is more efficient for higher molecular weight gases and least efficient for helium and hydrogen. At the bottom of the diffusion pump, the gas pressure is around 10^{-1} mbar (10 Pa), and the gas molecules are removed from the diffusion pump through the backing pipe by the rotary oil pump. In the meantime, the mercury or

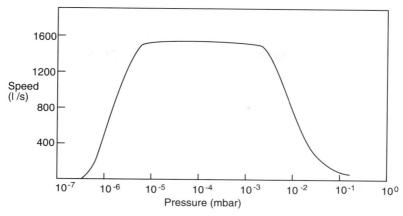

Figure 10-5 Pump speed compared to pressure for a typical diffusion pump.

oil vapors condense on the cool sides of the diffusion pump, apparently nearly completely liberating the entrained gas. The mercury or oils fall into the reservoir and are reheated in the reservoir.

The pumping speed of a diffusion pump (Fig. 10-5) is determined primarily by the area between the outer wall of the casing and the outer rim of the annular jets. As this area is increased, usually by increasing the diameter of the throat of the diffusion pump, the pumping speed also increases. Maximum pumping speeds vary from 15 l s^{-1} for a pump with a 2.5-cm throat to $30,000 \text{ l s}^{-1}$ for a pump with a 70-cm throat. Electron microscopes usually have diffusion pumps with about a 10-cm throat and a pumping speed of about 300 l s^{-1}.

Diffusion pumps can use mercury or oil as a pump fluid (Table 10-2). The construction of mercury and oil diffusion pumps is similar, although the fluids differ considerably in application.

TABLE 10-2 CHARACTERISTICS OF DIFFERENT FLUIDS USED IN DIFFUSION PUMPS

Fluid	Molecular weight	Approximate boiling point at 1 torr (°C)	Vapor pressure at 20°C (mbar)
Mercury	200.6	127	10^{-3}
Hydrocarbon oils			
Apiezon oil B	468	220	5×10^{-8}
Apiezon oil C	574	255	4×10^{-9}
Convoil 20	400	195	1×10^{-8}
Polyphenyl ether			
Santovac 5			1×10^{-10}
Silicone fluids	446	295	
Dow Corning 704	484	223	1×10^{-11}
Dow Corning 705	546	254	2×10^{-10}

Mercury. Mercury has fairly high vapor pressures of about 10^{-3} mbar (10^{-1} Pa), so a diffusion pump using mercury cannot be expected to obtain a vacuum better than this, unless a refrigerator trap is placed between the diffusion pump and the column of the scanning electron microscope. One problem with mercury is that it does not easily wet the cooled walls of the diffusion pump and does not flow as a sheet back down to the mercury reservoir at the bottom of the pump. Instead, the mercury stands upon the wall as many tiny drops, making point contacts with the cooled wall. These mercury drops do not easily coalesce into rivulets to stream down the wall to the mercury reservoir. Instead, the mercury drops mask the wall to newly arrived mercury vapors and do not themselves provide a well-cooled condensing surface. The rougher and/or oilier the surface, the larger the mercury drops and the worse the problem is. Therefore, the inner casing wall is finished to a high polish in manufacturing. Also, care is taken in operation of a mercury diffusion pump to make sure that oil does not backstream from the oil rotary pump into the mercury diffusion pump.

Oils (organic fluids). Almost all modern scanning electron microscopes with diffusion pumps use oils (organic fluids) as a fluid (Table 10-2). Oils used in diffusion pumps have molecular weights ranging from about 90 to 500 and vapor pressures ranging from 10^{-4} to 10^{-10} mbar (10^{-2} to 10^{-8} Pa) at room temperature. Most oils consist of a mixture of related molecules with similar vapor pressures. Oils can be broken (cracked or decomposed) into smaller molecules by excess heating. Generally, the smaller the molecule is, the higher the vapor pressure and the poorer the vacuum that can be attained in the diffusion pump. Therefore, the more the oil has decomposed, the poorer the vacuum that can be obtained. If the vapor pressure of the decomposition products is high enough (around 10^{-1} mbar [10 Pa]), they are removed from the diffusion pump by the rotary oil pump backing pump. Some of the decomposition products are of large molecular weight and lower vapor pressure than the original fluid. These decomposition products tend to accumulate in the boiler of the diffusion pump without evaporating, leading to the accumulation of tarlike products that eventually clog the jets. Silicon-based oils have a silicon–oxygen molecule backbone and have the advantage that they are resistant to decomposition under high temperatures. Three types of oils are in general use in diffusion pumps: (1) **hydrocarbon oils**, (2) **polyphenyl ether fluids**, and (3) **silicone fluids.**

Hydrocarbon Oils. These were the first organic fluids to be used in diffusion pumps and were refined from natural hydrocarbon oils by vacuum distillation. The hot oils will decompose in the presence of air. Commonly used hydrocarbon oils such as **Convoil 20, Apiezon Oil B**, and **Apiezon Oil C** all have vapor pressures ranging from 10^{-8} to 10^{-9} mbar (10^{-6} to 10^{-7} Pa) at 20°C.

Polyphenyl Ether Fluids. Polyphenyl ethers consist of chains of benzene radicals interbonded by oxygen. The compounds exhibit great thermal and chemical stability and do not easily decompose. The polyphenyl ethers have very low vapor pressures; the commonly used **Santovac 5** has a vapor pressure of 10^{-10} mbar (10^{-8} Pa) at 20°C. They have the disadvantage of costing more, but are commonly used in electron microscopes because of the high vacuums they can maintain.

Silicone Fluids. The silicone fluids are semiorganic with a molecular backbone built on silicon–oxygen linkages in place of the carbon–carbon linkages in organic fluids. Methyl and phenyl groups are attached to the silicon–oxygen backbone. The molecules have good chemical stability due to the great strength of the silicon–oxygen bond. Silicone fluids are so stable at working temperatures in diffusion pumps that atmospheric air can be admitted repeatedly without damage to the fluids or fouling of the pump. Two commonly used silicone fluids are **Dow Corning 704** and **Dow Corning 705** with vapor pressures of about 10^{-11} mbar (10^{-9} Pa) at 20°C. Both are phenylmethyl polysilicones. The phenyl content of the fluids reduces the volatility and increases thermal stability.

Malfunction of a diffusion pump. Malfunction of a diffusion pump is probably due to one of three problems:

1. The heater is not working properly.
2. There is insufficient or decomposed oil in the reservoir.
3. The rotary oil pump is not backing the diffusion pump properly.

Heater Malfunction. If the heater of the diffusion pump is below its proper temperature (usually because of too low wattage to the heater), the stream of mercury or oil will be too slow and not dense enough to compress the gas molecules effectively. If the heater is too hot, the stream of fluid molecules will be faster and denser and may lead to **backstreaming**, or the direct movement of mercury or oil into the column of the electron microscope. A minor amount of backstreaming will occur even in a properly operated diffusion pump when some of the oil or mercury vapor molecules coming out of the jets collide with the gas molecules in such a way that the oil or mercury molecules are directed upward toward the throat of the diffusion pump. Usually, some sort of a trap or baffle is placed between the column of the electron microscope and the diffusion pump to minimize backstreaming.

Insufficient or Decomposed Oil. Exposure of hot oil to the atmosphere will cause cracking and oxidation of the oil, leading to high vapor pressures and poor vacuums. If the cooling water is insufficient or stops, the oil will become hot and decompose. In modern electron microscopes, there are safety devices that have valve isolation systems preventing the introduction of air to the diffusion pump and devices that turn off the diffusion pump if there is insufficient cooling water.

Malfunctioning Rotary Oil Pump. A malfunctioning rotary oil pump will not reduce the backing pressure of the diffusion pump to 10^{-1} mbar (10 Pa) at the backing pipe. The accumulated gas near the base of the diffusion pump becomes too high and flows up along the wall of the diffusion pump, contaminating the high vacuum at the top of the pump.

Turbomolecular Pumps

The turbomolecular pump was developed by W. Becker in 1958 based on designs for a molecular pump by W. Gaede in 1913. The pumps are essentially high-vacuum gas turbines that are able to achieve high vacuums free of vapors from pumping fluids. A turbomolecular pump consists of a housing with a number of rows of stationary blades (stators) that project into the gaps formed by rows of blades on a rotor that is driven by a motor (Figs. 10-6, 10-7, and 10-8). The aluminum blades have a pitch of about 20°; the pitch is usually greater at the intake (high-vacuum side) and less at the low-vacuum exhaust. The pitch of the stator blades is a mirror image of the pitch of the rotor blades (Fig. 10-8).

Gas molecules from the electron microscope enter the pump at the low-pressure (high-vacuum) side, where they are knocked by the pitched blades toward the high-pressure (low-vacuum) end of the casing (Fig. 10-7). Here the accumulated gases are removed from the system by a rotary pump. Low molecular weight gases, such as hydrogen and helium, are not as effectively pumped as are high molecular weight gases (Fig. 10-9). The motor drives the rotor at about 50,000 rev min^{-1}, resulting in a compression ratio of 10^9 for nitrogen, 10^4 for helium, and 10^3 for hydrogen. A pump with a 6-in. throat has a constant pumping speed of about 240 l s^{-1} in a pressure range of 10^{-2} to 10^{-8} mbar (1 to 10^{-6} Pa). Beneath 10^{-2}

Figure 10-6 Turbomolecular pump that has had the rotor assembly removed from the casing. (Courtesy of Edwards High Vacuum International)

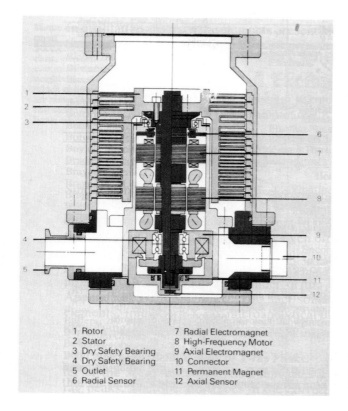

1 Rotor
2 Stator
3 Dry Safety Bearing
4 Dry Safety Bearing
5 Outlet
6 Radial Sensor
7 Radial Electromagnet
8 High-Frequency Motor
9 Axial Electromagnet
10 Connector
11 Permanent Magnet
12 Axial Sensor

Figure 10-7 Turbomolecular pump. (Courtesy of Edwards High Vacuum International)

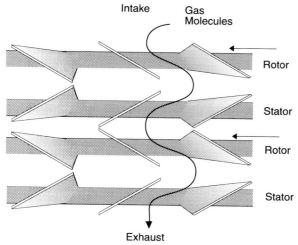

Intake

Gas Molecules

Rotor

Stator

Rotor

Stator

Exhaust

Figure 10-8 Passage of gas molecules through the turning rotor blades and stationary stator blades.

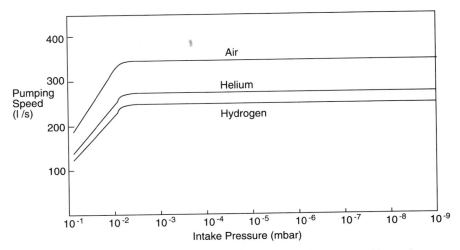

Figure 10-9 Relationship between intake pressure and pumping speed in a turbo-molecular pump.

mbar (1 Pa), there is no longer molecular flow of gas (viscous flow instead), and the pumping speed drops off.

Modern turbomolecular pumps have the motor inside the rotor and electromagnetic levitation of the rotor (Fig. 10-7). Magnetic levitation at the very high rotor speeds has the advantage of (1) eliminating the frictional contact associated with conventional bearings, resulting in a pump with little vibration or noise, (2) having no lubricant, so the high vacuum is completely free from hydrocarbon contaminants, and (3) magnetic bearings that are cool running, so little or no cooling is required. If a power failure occurs, a built-in battery maintains the electron magnetic levitation until the rotor stops.

Ion-getter Pumps

Getters are materials that remove gas by sorption in a vacuum system. Sorption is the uptake of gases by **absorption** (the penetration of a substance into the body of another substance) or **adsorption** (the surface retention of gas molecules, atoms, or ions by a solid or liquid). Ion-getter pumps used in electron microscopes have a pumping speed of about $100 \ l \ s^{-1}$ and work in the range of 10^{-3} to 10^{-8} mbar (10^{-1} to 10^{-6} Pa). Getters act as pumps since they hold gases in such a way that the gases are unable to contribute to the pressure in the system. Solid getters are normally elements with high melting points, with titanium usually the metal of choice.

Getter materials become saturated with gases and cease to act as pumps unless fresh getter surfaces are continually produced. Ion-getter pumps (getter-ion or sputter-ion pumps as they are also called) fall into two classes, those in which the fresh getter surface is produced by sputtering and those in which it is

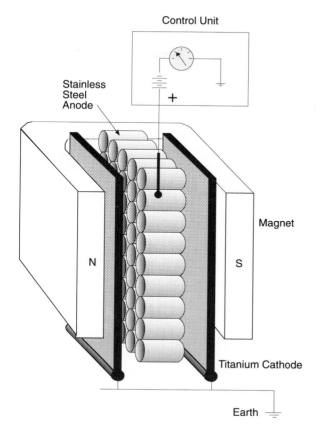

Figure 10-10   Ion-getter pump.

produced by evaporation. Most modern ion-getter pumps used in electron microscopes produce a fresh getter by sputtering. A commonly used design is based on
the work of L. D. Hall (1958) (Fig. 10-10). This ion-getter pump consists of an egg
crate anode made of stainless steel forming a number of cells within a titanium
cathode plate on each side. A magnet with a field of 1000 to 2000 gauss is placed
outside the titanium plates so that the magnetic lines of force pass through the
empty centers of the cells of the anode and are perpendicular to the surfaces of
the two titanium plates. Before the ion-getter pump is turned on, another type of
pump brings the pressure to at least 10^{-3} mbar (10^{-1} Pa). Once 10^{-3} mbar is
reached, the ion-getter pump is isolated from the fore pump. Commonly the
electron gun and upper part of the column are pumped by the ion-getter pump,
while the lower part of the column is pumped by the oil diffusion pump or turbomolecular pump. The two areas of the electron microscope are continuous through
the 100- to 400-μm aperture of the lower lens. The small size of the lower aperture
restricts the gases that can pass from one part of the column to the other, effectively maintaining the very high vacuum in the upper part of the column.
   Pumping is initiated in the ion-getter pump by placing a positive electric

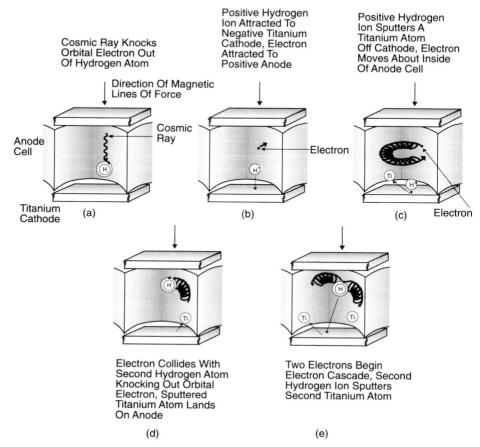

Figure 10-11   How a fresh getter layer of titanium is produced in an ion-getter pump.

potential of 3000 to 5000 V and 250 to 500 mA to the anode. The titanium cathodes are held at earth potential. Pumping action is initiated by a cosmic ray striking a gas molecule in the pump, creating a positive ion and a negative electron (Fig. 10-11). The negative electron is attracted to the nearest positive anode surface, but the strong magnetic field prevents the electrons from moving to the anode (because of the right-hand rule that determines the direction of force on an electron in a magnetic field; see Chapter 3). The electron moves in a helical path and is trapped in the cell between the two cathode plates while making a tortuous path to the anode. This greatly increases the path of the electron and also the probability of the electron colliding with a gas molecule. The electron strikes a gas molecule, knocking out an orbital electron from the gas molecule. The two electrons (the original electron plus the newly created electron) go on to collide with more gas molecules. The process is repeated, producing an electron cascade that results in an electron cloud between the two cathodes. The large number of electrons in the cloud results in a large number of collisions with any gas molecules in the area.

The positive ions produced from the collisions are attracted to the negative titanium cathode where they bombard the cathode and knock out individual titanium atoms in the process known as **sputtering**. The cathode is differentially sputtered, with maximum sputtering taking place opposite the open areas of the anode cells and little sputtering occurring opposite the solid lattice of the anode. The sputtered titanium atoms are electrically neutral and tend to travel in straight lines, building up a film of sputtered titanium over the anode, pump body, and the opposite cathode. This creates a fresh layer of getter material that is ready to remove gas molecules from the system by sorption. The removal of gas molecules by sorption occurs by two different methods, adsorption and absorption.

1. *Adsorption:* The clean titanium surface produced by sputtering acts as a getter and adsorbs gases such as nitrogen, oxygen, hydrogen, carbon monoxide, and carbon dioxide. Titanium has the ability to adsorb up to one atom of nitrogen from each atom of titanium by chemisorption forming nitrides. Oxygen reacts chemically with titanium, forming an oxide that must be buried with additional sputtered titanium before the surface can function as a getter again. Titanium adsorbs large amounts of hydrogen. The hydrogen is adsorbed on the surface, followed by diffusion into the interior under the influence of diffusion gradients. As hydrogen builds up on the surface of the cathode, concentration gradients are set up, forcing the hydrogen to diffuse farther into the titanium. At standard temperature and pressure, about 400 cm^3 of hydrogen can be sorped by 1 g of titantium. A certain amount of helium can also be pumped in this manner. Titanium reacts with water vapor to form titanium oxide and liberate hydrogen, which is adsorbed by a clean titanium surface. Inert gases such as argon (which makes up about 1% of atmospheric air) are not taken up by adsorption and must be buried by absorption.

2. *Absorption:* In absorption the energy of the ionized gas molecules is high enough (usually several kilo-electron volts) to drive them deeply into the titanium surface, where they are absorbed by ion implantation. This type of sorption is effective for all types of gases, including the rare and inert gases that are not removed from the system by adsorption. Unfortunately, these trapped gases will be removed from the titanium cathode as it is continually sputtered.

The forces binding and retaining the gases in adsorption and absorption may be physical (physical sorption) or chemical (chemisorption). **Physical sorption** results when ordinary intermolecular attractions (van der Waals forces with no transfer of electrons) hold the gases to the solid. Under high vacuum conditions, physical adsorption onto a surface rarely proceeds after the first complete layer

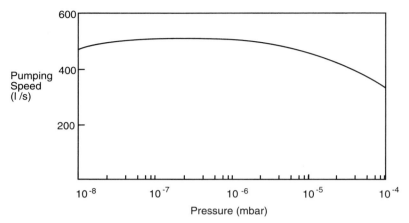

Figure 10-12 Relationship between pressure and pumping speed in a typical ion-getter pump.

(monolayer) of gases has been deposited. If a second layer is formed, it is weakly bound. Physical sorption to a surface is reversible. So desorption occurs when the temperature is raised or the pressure is reduced. **Chemisorption** relies on a chemical combination between the gas and the solid, with a resultant transfer of electrons between the atoms. The chemical reaction is usually permanent and irreversible (extreme heating may drive off the atoms, but usually not in the same molecular form that they were in when they were sorped onto the metal).

Ion-getter pumps used in electron microscopes have a pumping speed of about $100 \, l \, s^{-1}$ and work in the range of 10^{-3} to 10^{-8} mbar (10^{-1} to 10^{-6} Pa) (Fig. 10-12). Electron microscopes with thermal emission from a lanthanum hexaboride cathode or field emission from a tungsten cathode normally have an ion-getter pump because of the high-vacuum requirements of these types of electron emission (see Chapter 2). Using an ion-getter pump in an electron microscope with thermal emission from a tungsten cathode can routinely extend the life of the cathode to 200 h.

Unlike other types of pumps, ion-getter pumps have a lifetime that is determined by the amount of sputtering of the titanium cathodes. Eventually, the titanium cathodes are heavily sputtered away in the area of the open cells, at which time the cathodes have to be replaced with fresh titanium plates. The life of a pump depends on the pressure at which it is operated; a life of about 20,000 h can be expected for a commercial pump operated continually at 10^{-6} mbar (10^{-4} Pa). At poorer vacuums, there are more gas molecules and increased sputtering of the titanium cathodes occur, reducing the life of the ion-getter pump. The life of an ion-getter pump is largely determined by the accumulation of metal on the anodes, since the accumulated titanium begins to flake off, causing erratic pumping and failure of the pump. Certain chemicals such as Freon gas will quickly bind to (poison) all the clean titanium surface, resulting in diminished pumping. To clean a poisoned surface, argon is introduced into the pump, and a fresh surface of titanium is sputtered off the cathodes.

The ion-getter pump is also normally used as a vacuum gauge in electron microscopes. The number of electrons produced in the pump (discharge current) is proportional to the number of gas molecules in the pump. The discharge current therefore can be used as a measure of the gas pressure in the pump and is very accurate in the range of 10^{-4} to 10^{-8} mbar (10^{-2} to 10^{-6} Pa).

VACUUM GAUGES

Electron microscopes usually use three different types of systems for measuring vacuum:

1. 1 mbar (100 Pa) to 10^{-3} mbar (10^{-1} Pa): **Pirani (thermal conductivity) vacuum gauges** that measure the heat transfer by the conduction of gas molecules from a heated sensing wire.

2. 10^{-3} mbar (10^{-1} Pa) to 10^{-7} mbar (10^{-5} Pa): **Penning (cold cathode ionization) vacuum gauges** that measure the discharge current (production of positive ions and negative electrons) produced when generated electrons strike the gas molecules in the vacuum system of the electron microscope.

3. 10^{-7} mbar (10^{-5} Pa) to 10^{-9} mbar (10^{-7} Pa): This vacuum range is usually produced by an **ion-getter pump** whose discharge current is used as a measure of vacuum (see section on pumps).

Thermal Conductivity Vacuum Gauges

Vacuum in the range of 1 mbar (100 Pa) to 10^{-3} mbar (10^{-1} Pa) in an electron microscope is usually measured using a gauge that uses the thermal conductivity of the gas to indirectly measure the pressure. Thermal conductivity vacuum gauges are based on the principle that energy is dissipated from a heated surface in two ways: (1) by radiation, and (2) by conduction through the gas surrounding the surface (Fig. 10-13). **Radiation** carries heat energy away from the surface in the form of waves. Radiation is distinguished by the ability to transfer heat without any intervening medium and by the fact that the heat transfer occurs at the speed of light. **Conductance** relies on the gas molecules carrying away the heat energy. Gas molecules near a hot body acquire increased kinetic energy when they collide with the hot body. Between about 1 mbar (100 Pa) and 1 atm, the gas molecules rebound from the hot body, collide with other gas molecules, and impart the absorbed energy to the cooler (less energetic) molecules farther away. The presence of a hot body in the gas sets up a temperature gradient across the container because the gas in the immediate vicinity of the hot object acquires the temperature

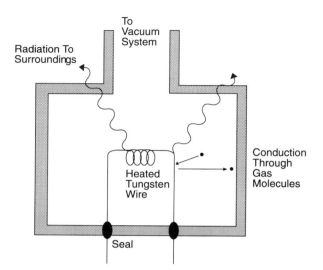

Figure 10-13 Heat transfer from a heated tungsten wire occurs by radiation and by conduction by gas molecules.

of the hot body, while the gas near the cool container wall is of the temperature of the wall. *At these gas pressures, the conductance of heat by the gas molecules is independent of gas pressure.* If the gas pressure is reduced further, the mean free path of the gas molecules approaches the dimensions of the container and, after striking the hot body, the gas molecules reach the container walls without striking other gas molecules on the way. Heat is conducted away from the hot body by the gas molecules without a temperature gradient being set up between the hot body and the cool container walls. *In this low gas pressure region (about 1 mbar [100 Pa] to about 10^{-4} mbar [10^{-2} Pa]), the thermal conductivity of the gas within the container is proportional to pressure and to the temperature difference between the hot body and the container walls.* At gas pressures beneath 10^{-4} mbar (10^{-2} Pa), there are so few gas molecules present that most of the heat is conducted away from the hot body by radiation, which is independent of gas pressure. The region of gas pressure, between about 1 mbar (100 Pa) and 10^{-4} mbar (10^{-2} Pa), where heat conductance is dependent on gas pressure, is the vacuum region effectively measured by thermal conductivity gauges. Measuring the rate of the transfer of heat energy at these gas pressures can, therefore, be used to indirectly measure the pressure of a gas in the system. Two basic types of thermal conductivity gauges are used to measure heat transfer in vacuums: **thermocouple gauges**, which measure the heat lost from a sensing wire heated by a constant current, and **resistance (Pirani) gauges**, which measure the resistance of a sensing wire. The latter type of gauges (Pirani gauges) are the most common types used in electron microscopes and will be discussed here.

Pirani (resistance) vacuum gauges (originally developed by M. Pirani in 1906) have a sensing wire (filament) heated by a constant voltage. With increasing gas pressure in the range of 10^{-4} mbar (10^{-2} Pa) to 1 mbar (100 Pa), the temperature of the wire decreases because of increasing conductance of heat away from the sensing filament by the gas. The electrical resistance of the sensing wire increases as the temperature of the wire increases. According to Ohm's law, if the potential (voltage) across the wire remains the same and the resistance increases (due to heating), then the current across the wire has to decrease, or

$$E = IR$$

voltage = current × resistance

volts = amperes × ohms

The change in current (amperes) across the sensing wire indirectly measures a change in gas pressure. The increased resistance of the sensing wire with increased temperature of the sensing wire and decreased gas pressure can be calibrated to known gas pressure. A measure of the current through the wire is then an indirect measure of the gas pressure.

The resistance of the sensing wire in the Pirani vacuum gauge is measured by using a Wheatstone bridge arrangement. In a Wheatstone bridge, there are four resistors arranged as in Fig. 10-14. The Wheatstone bridge is set up so that the resistance through resistors 2 and 3 is the same and the resistance through resistor 1 and the sensing wire in the vacuum system is about the same. Thus the sensing wire serves as a resistor in the system. The sensing wire, usually a 12-μm diameter tungsten filament heated to about 160°C, is usually used because it has a large change in electrical resistance in relation to change in temperature. All the resis-

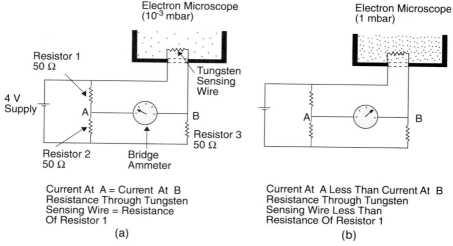

Figure 10-14 Circuitry of a Pirani gauge operating at two different gas pressures in an electron microscope. (a) The electron microscope is under a vacuum of 10^{-3} mbar; (b) the electron microscope is under a vacuum of 1 mbar.

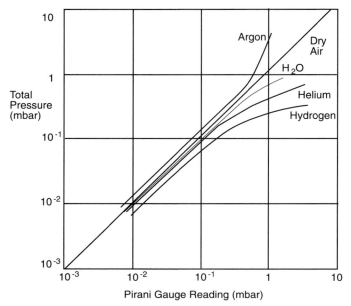

Figure 10-15 Relationship between total pressure and the reading of a Pirani gauge for different gases.

tances are usually about 50 ohms (Ω) and the power supply is about 4V. The Pirani gauge is calibrated at about 10^{-3} to 10^{-4} mbar (10^{-1} to 10^{-2} Pa) by setting the resistance of resistor 1 so that it has the same resistance as the sensing wire at this gas pressure (Fig. 10-14). Therefore, the currents at points A and B are the same and the current through the bridge ammeter is zero. Gas pressure measurement is now a function of resistance across the sensing wire in the vacuum system to be measured. If the gas pressure in the vacuum system increases, the conductance of heat from the sensing wire increases, the temperature of the sensing wire decreases, there is decreased resistance through the sensing wire, and the current across the sensing wire increases. The current at B is now more than at A, and current flows through the bridge ammeter between A and B (Fig. 10-14). The deflection of the needle on the bridge ammeter is used to indicate the vacuum in the system.

Pirani gauges can be built so that either the temperature, current, or voltage is maintained constant across the sensing wire. The design shown here, in which the voltage is maintained constant across the sensing wire, is the most common design, primarily because of the simplicity of the constant-voltage circuitry and because the gas pressure is approximately a linear function of the current across the sensing filament over a range of pressures (Fig. 10-15).

Thermal conductivity gauges measure the total pressure in a system; that is, they measure the pressure of vapors as well as the pressure of permanent gases.

The gauges read continuously and remotely, and the circuitry is not complex. The gauge tube elements are usually not harmed by operation at atmospheric pressure, although the calibration will change if the wire surface becomes coated with charred oil or any similar substance.

Penning (Cold Cathode Ionization) Gauges

Penning vacuum gauges (originally developed by F. M. Penning in 1937) are used in electron microscopes to measure vacuums in the 10^{-3} to 10^{-7} mbar (10^{-1} to 10^{-5} Pa) range. A Penning gauge functions similarly to that of an ion-getter pump (see section on pumps). A stainless steel anode (commonly in the shape of a ring) is positioned between two aluminum or stainless steel cathode plates, which are between the north and south poles of a magnet with a strength of about 400 gauss (Fig. 10-16). The cathode plates are usually at ground, while the anode ring is around $+2000$ V through a direct current supply. At this point, no electrons flow between the anode and the cathodes, and the discharge current through the ammeter is zero. A cosmic ray will usually initiate the action of the gauge by striking an atom in the cold cathode, knocking an orbital electron out of the atom. The released orbital electron attempts to move toward the positive anode. However, the lines of magnetic force of the magnet push the electron (because of the right-hand-thumb rule; see Chapter 3) in a direction perpendicular to the lines of magnetic force and perpendicular to the direction of the anode. As a result of these two forces, the electrostatic attraction of the electron toward the anode and the pushing of the electron by the magnetic field at a direction perpendicular to the direction of the anode, the released electron moves in a circuitous spiral path toward the

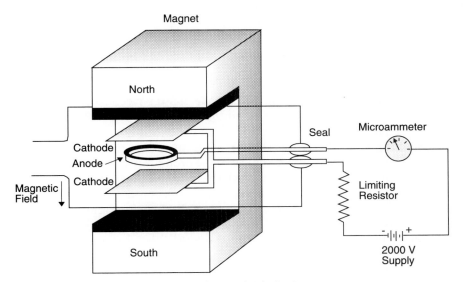

Figure 10-16   Structure of a Penning ionization vacuum gauge.

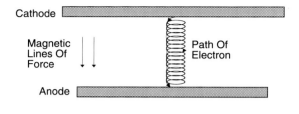

Cathode

Magnetic
Lines Of
Force

Path Of
Electron

Anode

Cathode

Figure 10-17 Path of an electron produced at the cathode in a Penning ionization vacuum gauge. The electron spirals along the magnetic lines of force.

anode along the lines of magnetic force (Fig. 10-17). The electrons shoot through the anode ring, are repelled by the opposite cathode plate, spiral back through the anode ring, and so on, several times before finally being captured by the anode. The total length of the path traveled by the electron is many hundred times the direct distance between the anode and cathodes. The chance that this electron will strike a gas molecule in the system is greatly increased by the long, tortuous path the electron takes before it is finally captured by the anode. Usually, the electron will strike a gas molecule before being captured by the anode, knocking an orbital electron out of the gas molecule and creating a positive ion and a second electron. Both electrons continue spiraling toward the anode. The positive ion, because of its larger mass, is less affected by the magnetic field and travels to the cathode, striking the cathode and setting up a discharge current that is registered on the microammeter. On striking the cathode, the positive ion can sputter the cathode, knocking out an atom and/or electrons. The electrons go on to interact with more gas molecules. There are such a large number of collisions between spiraling electrons and gas molecules in this system that the discharge current does not need to be amplified and can be read directly on the microammeter, which is calibrated in pressure units. The magnitude of the total discharge current, which is the sum of the positive ion current to the cathode and the electron current leaving the same cathode, is used as a measure of the gas present (Fig. 10-18).

The Penning gauge construction described does not work well at gas pressures beneath 10^{-5} mbar (10^{-3} Pa) because of the inability to initiate a discharge current. An electron knocked out of the orbit of an atom in the cathode by a cosmic ray fails to start an electron discharge because there are so few gas molecules at these low pressures. The chance of a collision between an electron and a gas molecule is so small that a continuous discharge current cannot be maintained. Electron microscopes use a modification of the original design of the Penning gauge that increases the length of the path of the electron, leading to an increased chance of collision between an electron and a gas molecule and the ability to maintain a continuous, self-sustaining discharge current. These modified Penning gauges can read down to at least 10^{-7} mbar (10^{-5} Pa).

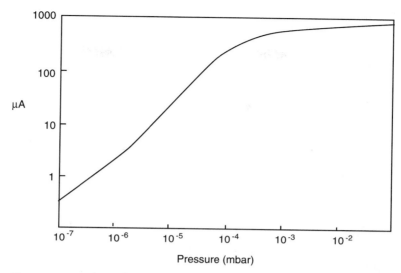

Figure 10-18 Relationship between discharge current and pressure in a Penning vacuum gauge.

Contamination of the Penning gauge occurs with use, due to the dissociation of hydrocarbon or silicone vapors during bombardment by electrons in the discharge process. These vapors condense on the parts of the vacuum gauge, forming conducting or insulating layers. These layers result in erroneous readings because of the inability of the positive ions and electrons to penetrate the layers and produce a readable current on the cathode. Gauge heads are normally demountable and sufficiently rugged to be cleaned with acids, solvents, and abrasives. Failure to reassemble the vacuum gauge correctly (that is, the wrong magnetic field to anode deposition) will prevent a discharge being struck due to insufficiently long electron path lengths.

A Penning vacuum gauge also acts as an ion-getter pump. A clean surface is produced by the positive ions that sputter the cathode material into the pump. These clean atoms have the ability to adsorb gas molecules and remove the gas molecules from the system. Consequently, it is important that the Penning gauge be connected to the vacuum system to be measured with a large-diameter opening; otherwise, the pumping action of the gauge will lead to erroneously low vacuum readings.

The advantages of the Penning gauge are (1) simple electronic circuitry, (2) rugged construction, (3) exposure to the atmosphere will cause no damage, (4) high resistance to mechanical shock, (5) long life, and (6) easy cleaning. The disadvantages are that (1) it is not as accurate as some other high-vacuum gauges, and (2) a strong magnet is required, which can affect the electron beam if the gauge is not properly shielded.

SEALS

Electron microscopes are composed of a number of parts, such as the specimen door, the electron gun chamber, the column itself, and the vacuum pumps. These parts are joined together by vacuum seals that are either (1) elastomer O-rings or (2) metal gaskets. A seal suitable for high vacuum must be made of a material with very low vapor pressure at room temperature and must not have any components that will vaporize at the pressures and temperatures to which the joint is subjected. In addition, the material must also have a very low permeability; that is, atmospheric gas must not pass through the seal material to contaminate the high vacuum on the other side of the seal. Elastomer O-rings are best limited to systems requiring a vacuum poorer than 10^{-8} mbar. For better vacuums, metal gaskets should be used, even though they are more troublesome to make.

O-rings

An O-ring is a molded, ring-shaped gasket with a circular cross section, made in a wide range of dimensions and of any of several elastomers (a plastic or synthetic rubber having some of the properties of natural rubber; they have the inherent ability to accept and recover from extreme deformation), depending on its intended use. O-rings are commercially available in a wide range of sizes, varying from about ⅛ in. to about 18 in. Larger O-rings can be made from extruded cord cut to the desired length and vulcanized (heated) or cemented, preferably with a 45° splice.

For sealing parts of an electron microscope, an O-ring is inserted in a groove cut in one part of the electron microscope, the other part having a smooth surface (Fig. 10-19). A square or wedge-shaped groove can be used. However, whatever the shape of the groove, its depth should not exceed two-thirds the cord diameter of the O-ring. When the parts of the microscope are forced together, metal to metal, the O-ring is sufficiently distorted to provide a good seal, but does not fill

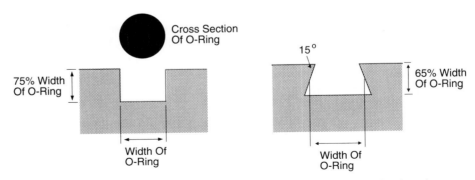

Figure 10-19   Dimensions of grooves for O-rings with a square side and a slanted side.

completely the volume available in the groove. It is important that the groove surfaces be ground and polished so that no residual tool marks or scratches occur at right angles to the groove. Scratches parallel to the length of the groove do not, in general, produce leaks and can be tolerated, provided their magnitude is not too great.

O-rings can be made of various elastomers:

1. *Viton A*, a copolymer of perfluoropropylene with vinylidene fluoride: Viton has a very low outgassing rate after it has been baked out at approximately 220°C in a vacuum chamber to remove any unreacted plasticizers, catalysts, and water vapor. The low outgassing rate and vapor pressure allow vacuums in the range of 10^{-9} mbar (10^{-7} Pa) to be attained. Viton A will not decompose until 300° C. The low outgassing rate and high chemical stability make Viton A the choice for O-rings in areas of the electron microscope where high vacuums are required.

2. *Nitrile*, a copolymer of butadiene and acrylonitrile: Commercially marketed as Buna N, Perbunen, and Hycar, has a maximum temperature of 80°C, and a vacuum of 10^{-5} mbar (10^{-3} Pa) can be attained in systems using O-rings made of nitrile.

3. *Neoprene*, a chloroprene polymer: Neoprene has a vapor pressure of 10^{-5} mbar (10^{-3} Pa). O-rings made of neoprene and nitrile can be used in low-vacuum areas of the electron microscope (such as around oil rotary pumps).

4. *Butyl rubber*, a copolymer of isobutylene and isoprene: Butyl rubber (used in automobile tires) has a very low permeability to gases, but unfortunately it has a very high outgassing rate, so it is not widely used in electron microscopes.

O-rings should be cleaned using a small amount of acetone or alcohol to remove all dirt and foreign material. A very thin coat of vacuum grease is applied by drawing the O-ring through fingers slightly coated with vacuum grease. The vacuum grease is not the sealing material, but simply a lubricant to enable the O-ring to completely fill the groove in which it is placed and compressed. *The vacuum grease has no other function and cannot of itself provide a vacuum seal. Many people consider the vacuum grease to be the sealing agent when it is not.* Apiezon grease L and M are commonly used vacuum greases. Apiezon grease L and M are hydrocarbon greases with a maximum usable temperature of 30°C. Apiezon grease L has a very low vapor pressure, around 10^{-11} mbar (10^{-9} Pa), but is very viscous. If poorer vapor pressures can be tolerated (around 10^{-9} mbar [10^{-7} Pa]), less viscous Apiezon M can be used. Braycote Micronic 803 is a vacuum grease with a 10^{-10} mbar (10^{-8} Pa) vapor pressure and is stable in the range of $-23°$ to 260°C. The Dow Corning silicone greases are not suitable for electron microscopes because of their high vapor pressure (10^{-5} mbar [10^{-3} Pa]), even though they are stable at high temperatures.

Metal Gaskets

Metal gaskets are commonly used in areas of the electron microscope that require very low gas pressures, such as in the gun chamber. These gaskets are made from a metal of high ductility (easily molded or drawn) and low vapor pressure and with a melting point higher than the maximum temperature to which the system is raised. Gaskets have been made from the following metals (their melting points are given in parentheses):

Indium	(156°C)
Tin	(232°C)
Aluminum	(659°C)
Copper	(1084°C)
Nickel	(1452°C)
Gold	(1063°C)

Indium is the most commonly used metal gasket material in electron microscopes. Because of its low melting point (156°C), indium cannot be used in systems requiring a high temperature. This is not a problem in electron microscopes, which are usually at room temperature. At room temperature, the vapor pressure of indium is so low that it is not possible to measure it (at 500°C the vapor pressure is 10^{-8} mbar [10^{-6} Pa]). Indium seals are very easily made by compressing a loop of $\frac{1}{32}$ or $\frac{1}{16}$ in. thick wire between two flat flanges of stainless steel, until the thickness of the indium is reduced to 0.005 to 0.010 in. It is not even necessary to melt the ends together to form a complete loop. If enough indium is used to allow an overlap at the ends (Fig. 10-20), they will cold-weld during compression since the metal is so soft. Indium metal gaskets are used only once and have to be scraped off the metal surfaces after use, during which it is important not to damage the polished surfaces or drop any indium down the column of the electron microscope. High pressures are needed on the joints to effect a tight seal; often, twice as many bolts as would be required for an elastomer O-ring are used. In addition, the sealing surface must have a high polish as the indium will not effectively seal across a large scratch.

Figure 10-20 Indium wire overlapped and ready for cold welding with pressure to form a metal gasket.

VALVES

All vacuum systems require valves to control the rate and direction of gas flow and to isolate the components of the system. A vacuum valve generally consists of the following parts (Fig. 10-21):

Body or **housing** of the valve

Port or opening within the valve body through which the gases pass on their way to the vacuum pump

Disc containing an elastomer O-ring, which is pushed up against the **seat** to stay the flow of gases

Stem that supports and manipulates the disc

Bonnet that supports the stem

Types of Valves Used in Electron Microscopes

A large number of different designs of vacuum valves are used in electron microscopes. However, the most common types are (1) **disc valves**, (2) **gate valves**, and (3) **butterfly valves**.

1. **Disc valve:** The disc valve (Fig. 10-21) consists of a disc mounted on a movable stem. The valve is closed when a pneumatic or solenoid system drives

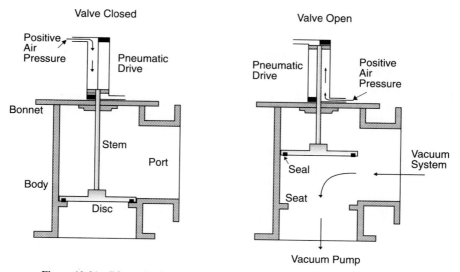

Figure 10-21   Disc valve in the open and closed configuration. The valve is driven by a pneumatic drive.

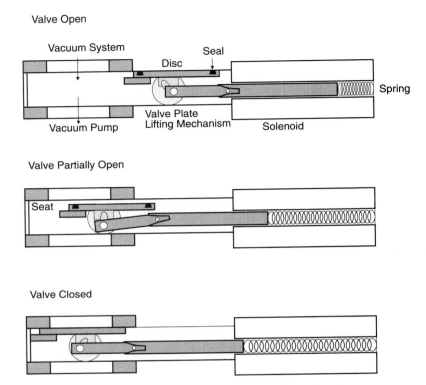

Valve Open

Vacuum System

Disc   Seal

Spring

Vacuum Pump   Valve Plate
               Lifting Mechanism   Solenoid

Valve Partially Open

Seat

Valve Closed

Figure 10-22   Closing of a gate valve by a solenoid.

the disc into the seat. The disc has an O-ring that makes a seal with the seat. When the valve is open, the disc is in the port of the valve, offering some impedence to the passage of gas molecules.

2. **Gate valve:** A gate valve has a relatively thin disc that is mounted on a carriage that forces the O-ring in the disc into the seat of the valve when the stem is extended (Fig. 10-22). When the stem is retracted, the carriage drops down (by a mechanism such as a eucentrically mounted wheel) and the disc is pulled down and away from the valve port, providing for unimpeded passage of the gas molecules from the vacuum system into the vacuum pump. Gate valves are relatively thin and therefore offer a high conductance of gas molecules through the valve.

3. **Butterfly valve:** The butterfly valve is similar to dampers used in stove pipes. The valve consists of a relatively thin circular disc with an O-ring around the outside of the disc (Fig. 10-23). The O-ring of the disc fits into the seat of the valve. The valve is opened by tilting the disc 90°. The butterfly valve is the most compact valve that can be manufactured for the port size produced (not

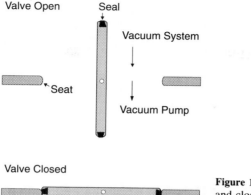

Figure 10-23 Butterfly valve opened and closed by the rotation of the disc.

considering the space that must be provided into which the disc can swing), although in the open position the disc offers some impedence to gas flow.

Vacuum valves are usually opened or closed in modern electron microscopes with either **pneumatic cylinders** or **solenoids**. A pneumatic cylinder consists of two chambers, one of which is used to open and the other, to close the vacuum valve (Fig. 10-21). A switch controls the positive air pressure into one of the two chambers. A solenoid pulls the stem toward the solenoid when current is passed through the solenoid. The magnetic field that is generated pulls the stem into the solenoid and opens the valve (Fig. 10-22). The valve is closed when the current is turned off and the spring in the solenoid pushes the stem back into the valve, seating the disc.

11

Specimen Preparation

The column of the scanning electron microscope is maintained at a high vacuum so that the electron beam has unimpeded passage from the electron-emitting cathode to the specimen. Therefore, specimens examined in the scanning electron microscope cannot have volatile compounds. Specimens that are dry and do not contain volatile components are prepared for scanning electron microscopy with minimal effort. Specimens containing volatile components are more difficult to prepare because it is necessary to remove the volatiles without significantly altering the specimen structure before the specimen can be examined in the scanning electron microscope.

DRY SPECIMENS

Specimens that do not contain significant amounts of volatile components are easy to prepare for scanning electron microscopy. The specimen preparation involves mostly common sense. It is the surface of the specimen that is examined in the scanning electron microscope. Therefore, the investigator should ensure that the specimen is clean. It may be necessary to blow off dust particles with clean nitrogen gas or compressed air. If the surface is oily, it may be necessary to wash the specimen with an organic solvent. Once a clean surface has been obtained, the specimen is mounted and coated for viewing.

Volatilization of Specimen Components during Viewing.   Some heating of the specimen occurs from interaction of the electron beam with the specimen. Some specimens contain components that volatilize when the specimen is heated. Specimen heating will volatilize components such as plastics or other long-chain hydrocarbons (for example, rubber). These volatiles condense down on the colder parts of the specimen and specimen chamber, which means everything except the portion of the specimen being viewed. This results in a coating of contaminating material in the chamber. Normally, this is not a significant problem. However, if this type of specimen is continuously viewed, enough of a coating can build up on the scintillator of the detector so that there is a reduction in the detection efficiency of the system. If it is necessary to view this type of specimen, it would be advisable to coat the specimen with a relatively thick layer of carbon or metal to reduce the volatilization.

SPECIMENS CONTAINING VOLATILE COMPONENTS

With physical specimens, such as an oil-bearing rock, it is usually relatively easy to remove the volatiles with solvents. Subsequent air drying will usually result in little or no alteration of the structure of the specimen.

However, biological specimens (usually containing over 90% water) are another matter. Biological specimens are very difficult to prepare for scanning electron microscopy; the specimen generally suffers some shrinkage and distortion during preparation, even with the best methods available.

Two basic methods are usually used to prepare biological specimens:

1. Chemical fixation, followed by air drying or critical point drying
2. Rapid freezing, followed by freeze drying at low temperatures and high vacuum

Chemical Fixation

Fixatives can be divided into noncoagulant or coagulant fixatives. **Noncoagulant fixatives** stabilize macromolecules without much structural distortion from their living natural state. There is little dissociation of proteins from water, and the proteins retain at least some of their reactive groups. **Coagulant fixatives** (such as ethanol) result in gross precipitation of the cellular macromolecules, so the cellular components are distorted. Noncoagulant fixatives, such as glutaraldehyde, formaldehyde, and osmium tetroxide are used to fix biological tissue. Once the tissue is fixed in a noncoagulant fixative, it will not be significantly affected by exposure to a coagulant fixative such as ethanol.

The main aim of chemical fixation is to crosslink the macromolecules of the cells into a rigid reticulate network that holds the size and shape of the cells in a condition as near to natural as possible. Chemical fixation also renders the macromolecules insoluble so that they are not extracted in subsequent dehydration steps. The four types of important macromolecules in cells are the following:

- Proteins
- Polysaccharides
- Lipids
- Nucleic acids

Chemical fixation of cells is usually carried out by a **primary fixation** with a buffered glutaraldehyde or a glutaraldehyde–formaldehyde mixture, followed by a **postfixation** with osmium tetroxide. The aldehyde fixation (glutar**aldehyde** or form**aldehyde**) fixes the proteins and a large number of the polysaccharides. Osmium tetroxide fixes the unsaturated lipids. In transmission electron microscopy, the nucleic acids are fixed with uranium. However, the uranium step is usually omitted in the scanning electron microscope since the nucleic acids make up such a small proportion of the cells.

It is essential that the size and shape of the cells after fixation be as near as possible to that in the natural condition. Therefore, the tonicity, pH, temperature, and oxygen content of the fixative solutions have to be taken into consideration to make sure that the cells do not shrink or swell during the procedure.

Glutaraldehyde fixes proteins and some polysaccharides. The living cell is basically a **sol** composed of a solid phase (proteins, polysaccharides, lipids, nucleic acids, and the like) suspended in a liquid phase (water). While the protoplasm is in a sol state, the shape of the cell is determined by the shape of the container and its supports. A solitary cell in blood plasma has the cell shape determined by the microfilaments and microtubules that support the plasma membrane. The shape of a cell in tissues is often determined by the packing of the other cells and connective tissue. In the scanning electron microscope, it is the shape of the cells and their constituents that are examined. Therefore, it is essential that the shape of cells and tissues not be changed during fixation.

Glutaraldehyde is an excellent fixative because it converts the sol state of the living cell into a **gel** state in the fixed cell without significantly altering the shape of the cell. Glutaraldehyde fixes the protoplasm by forming methylene bonds ($-CH_2-$) between protein molecules to transform the protoplasm into a gel, a network structure loaded with fluid. The crosslinked proteins form the gel network that holds the other cell constituents suspended in an aqueous medium. *The gelled protoplasm is elastic and has a form of its own, so the shape of the cell is not significantly affected by subsequent changes outside the cell.*

Glutaraldehyde ($C_5H_8O_2$) is a 5-carbon dialdehyde (molecular weight of 100.12) (Fig. 11-1). At room temperature, glutaraldehyde undergoes aldol condensation (Fig. 11-2). Initially, two 5-carbon glutaraldehyde molecules condense to a 10-carbon molecule, followed by condensation of an additional 5-carbon glutaral-

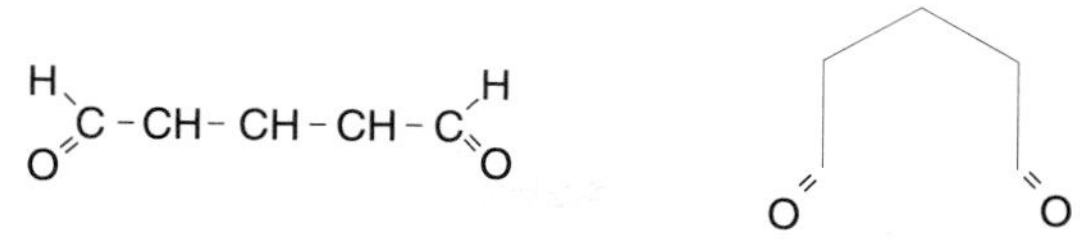

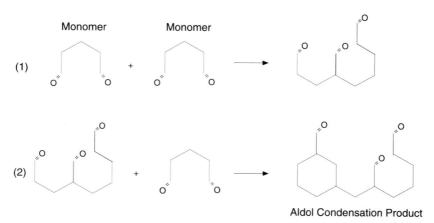

Figure 11-1 Structure of glutaraldehyde, actual and diagrammatic.

dehyde to yield a 15-carbon aldol. It is this aldol that reacts with sulfhydryl groups and nitrogen atoms of amino groups of proteins during fixation (Fig. 11-3). The amino acid residues of lysine, tyrosine, tryptophan, histidine, and phenylalanine, along with sulfhydryl groups, react with the glutaraldehyde aldol (Fig. 11-4). The crosslinking of proteins is so stable that the exposure of the tissue to boiling water or dilute acid at 20°C fails to remove a significant portion of the fixed proteins.

Lipids are not fixed by glutaraldehyde. However, lipoproteins can be fixed by virtue of the crosslinking of the protein portion of the molecule. Polysaccharides, such as glycogen, are partially fixed by glutaraldehyde. Glutaraldehyde travels into the tissue fairly rapidly; after 1 h at room temperature, a 4% glutaraldehyde solution travels about 0.4 mm into liver. Primary fixation of tissue with glutaraldehyde is usually performed by a well-oxygenated, buffered solution at the temperature at which the organisms or cells are grown.

Formaldehyde travels into the tissue faster than glutaraldehyde. Formaldehyde is rarely used alone as a fixative for electron microscopy because the crosslinks formed between macromolecules are reversible by such mild conditions as a water wash. This results in tissue that is poorly fixed, with a large amount of extraction of the cell constituents.

Formaldehyde, however, is frequently used in a mixture with glutaraldehyde (for example, as in Karnovsky's fixative, 4% glutaraldehyde, 1% formaldehyde in 0.1 M cacodylate buffer, pH 7.2) as a primary fixative. The penetration rate of

Figure 11-2 Glutaraldehyde monomers react to produce an aldol condensation product, which is the principal form of glutaraldehyde that reacts with proteins.

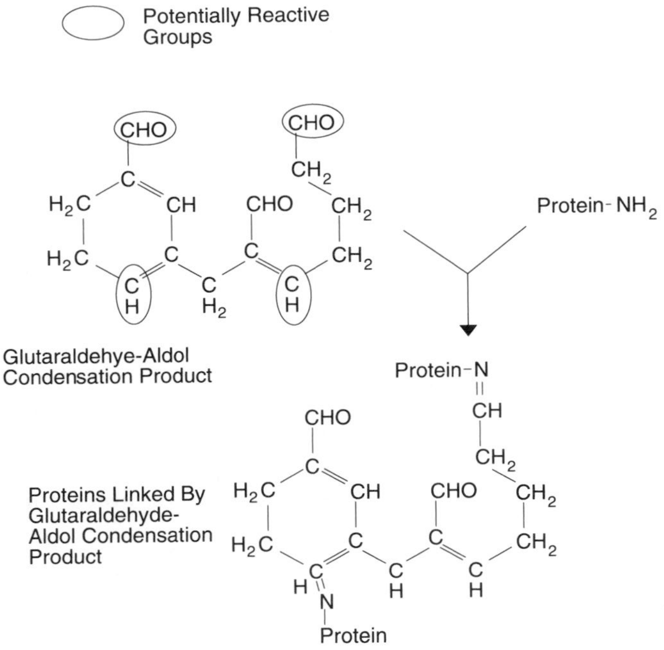

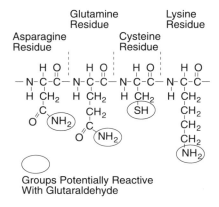

Figure 11-3   Reaction of the aldol condensation product of glutaraldehyde with proteins.

aldehyde fixative into tissue is inversely proportional to the molecular weight of the aldehyde. Formaldehyde has a molecular weight of 30.03 and is much smaller than glutaraldehyde (molecular weight of 100.12). Therefore, when formaldehyde is used in a mixture with glutaraldehyde, the formaldehyde penetrates more rapidly, producing temporary crosslinks that are stabilized on the arrival of the glutaraldehyde molecules. This results in better fixation, since the quicker stabilization of the cell constituents minimizes autolytic changes associated with cell death.

Figure 11-4   Portion of a protein molecule showing those groups that have the potential to react with glutaraldehyde or formaldehyde.

1. $\text{Protein-NH}_2 \;+\; \text{CH}_2\text{O} \longrightarrow \text{Protein-NH-CH}_2\text{OH}$

               Formaldehyde   Amino Methyl Group

2. $\text{Protein-NH-CH}_2\text{OH} \;+\; \text{Protein-NH}_2 \longrightarrow \text{Protein-NH-CH}_2\text{-NH-Protein}$

   Protein With Amino   Crosslinked Protein Molecules
   Methyl Group

Figure 11-5 Reaction of formaldehyde with proteins. In the first step the NH_2 group of a protein reacts with formaldehyde to form an amino methyl group on the protein. In the second step the protein with the amino methyl group reacts with a second protein molecule, resulting in crosslinking of the two protein molecules.

Formaldehyde (CH_2O) has a single aldehyde group (it is a monoaldehyde). **Formalin** is an aqueous solution of formaldehyde gas. Commercial preparations of formalin have a small amount of methanol added as a preservative and some formic acid. Such formalin solutions should not be used for preparation of specimens for electron microscopy because the methanol is a coagulant fixative causing precipitation of macromolecules and poor preservation of structure. Electron microscopists prepare formaldehyde from the solid polymer **paraformaldehyde** $(\text{CH}_2\text{O})_n$ by dissolving paraformaldehyde powder in a slightly alkaline solution at 60°C for 1 h. The solution is stable for about 2 weeks at 4°C in the dark.

Formaldehyde reacts with a number of different groups of proteins and polysaccharides. One of the most common reactions is with the amino groups of proteins to form amino methyl groups (Fig. 11-5). The amino methyl group condenses with functional groups such as phenols, imidazoles, or indoles on other proteins to form methylene bridges (protein–CH_2–protein), crosslinking and fixing the molecules. As mentioned before, the reactions are reversible and the crosslinks can be removed by washing with water. This is why formaldehyde is a poor fixative and requires stabilization of the crosslinks by glutaraldehyde.

Formaldehyde will fix glycogen and mucoproteins and prevent their extraction from tissues, providing the duration of fixation is not too long.

Osmium tetroxide fixes unsaturated lipids. Osmium tetroxide (OsO_4) is usually used after the tissue has been fixed in an aldehyde (glutaraldehyde or a glutaraldehyde–formaldehyde mixture). The aldehyde fixation is referred to as **primary fixation**, while the osmium fixation is referred to as **postfixation**.

Osmium tetroxide reacts primarily with unsaturated lipids and does not react with saturated lipids. Unsaturated lipids contain one or more double bonds, while saturated lipids do not contain double bonds (Table 11-1). Lipids occur in cells principally in the membranes. The three major kinds of membrane lipids are **phospholipids, glycolipids,** and **cholesterol** (Fig. 11-6). The phospholipids and glycolipids contain fatty acid residues that may be unsaturated and contain double bonds. Cholesterol does not contain double bonds.

TABLE 11-1 SOME SATURATED AND UNSATURATED FATTY ACIDS

	Saturated fatty acids
Lauric	$CH_3(CH_2)_{10}COOH$
Myristic	$CH_3(CH_2)_{12}COOH$
Palmitic	$CH_3(CH_2)_{14}COOH$
Stearic	$CH_3(CH_2)_{16}COOH$
Arachidic	$CH_3(CH_2)_{18}COOH$
Lignoceric	$CH_3(CH_2)_{22}COOH$

	Unsaturated fatty acids
Palmitoleic	$CH_3(CH_2)_5CH{=}CH(CH_2)_7COOH$
Oleic	$CH_3(CH_2)_7CH{=}CH(CH_2)_7COOH$
Linoleic	$CH_3(CH_2)_4CH{=}CHCH_2CH{=}CH(CH_2)_7COOH$
Linolenic	$CH_3CH_2CH{=}CHCH_2CH{=}CHCH_2CH{=}CH(CH_2)_7COOH$
Arachidonic	$CH_3(CH_2)_4CH{=}CHCH_2CH{=}CHCH_2CH{=}CHCH_2CH{=}CH(CH_2)_3COOH$

Osmium tetroxide reacts with double bonds of the unsaturated fatty acids.

Osmium tetroxide reacts with the double bonds of unsaturated lipids to form crosslinks between lipid molecules and therefore acts as a noncoagulant fixative. Osmium tetroxide first oxidizes a double bond in an unsaturated lipid, forming a cyclic osmic acid monoester (Fig. 11-7). The cyclic osmic acid monoester reacts with two molecules of water to yield a diol and osmic acid. Another cyclic osmic acid monoester reacts with the diol, forming a very stable diester that crosslinks two molecules of lipids, thus fixing the lipids in the cells.

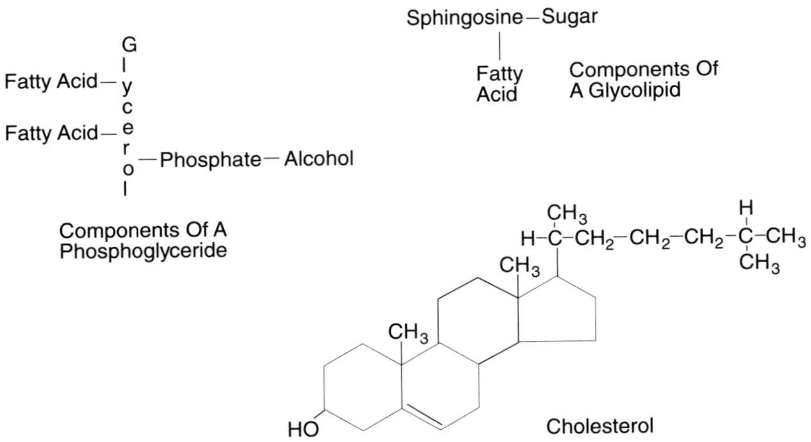

Figure 11-6 The three types of lipids in biological membranes: phosphoglycerides, glycolipids, and cholesterol. Phospholipids and glycolipids can contain unsaturated fatty acids and can, therefore, be crosslinked by osmium. Cholesterol does not contain fatty acids.

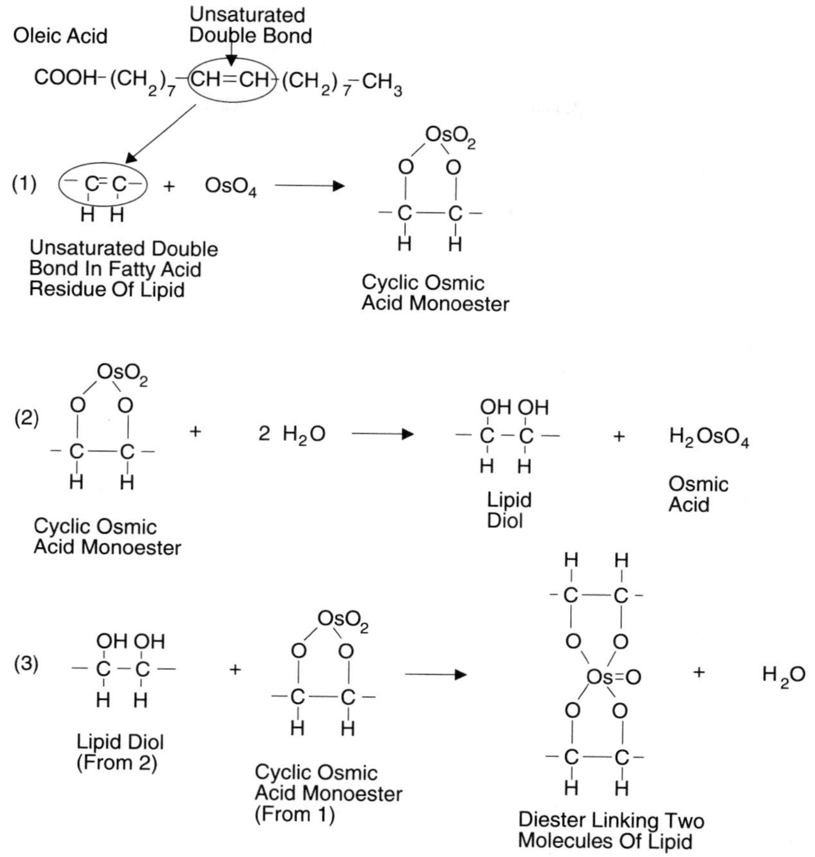

Figure 11-7 Reactions of osmium tetroxide with unsaturated fatty acids.

Osmic acid is usually used as a 1% to 2% buffered solution (usually around pH 7.2) for 1 to 2 h. Most investigators carry out osmium fixation at 4°C to minimize autolytic changes associated with cell death. After 1 h, osmium penetrates into about 0.6 mm of biological tissue.

Osmium tetroxide (molecular weight of 254.2) melts at 41°C, boils at 131°C, and has an acid, chlorinelike odor. It dissolves slowly in water (solubility of 7.24 g/100 cc^3 at 25°C) and in acetone, benzene, paraffin oil, and saturated lipids. *Osmium tetroxide is volatile and poisonous*; its vapor kills the epithelial cells of the eyes, nose, and mouth. It should be handled with gloves in a ventilated hood.

Osmium tetroxide is normally prepared from yellowish crystals purchased in a sealed tube made of glass. The osmium tetroxide is dissolved in the buffer solution contained in a glass-stoppered bottle. The glass-stoppered bottle is cleaned with concentrated nitric acid before use to remove all the organic matter and then thoroughly washed with distilled water to eliminate the acid. Small amounts of organic matter will reduce osmium tetroxide to a black, hydrated dioxide, which is a poor fixative. The clean, glass ampul containing the osmium tetroxide is placed in the buffer in the bottle, and the ampul is broken with a heavy, clean, glass rod. The solution should be left overnight for the osmium tetroxide to dissolve in the buffer.

Fixation with glutaraldehyde–osmium tetroxide mixtures prevents the formation of membrane blisters.

One problem with glutaraldehyde primary fixation, followed by osmium tetroxide postfixation, or with glutaraldehyde fixation alone, is the formation of **membrane blisters**, which are artifacts of the fixation process and not natural structures. In the scanning electron microscope, these blisters are seen as mounds of baglike extensions of the plasma membrane. These blisters form during the time that the cells are in the glutaraldehyde solution. The lipid molecules in the membranes are in a fluid form and are in continuous flow. The glutaraldehyde solution fixes the proteins in the cell, but does not fix the lipids, unless the lipids are attached to proteins. The lipids flow while the cells are in the glutaraldehyde fix, with the plasma membrane being drawn out into blisters.

To complicate the situation, **blebs** occur naturally in the plasma membrane of some living cells (Fig. 11-8). These blebs are real structures, usually containing ribosome-rich cytoplasm, and are not to be confused with the artifactual blisters, even though they look the same in the scanning electron microscope. Fixation with a mixture of buffered glutaraldehyde and osmium tetroxide can be used to differentiate between blebs and blisters. This immediate exposure to osmium tetroxide in the mixture fixes the lipids quickly and prevents the lipids from flowing during fixation, stopping the formation of blisters. The naturally occurring blebs in the plasma membrane are preserved.

A problem with fixing with a mixture of glutaraldehyde and osmium tetroxide is that the chemicals react with each other. Mixing the two chemicals at room temperature soon results in a solution that is an ineffective fixative. To be an

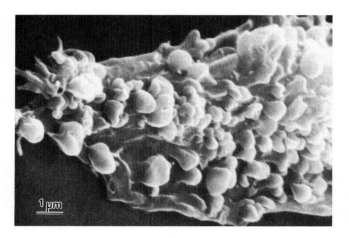

Figure 11-8 Blebs in the plasma membrane of a cultured Chinese hamster ovary cell. These blebs are real structures, whereas similar appearing blisters are artifacts of fixation.

effective fixative, the glutaraldehyde and osmium tetroxide have to be prepared as two separate solutions at 4°C. The two solutions are mixed at 4°C just before use, and the tissue is fixed in the mixture at 4°C to retard reaction of the glutaraldehyde with the osmium tetroxide. Such a fixation procedure quickly stabilizes the naturally occurring blebs and prevents the formation of membrane blisters.

Tonicity, pH, temperature, and oxygen content of the fixative solution. During primary fixations, it is usually desirable for the fixative solution to have a pH, temperature, and oxygen content very close to that of the tissue in vivo. The tonicity of the fixative solution, however, should have an osmotic value that is slightly hypertonic to the physiological condition of the tissue.

Tonicity. Experiments with the tonicity of the fixative solution have shown that **hypotonic** (of lower osmotic value that the cellular protoplasm; Table 11-2) or **isotonic** (of the same osmotic value as the cellular protoplasm) fixative solutions result in swelling of the cells during fixation. Fixative solutions that are strongly **hypertonic** (of higher osmotic value than the cellular protoplasm) result in shrinkage of the cells. *The best fixation, resulting in no shrinking or swelling of the cells, occurs in fixative solutions that are somewhat hypertonic.* A fixative solution of 2.5% glutaraldehyde in 0.1 *M* phosphate buffer, pH 7.2, has an osmolarity of around 500 milliosmols and is a commonly used primary fixation solution that is slightly hypertonic to most tissues. The tonicity of the primary fixative solution is the most critical, with the tonicity of the following solutions being less critical to maintenance and the size and shape of the cells. The tonicity of fixative solutions is determined by the fixative, the buffer, and any other added compounds (for example, sodium chloride and sucrose; Table 11-3).

pH. The pH of the fixative solution should be the same as the physiological

TABLE 11-2 OSMOTIC CONCENTRATIONS OF THE BLOOD OF SOME ANIMALS

Class	Species	Osmolarity of blood (mOsmol 1^{-1})
Mammalia	*Homo sapiens*	285
	Rattus albinicus	310
Aves	*Gallus domesticus*	299
Reptilia	*Emys orbicularis*	237
Amphibia	*Rana esculenta*	237
Insecta	*Mantis religiosa*	477
	Aedes aerypti (larva)	266
Arachnida	*Euscorpius italicus*	607
Crustacea	*Oniscus* sp.	560
Annelida	*Pheretima posthuma*	250
Gastropoda	*Helix pomatia*	228

From W. T. W. Potts and G. Parry, *Osmotic and Ionic Regulation in Animals*, Macmillan, New York, 1964, 423 p.

TABLE 11-3 OSMOLARITY OF SOME COMMONLY USED BUFFERS AND FIXATIVE SOLUTIONS

Substance	Osmolarity (mOsmol l^{-1})
0.05 M Phosphate buffer, pH 7.2	145
0.1 M Phosphate buffer, pH 7.2	230
0.2 M Phosphate buffer, pH 7.2	420
1.2% Glutaraldehyde in 0.1 M phosphate buffer, pH 7.2	370
2.3% Glutaraldehyde in 0.1 M phosphate buffer, pH 7.2	490
4.0% Glutaraldehyde in 0.1 M phosphate buffer, pH 7.2	685
2.5% Glutaraldehyde in H_2O	270
2.0% Formaldehyde in H_2O	215
0.05 M Sodium cacodylate buffer, pH 7.4	110
0.1 M Sodium cacodylate buffer, pH 7.4	195
0.2 M Sodium cacodylate buffer, pH 7.4	390
2.5% Glutaraldehyde plus 2% formaldehyde in 0.05 M cacodylate buffer, pH 7.4	480
2.5% Glutaraldehyde plus 2% formaldehyde in 0.1 M cacodylate buffer, pH 7.4	550

pH of the tissue. Generally, fixative solutions are prepared between pH 7.0 and pH 7.4, which is within the pH range of most tissues. Buffers are used to stabilize the pH at a certain value. A buffer solution contains a weak acid and its salt or a weak base and its salt. Buffers can resist changes in hydrogen ion concentration when small amounts of a strong acid or base are added to the buffer. Phosphate and cacodylate are the most commonly used buffers. Phosphate buffer is inexpensive and phosphate ions are common in most cells, so it is a natural buffer. Cacodylate is more expensive and it is toxic (it contains arsenic).

Temperature.   The primary fixation is usually at least started at the physiological temperature of the organism. Some investigators prefer to cool the tissue and fixative solution to 4°C after approximately 15 min. A disadvantage is that cooling causes a slower penetration of the fixative into the tissue. An advantage is that cooling reduces the rate of autolytic changes in the cell associated with cell death.

Oxygen.   Fixatives penetrate into tissues at a rate of about 1 to 2 μm per second. Tissue continues to respire while the primary fixative is penetrating into the tissue. Oxygen is used up during this period of time. Anaerobic conditions can cause swelling and distortion of mitochondria. The primary fixative, therefore, should be well oxygenated by bubbling air or oxygen through the solution before it is used.

Methods of delivering the primary fixative to the tissue.   There are three methods by which the primary fixative (usually glutaraldehyde) can be delivered to the tissue: (1) **immersion**, (2) **injection**, or (3) **perfusion**.

Immersion Fixation.   The tissue is immersed in the fixative in this method. This results in excellent fixation of small organisms such as bacteria, algae, and protozoa. However, it usually takes a long time for the fixative to penetrate into the center of large organisms or pieces of tissue, with resulting poor fixation in these areas, partially due to anoxic conditions arising during penetration of the fixative. The rate of penetration of the fixative can be accelerated by placing the cells or tissues under a mild vacuum during fixation.

Some tissue in animals, such as the lens of the eye, have to be fixed by immersion fixation because they have no or little vascular tissue associated with them, ruling out perfusion as a method of fixation. In such cases the tissue is dissected out as quickly as possible and immersed in the fixative. Plant tissues are fixed by immersion since perfusion or injection of the fixative is impractical.

Injection Fixation.   The primary fixative can be injected directly into the tissue. After injection of the fixative, the tissue is cut up into small blocks and immersed in the fixative. This is not a widely used method since some mechanical damage is induced by the pressure of the fixative as it is injected into the tissue.

Perfusion Fixation.   In perfusion fixation, the primary fixative is delivered to the tissue through the vascular system (arteries and veins) of the organism. This is the most effective way of delivering the fixative in animals, since most cells in the body are usually only a few cell diameters away from a part of the vascular system. Whole-body perfusions are usually performed on small animals by delivering the fixative through the left ventricle of the heart, to the organs of the body through the vascular system (Fig. 11-9). For large animals, where an inordinate amount of fixative solution is required for whole-body perfusion, the organ of interest is usually perfused by cutting the artery to the organ and delivering the fixative solution to the artery.

A perfusion device consists of two containers with tubes connected by a three-way valve (Fig. 11-10). A single tube carries the solution from the three-way valve to a cannula (small tube, often with a flared end) or a large-diameter needle (about no. 16). One of the containers has the solution used to flush the blood out of the vascular system of the animal. Usually, the solutions are fed by gravity by raising the containers to a height of 120 to 150 cm (0.117 to 0.145 atm of pressure; 10.3 m of water is 1 atm) above the animal if the solution is delivered through an artery, or 20 to 30 cm if the solutions are delivered through a vein. The blood pressure of humans and rats is about 120/80 mm of mercury (0.160 to 0.105 atm; 760 mm of mercury is 1 atm). The pressure should be enough to displace the blood, but not much higher than the pressure occurring naturally in the vascular system. Pumps can be used to produce the needed pressurization of the fixative solution (instead of gravity).

A flush or preperfusion solution is used to wash the blood out of the vascular system before delivering the fixative solution. The fixative solution will cause coagulation of the blood as the fixative enters into the vascular system, blocking the vascular system and preventing further passage of the fixative. The flush solu-

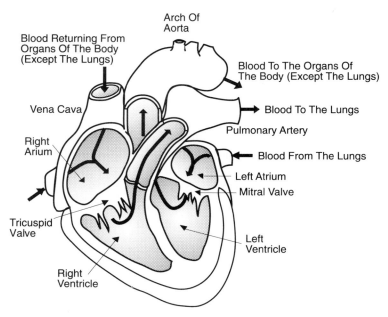

Figure 11-9   Blood flow through a mammalian heart. Blood from the right ventricle passes to the lungs, while blood from the left ventricle passes to the rest of the organs of the body.

tion is usually phosphate-buffered saline, which is approximately the same pH and osmolarity of blood. The flush will also commonly have an anticoagulant such as heparin (0.25% w/v) and a dilator such as procaine (0.5% w/v). The anticoagulant will help to prevent clotting of the blood in the vascular system. The dilator will prevent vasoconstriction (narrowing) of the arteries. The fixative is a powerful vasoconstrictor, so it is important to dilate the vascular system as much as possible during the flush. It is also important that the flush solution be well oxygenated. The cells are alive during the flush. If the flush solution does not contain oxygen, the cells will be subjected to anaerobic conditions with attendant changes in morphology. The fixative solution is oxygenated by bubbling oxygen or air through it for 5 min. The flush and the fixative solution should be delivered at the physiological temperature of the animal.

Immediately prior to perfusion, any gas bubbles in the tubing are removed by passing fixative and then flush solution through the tubes. If gas bubbles are in the tubing, they will pass into the vascular system, blocking the vascular system and preventing passage of the fixative.

The animal is placed under anesthesia and a central incision is made in the abdominal cavity up to the diaphragm. At the diaphragm, a V-shaped incision is made toward the front legs, and the rib cage is reflexed with a hemostat. The cannula or needle of the perfusion device is inserted into the left ventricle of the

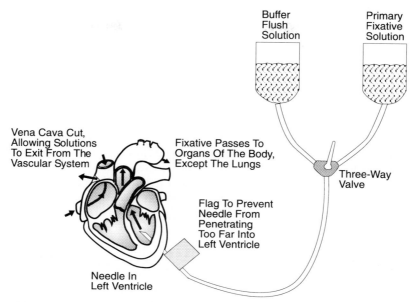

Buffer
Flush
Solution

Primary
Fixative
Solution

Vena Cava Cut,
Allowing Solutions
To Exit From The
Vascular System

Fixative Passes To
Organs Of The Body,
Except The Lungs

Three-Way
Valve

Flag To Prevent
Needle From
Penetrating
Too Far Into
Left Ventricle

Needle In
Left Ventricle

Figure 11-10 Perfusion through the left ventricle. The needle is embedded in the left ventricle. The vena cava or right atrium is cut. The flush and fixative pass through the left ventricle, through the organs of the body (except the lungs), and exit through the cut vena cava or right atrium.

heart (Fig. 11-10). A flag is commonly placed on the cannula or needle to ensure that the needle does not penetrate too far into the left ventricle. The right ventricle will be perforated if a needle is inserted too far, resulting in perfusion of the lungs through the pulmonary artery. The perfusion begins by allowing the flush to pass through the left ventricle, to the aorta and organs of the body (except the lungs), and out the snipped right atrium or vena cava. The flush is continued until the solution coming out the cut right atrium or vena cava is clear. The three-way valve in the perfusion device is rotated so that the fixative is introduced into the animal. Usually, about 250 ml of fixative is sufficient to fix a rat, while 100 ml is sufficient for a frog. After perfusion, the tissue of interest is removed and placed in the fixative solution for 1 h before processing further. Some investigators like to cool the solution to 4°C during this period and carry out the remaining fixation steps at this temperature.

Dehydration

The aim of dehydration is to replace the water in the specimen with a fluid that is miscible with water and the transitional fluid in critical point drying. If air drying is to be used, the aim is to replace the water with a fluid that will produce low surface tension forces during air drying. In most cases, ethanol or acetone is used to dehydrate tissue. Acetone has the potential to extract more of the constituents of the cells than does ethanol. However, acetone is more miscible in carbon dioxide, the commonly used transitional fluid in critical point drying. There is also evidence that acetone does not cause as much shrinkage of the tissue during dehydration as does ethanol. Whichever chemical is used, it must be anhydrous (not containing water). Both chemicals are hygroscopic, taking up water when exposed to the atmosphere.

The specimen is washed with buffer after osmication and then dehydrated through a graded series of acetone or ethanol, usually 50%, 75%, 90%, 95%, to 100%. Ten minutes should be long enough in each solution if the specimens are 1 mm or less thick; 15 to 20 min should be alloted for 1- to 3-mm-thick specimens. There may be a fine black precipitate in the first dehydration step if the dehydration fluid is ethanol. This results when osmium tetroxide coming out of the tissue is reduced. After the first 100% change, the tissue is placed in a 1-h change of 100%. After this step, all the water should have been replaced by the dehydration fluid. The changes of dehydration fluid should be as short as possible to minimize extraction and shrinking of the specimen, yet long enough to remove the water from the specimen. At the same time, too rapid dehydration may cause rapid osmotic changes in the cells, possibly resulting in morphological damages. Most cells maintain their shape and volume during dehydration until the 70% acetone or 70% ethanol stage is reached. During this stage, the bulk water (such as that in vacuoles) in the cells is replaced by the dehydration fluid. The water that is bound to hydration shells of macromolecules is removed during passage from 70% to 100% acetone or ethanol. Removing water from hydration shells results in more serious alterations of fine structure because the water is strongly bound to the cell macromolecules. Cells and tissues commonly shrink 10% to 20% during the transition from 70% to 100% dehydration fluid.

Specimen Drying

The column of the scanning electron microscope is maintained under a high vacuum to allow the electron beam unimpeded passage down the column to interact with the specimen. This high vacuum implies that the specimen itself cannot contain volatile components. Volatiles would evaporate rapidly in the high vacuum, destroying the structure of the specimen by the forces associated with surface tension and explosive evaporation of the volatile component. Consequently, the specimen must be dried before it is placed in the high vacuum of the scanning electron microscope.

Specimens containing volatile components are subjected to two types of forces during drying: (1) volume changes and (2) surface tension.

Volume Changes. Liquids, such as water, are able to support structures by their own mass. Thus a plastic bag full of water will have a roughly spherical shape due to the supporting mass of water in the bag. Removal of water from the bag

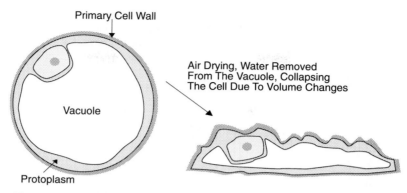

Primary Cell Wall

Vacuole

Protoplasm

Air Drying, Water Removed
From The Vacuole, Collapsing
The Cell Due To Volume Changes

Figure 11-11   Volume changes causing distortion of morphology during drying. The aqueous solution in the large vacuole supports the morphology of the cell. The cell becomes severely distorted during drying when the water is lost from the vacuole.

causes the bag to change shape and become shriveled and shrunken. These volume changes produce forces that are perpendicular to the surface, pulling the structures inward. Similar changes occur in cells with large vacuoles. Plant cells commonly have large vacuoles (Fig. 11-11). Placing a plant cell in a hypertonic solution (a solution with a higher osmotic pressure than the aqueous medium in the cell vacuole) results in water flowing out of the vacuole. The vacuole and the cell shrink, resulting in distortion of the shape of the cell. This is similar to what happens during drying when the water is removed from the cell with the attendant morphological changes.

Surface Tension.   Surface tension forces result from the attraction of liquid molecules to each other, producing a force that tends to minimize the area of the surface of the liquid. A water droplet in gravity-free space has the shape of a sphere due to the surface tension forces minimizing the surface area. At the dimensions of a water droplet and smaller, surface tension forces are important in the production of morphological distortions of specimens during drying.

A molecule in the interior of a liquid has no directional forces of long duration placed on it, even though the molecule may undergo random displacements because of collisions with other molecules (Fig. 11-12). The molecule is surrounded by other molecules, all with the same attractant force for other molecules. At the surface of the liquid, however, there are different circumstances. Beyond the free surface of the liquid, there are not any similar molecules to counteract the attractive forces exerted by the molecules inside the surface. This results in a net attraction of the molecules at the surface toward the interior of the liquid. A molecule at the surface has a greater potential energy than a molecule in the interior of the liquid. This is because work has been expended to bring the molecule from the interior to the surface. Thus the potential energy at the surface is proportional to

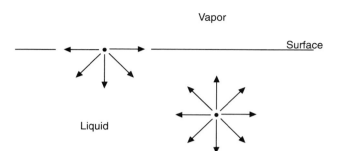

Vapor

Surface

Liquid

Figure 11-12 A molecule within a liquid has forces applied equally to it. A water molecule at the surface has forces applied internally from the liquid, which act to draw the molecule into the liquid.

the surface area. An increase in the surface area of the liquid requires an amount of work equal to the energy necessary to bring molecules from the interior to the surface. Like a stretched membrane, the surface will assume a shape of minimum area because this is the shape having minimum potential energy, the condition for a stable equilibrium.

The surface tension of a liquid is defined as the **surface potential energy**, which is the same as the work done per unit increase in surface area. This surface energy is contained in a surface layer just a few molecular diameters thick. Most liquids have surface tension forces of 20 to 40 dyn/cm² at room temperature, but water has the exceptionally high value of 72.75 dyn/cm² at 20°C (68°F).

Surface tension forces are thus parallel to the surface and attempt to contract a membrane laterally. Water molecules are an integral part of most cellular macromolecules. When cells dry, the cellular macromolecules are drawn up along the receding water meniscus and subjected to the enormous stresses of interfacial tensions that are set up (Fig. 11-13). This water meniscus is the same as that maintained by interfacial tension at any liquid surface, such as that in a graduated

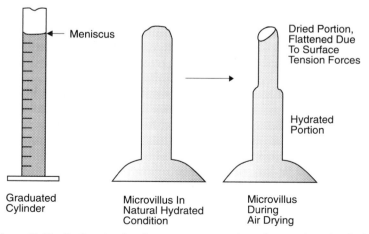

Meniscus

Dried Portion, Flattened Due To Surface Tension Forces

Hydrated Portion

Graduated Cylinder

Microvillus In Natural Hydrated Condition

Microvillus During Air Drying

Figure 11-13 Surface tension forces create a meniscus in a graduated cylinder holding a liquid. The surface tension forces involved in creating a meniscus result in flattening of a microvillus during air drying.

cylinder. **Interfacial tension** is the pressure difference between two sides of the liquid meniscus caused by the surface tension of the liquid.

It is the **stress** created by surface tension that causes the damage to biological structures. The stress forces caused by surface tension on a large structure are minimal. However, as the structure becomes smaller and smaller, the stress forces set up by surface tension increase dramatically. *The surface tension forces produce a stress (force/area) in the specimen that is proportional to the reciprocal of the radius (1/r) and that rises without limit as the structure becomes smaller and smaller.* The stress forces across a small structure such as a microvillus are about 2.8×10^7 dyn/cm^2 (28 atms, 412 psi). All biological specimens, with the possible exception of bone or wood, are severely distorted by such stress.

Air drying.   Air drying specimens is the simplest method of drying specimens. Unfortunately, the surface tension forces associated with the evaporation of water are so great that, except with the toughest specimens, an unacceptable amount of specimen distortion and shrinkage occurs during drying. Delicate specimens are flattened and twisted and commonly lose up to 50% of their volume during air drying.

An alternative is to replace water (with a surface tension of 72.75 dyn/cm) in the specimen with a solution of lower surface tension, such as ethanol (23 dyn/cm) or acetone (24 dyn/cm). Infiltrating a fixed sample with ethanol or acetone, followed by air drying, will reduce the surface tension forces by about one-third. While these reduced surface tension forces may result in little distortion of rugged specimens, they will still severely distort delicate specimens.

Critical point drying.   In critical point drying, a fluid in the specimen passes from the liquid phase through the critical point to the gas phase. The **critical point** is that combination of temperature and pressure at which the densities of the vapor and liquid phases are the same, so they intermingle and cannot be distinguished. *At the critical point, there are no surface tension forces associated with passing from the liquid to the gas phase and therefore no distortion of the specimen due to surface tension forces.* Although there is a little distortion, a specimen of vertebrate tissue will commonly have shrunk by one-third at the end of critical point drying, although some of the shrinkage will have occurred during dehydration. The shrinkage occurs because the dehydration and transitional fluid have the ability to extract some of the tissue contents.

Understanding critical point drying involves the use of the following terms. A **fluid** covers both liquid and gas phases; a fluid can flow, as can these two phases. A **liquid** is a fluid with a definite volume at any given temperature, whereas a **gas** is a fluid that can expand indefinitely to fill all available space. A **vapor** is a gas at low enough temperature and a high enough pressure so that a solid or liquid phase also exists, for example, carbon dioxide gas above dry ice or liquid carbon dioxide. A gas does not coexist with a solid or liquid phase, for example, carbon dioxide gas at room temperature and pressure.

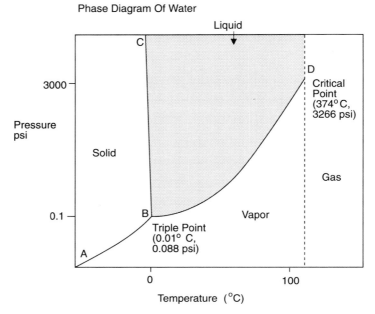

Figure 11-14   Phase diagram of water.

Before going into a more detailed explanation of critical point drying, the basics of phase diagrams will be reviewed.

Phase Diagrams.   Water is able to exist as a solid, liquid, or gas depending on the temperature and pressure to which it is subjected (Fig. 11-14). The **triple point** is the temperature and pressure at which there is an equilibrium between the solid, liquid, and gas phases. For water the triple point is 0.01°C and 6.1 mbar (0.0061 atm, 0.088 psi). Every known chemical substance, except helium, has its own characteristic triple point, which is controlled by the balance of intermolecular forces in the solid, liquid, and gas phases.

Line BC in Fig. 11-14, extending upward from the triple point, is the solid–liquid equilibrium or melting point line. At 1 atm (760 mbar, 14.7 psi), ice melts at 0°C. Water is very unusual in that the melting point decreases with pressure. Almost all the substances have melting points that increase with pressure (for example, carbon dioxide, Fig. 11-15).

Line AB in Fig. 11-14 is the sublimation curve; along this line, ice and its water vapor are in equilibrium. Line BD is the evaporation curve for water; along this line, liquid and vapor are in equilibrium. At 100°C and 1 atm, liquid water boils into gas.

A phase diagram is important in that it defines regions of temperature and pressure where only a single phase can exist. Thus, at 1 atm, water is liquid between 0° and 100°C; below 0°C it is a solid, and above 100°C it is a gas.

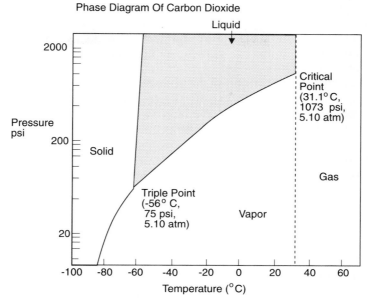

Figure 11-15   Phase diagram of carbon dioxide.

The evaporation line BD does not continue on indefinitely as the temperature and pressure increase. It eventually reaches the **critical point** (374°C and 217.7 atm [3266 psi] for water) at which the density for the liquid phase is equal to that of the gas phase, and surface tension forces are equal to zero. For example, if a fluid in a sealed container is heated, two things happen (Fig. 11-16). First, the liquid

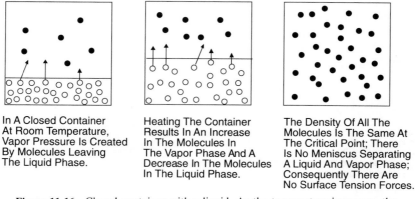

In A Closed Container At Room Temperature, Vapor Pressure Is Created By Molecules Leaving The Liquid Phase.

Heating The Container Results In An Increase In The Molecules In The Vapor Phase And A Decrease In The Molecules In The Liquid Phase.

The Density Of All The Molecules Is The Same At The Critical Point; There Is No Meniscus Separating A Liquid And Vapor Phase; Consequently There Are No Surface Tension Forces.

Figure 11-16   Closed container with a liquid. As the temperature increases, the density of the liquid decreases and the density of the vapor increases. At the critical point for the substance, the density of all the molecules is the same.

will expand and evaporate. Second, the gas will contract and begin to condense. Thus the density of the gas phase increases as the density of the liquid phase decreases. As the temperature and density approach the critical point, the densities of the liquid and gas phases will approach each other. At this point, **critical opalescence** occurs caused by the formation of areas of different density as the liquid and gas phases begin to be indistinguishable at the critical point. These areas of different density scatter light, giving rise to the opalescence. When the temperature and pressure have increased to the critical point, the density of the liquid phase will become the same as the vapor phase, the meniscus distinguishing the liquid phase and the gas phase will disappear, and the surface tension forces are zero. At temperatures above the critical point, no amount of pressure can condense the gas into liquid, and the substance can exist only as a gas.

Transitional Fluids.   From the standpoint of drying specimens for scanning electron microscopy, the critical point is the place on the phase diagram that is important, because it is here that the transition from the liquid to the gas phase can be made with zero surface tension forces. A specimen can be dried with little if any surface tension forces by placing the specimen in a liquid, raising the temperature and pressure through the critical point to the gas phase, and then slowly lowering the pressure to 1 atm while the temperature is kept above the critical point (so that the molecules remain in the gas phase).

The liquid that is used to pass through the critical point is called the **transitional fluid**. Unfortunately, water is not a practical transitional fluid because its critical temperature and pressure are too high (374°C and 217.7 atm [3266 psi]). Any biological specimen would be cooked and pressurized into an unrecognizable mass at this temperature and pressure. The most commonly used transitional fluid is carbon dioxide because it is economical, does not damage the environment, and has a reasonable critical temperature and pressure (31.1°C and 72.9 atm [1073 psi]) (Fig. 11-15). Freon 13 ($CClF_3$), 23 (CHF_3) and 113 ($CCl_2F\text{-}CClF_2$) are also used by some investigators (Table 11-4). Freon is the du Pont trademark for a series of unreactive halogenated hydrocarbons. Identical compounds are manufactured by other companies under their own trade names, such as Genetron, Arcton, and Frigon. The number after each trade name is the same for the identical chemical; for example, Freon 13 and Genetron 13 are the same compound ($CClF_3$). The

TABLE 11-4 CRITICAL TEMPERATURES AND PRESSURES OF COMMONLY USED TRANSITIONAL FLUIDS

Compound	Critical point, °C	Critical pressure, psi	Critical pressure, atm
Carbon dioxide	31.1	1073	72.9
Freon 13	28.9	561	38.2
Freon 23	25.9	701	47.7
Freon 113 (TF)	214.1	495	33.7

Freons (commonly used as refrigerants) are not flammable, have lower critical pressures than carbon dioxide, and therefore have less potential to cause explosive bursting of the critical point dryer. However, the Freons are considerably more expensive than carbon dioxide. Also, halogenated hydrocarbons deplete the ozone in the atmosphere and will be phased out. Other fluids (R-134, R-124, R-125, R-152a) that are more environmentally friendly will soon replace the older halogenated hydrocarbons.

Dehydration Fluids. Acetone and ethanol, as well as the lower alcohols (methyl, propyl, and butyl), are miscible with carbon dioxide and Freon. Amyl acetate is also sometimes used. The common method is to dehydrate in a graded series of ethanols or acetones and place the specimen into liquid carbon dioxide or Freon under pressure in the critical point dryer. The investigator should make sure that the specimens never dry out during any of the steps; otherwise, the specimen will be subjected to strong surface tension forces, and the benefits of critical point drying will be negated.

Critical point dryer. The critical point dryer is a bomb or sturdy metal container that can withstand high pressures. A bomb is defined as a chamber in which a process takes place at elevated pressure and often elevated temperature. Contrary to the military term, a bomb is not meant to explode, although it has the potential for doing so, particularly if it is operated improperly. There are a number of different designs for critical point dryers, although they are all basically the same. The instrument manufactured by Polaron will be described here (Fig. 11-17). The Polaron critical point dryer has a cylindrical pressure chamber. At one end of the pressure chamber is a 2-cm-thick **window** covered with a Plexiglass cover for viewing the process. At the other end of the pressure chamber is a removable **door** for placing a trough with the specimen in the pressure chamber. The pressure chamber has four valves: (1) The **safety valve** is built into the support pillar. It opens at 1850 psi to bring the pressure in the chamber to 1 atm. The valve has to be resealed before the device is used again. (2) The **drain valve** is built into the bottom so that the dehydrating and transitional fluid can be flushed from the pressure chamber. (3) A **vent valve** is in the top to vent gases off from the pressure chamber. (4) An **input valve** is used to introduce the transitional fluid into the pressure chamber. The input valve is connected, by a high-pressure metal hose, to the pressurized cylinder containing the transitional fluid. If carbon dioxide is used as a transitional fluid, the metal hose is connected to a siphon cylinder containing anhydrous carbon dioxide (a supply cylinder with a siphon tube from the bottom of the cylinder to the exit valve so that liquid carbon dioxide is forced out by the gas pressure at the top of the cylinder). The temperature of the critical point dryer is controlled by a water jacket through which either hot or cold water is passed. A **pressure** and **temperature gauge** are inserted into the apparatus.

Critical point drying technique.

Setup. The critical point dryer is cooled to about 5°C by running cold water

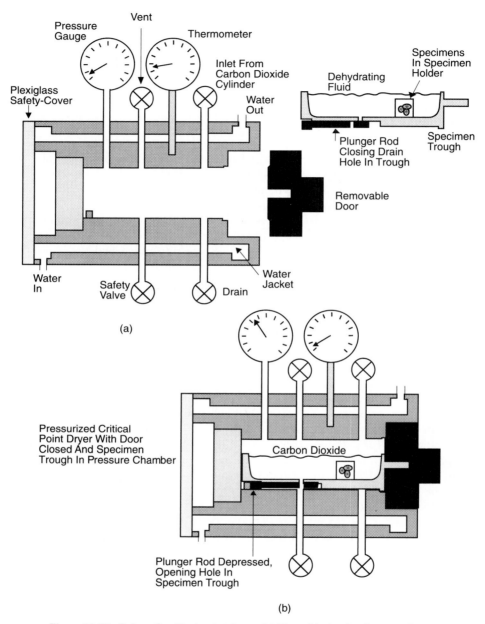

Figure 11-17 Polaron® critical point dryer. (a) The critical point dryer ready to be loaded with the pressure chamber exposed to the atmosphere. The door is off and the specimen trough has a specimen holder in the dehydration fluid. The plunger-rod of the trough plugs the hole in the bottom of the specimen trough. (b) The critical point dryer loaded and pressurized. The stud in the front of the pressure chamber has depressed the plunger rod, opening the hole in the trough and allowing the dehydrating fluid to drain out of the trough. The inlet valve is open to the carbon dioxide supply cylinder.

through the water jacket. The colder temperature helps to keep carbon dioxide in the liquid state in the pressure chamber, especially during the flushes when the pressure in the chamber can fall to the level where the specimen may be exposed, resulting in surface tension forces being applied to the specimen. The specimen in the dehydration fluid (usually ethanol or acetone) is placed in a fine wire screen or porous polyethylene specimen holder immersed in the dehydration fluid. Four or five specimen holders immersed in the dehydration fluid can be accommodated in the trough.

Loading.   The door is removed from the rear of the pressure chamber, the trough containing the specimen holders immersed in the dehydration fluid is inserted into the pressure chamber, and the door is closed (Fig. 11-18, point A). The trough containing the specimen holders has a small hole in the bottom that is closed by a plunger rod when the trough is outside the pressure chamber. Inserting the trough in the pressure chamber results in a peg in the front of the chamber pushing on the plunger in the trough, opening the small hole in the trough. The

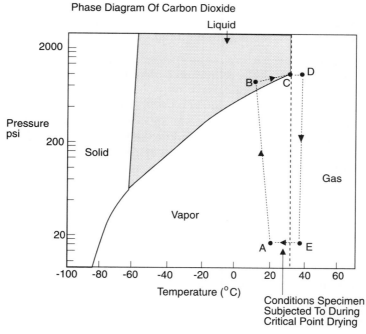

Figure 11-18   Phase diagram of carbon dioxide showing conditions that a specimen is subjected to during critical point drying. (A) The specimen in the dehydration fluid at room temperature and pressure. (B) The specimen in the critical point dryer with full cylinder pressure of carbon dioxide in the pressure chamber. (C) The specimen at the critical point. (D) The specimen above the critical temperature so that all the carbon dioxide is a gas. (E) The dry specimen at room pressure ready to be removed from the critical point dryer.

dehydration fluid begins to slowly drain out of the trough into the pressure chamber. At the same time, carbon dioxide is introduced into the pressure chamber from the pressurized supply cylinder. The carbon dioxide is introduced slowly over a period of 3 to 4 min until full cylinder pressure is reached (usually 800 to 900 psi). Thus, while the dehydration fluid is draining out of the specimen trough, liquid carbon dioxide is filling the pressure chamber. The pressure chamber becomes full of liquid, the heavier dehydration fluid settles on the bottom, and the lighter carbon dioxide is on the top, with a visible interface between them.

Flushing out the Dehydrating Fluid.   The dehydration fluid plus some carbon dioxide is drained out of the bottom of the critical point dryer through the drain valve while fresh carbon dioxide flows in through the inlet valve. While draining the pressure chamber, it is important to look through the viewing port to ensure that the level of carbon dioxide does not drop beneath the level of the specimen. This can happen if the fluid is drained out so fast that the inflowing carbon dioxide cannot maintain the pressure in the chamber. A flush will normally continue for 1 to 2 mins and is then stopped for 5 min to allow the carbon dioxide to replace the dehydration fluid in the specimen. This process (this flush plus the 5-min break) is usually repeated about five times, during which time the transitional fluid will have replaced all the dehydration fluid (Fig. 11-18, point B).

Going through or around the Critical Point.   All the valves to the critical point dryer are closed, so the critical point dryer is a closed bomb. Hot water is used to raise the temperature and pressure to the critical point of carbon dioxide (31.1°C and 1073 psi), where the liquid carbon dioxide passes to the gas phase without surface tension force applied to the specimen (Fig. 11-18, point C). Seldom do the temperature and pressure coincide exactly at the critical point. Instead "going around the critical point" usually occurs when either the temperature exceeds the critical temperature (31.1°C) before the pressure reaches the critical pressure (1073 psi), or vice versa. The surface tension forces near the critical point are so low that going around the critical point usually results in minimal surface tension forces being applied to the specimen.

Venting.   With the temperature above the critical temperature (the carbon dioxide remains in the gas phase) (Fig. 11-18, point D), the pressure is lowered to 1 atm by venting carbon dioxide over 5 min (Fig. 11-18, point E). The dry specimen is removed from the device and stored in a desiccator or mounted for viewing.

Safety.   While using the critical point dryer, it is important to remember that very high pressures are involved. The operator should use common sense by making sure the critical point dryer is functioning properly and that unreasonably high pressures are not produced, particularly when heating the device to bring the specimen to the critical point.

Fixation by Rapid Freezing Followed by Freeze Drying

Fixation of hydrated specimens by rapid freezing followed by dehydration by freeze drying is usually the best way to prepare tissue for scanning electron microscopy. However, relatively sophisticated equipment is required to maintain the specimen at high vacuum and low temperature during the freeze drying, making this method the most expensive. The method gives best preservation of tissue for the following reasons.

1. *Fixation by rapid freezing occurs much more quickly than fixation with chemicals.* Penetration of a chemical fixative into a specimen occurs by diffusion of the fixative into the specimen, a relatively slow process occurring at about 1 to 2 µm/s. Once the fixative has arrived at a particular place in the specimen, it takes a few seconds to a minute to crosslink the macromolecules of the cell. On the other hand, fixation by rapid freezing occurs in milliseconds, considerably reducing the chance of artifactual changes occurring in the specimen.

2. *Fixation by rapid freezing, followed by freeze drying, immobilizes molecules in place.* In chemical fixation, some of the larger molecules and most of the smaller molecules are not crosslinked by the standard glutaraldehyde/osmium tetroxide fixation. The molecules that are not fixed are washed out of the specimen or moved in the specimen during the fixation and dehydration steps.

3. *Fixation by rapid freezing, followed by freeze drying, results in minimal shrinkage and distortion of the specimen.* This is the most important aspect as far as scanning electron microscopy is concerned, since it is the size and shape of objects that are viewed in the scanning electron microscope. Rapid freezing immobilizes the specimen in approximately the same shape and size as the natural condition. Subsequent freeze drying can result in shrinkage of up to 20% of the specimen volume. However, this is the smallest amount of volume shrinkage of the methods available for specimen preparation (Fig. 11-19).

Basics of liquid to solid transitions.   Liquids become solids by the removal of heat. Lowering the temperature of a liquid causes the molecules to slow down. The liquid begins to freeze when the temperature is lowered enough so that the attractive forces between molecules cause the slower moving molecules to be held in rigid position. At the solidification point, the temperature will usually remain the same until all the liquid has been transformed into the solid phase. This is because it is necessary to continually remove heat from the liquid molecules if they are to lose enough kinetic energy to be added to the solid phase. The amount of heat that must be removed to freeze a liquid is called the **heat of crystallization** (Fig. 11-20). The solid molecules thus have less energy than the liquid molecules. If the molecules are frozen in a well-ordered lattice, a **crystalline solid** is produced.

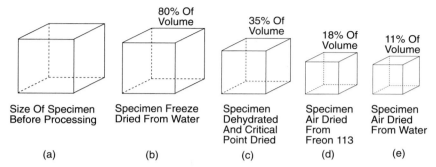

Figure 11-19 (a, b) Minimal shrinkage of a specimen results from rapid freezing followed by freeze drying. Specimens chemically fixed and critical point dried (c) have less shrinkage than specimens that are air dried (d, e). (*Redrawn from* A. W. Robards and U. B. Sleyter, Low temperature methods in biological electron microscopy, in *Practical Methods in Electron Microscopy* (A. M. Glauert, ed.), Vol. 10, Elsevier Publishing Co., New York, 1985.)

If the molecules are frozen in a random fashion, an **amorphous solid** is formed. Once all the molecules are in a solid form, the temperature of the substance can decline farther.

Melting of a solid to a liquid results in the release of **heat of fusion**. The temperature of the substance remains the same until all the solid is converted to liquid, since the energy from the heat of fusion is used to melt the remaining solid parts to liquid.

Formation of ice.   Some liquids, such as water, do not have a smooth transition from the liquid into the solid state, as just described. Instead, the liquid undergoes **supercooling**, with the temperature dropping below the freezing point of the substance, until crystallization begins to form solid particles (Fig. 11-20).

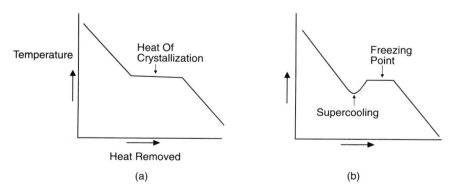

Figure 11-20 Solidification curves for (a) a substance that does not exhibit supercooling and (b) a substance that does exhibit supercooling.

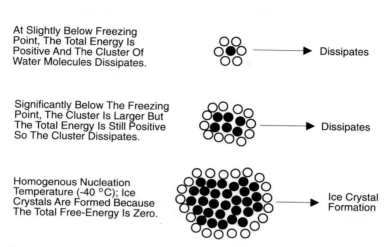

○   Water Molecule With Volume Free-Energy (Negative)

●   Water Molecule With Surface Free-Energy (Positive)

At Slightly Below Freezing
Point, The Total Energy Is
Positive And The Cluster Of   ⟶ Dissipates
Water Molecules Dissipates.

Significantly Below The Freezing
Point, The Cluster Is Larger But
The Total Energy Is Still Positive   ⟶ Dissipates
So The Cluster Dissipates.

Homogenous Nucleation
Temperature (-40 °C); Ice
Crystals Are Formed Because   ⟶ Ice Crystal
The Total Free-Energy Is Zero.   Formation

Figure 11-21 Two-dimensional view of nucleation of ice crystals. Clusters of water molecules formed at temperatures near the freezing point are small and contain more molecules with surface free energy, making the total energy of the cluster negative and resulting in dissipation of the cluster. At a colder temperature, the ice clusters are larger with more molecules with volume free energy, resulting in the total energy of the cluster being positive. Nucleation of ice crystal formation then begins.

Once the crystals begin to grow by the accumulation of additional molecules, potential energy is released as kinetic energy of the heat of fusion. This increases the kinetic energy of the liquid and solid molecules of the substance, with a resulting increase in temperature up to the freezing point. Further removal of heat eventually leads to complete conversion of liquid to solid.

Supercooling.   Supercooling of water can be explained through the classical nucleation theory. As the temperature of water is lowered to the freezing point, some of the water molecules lose enough energy to aggregate together as clusters (the microstructure of which is not necessarily crystalline) (Fig. 11-21). The total energy of each cluster is divided into two components: (1) the volume free energy, and (2) the surface free energy. The **volume free energy** is negative (lower energy) and arises from interactions of the water molecules in the interior of the cluster. The **surface free energy** is positive (higher energy) and arises from water molecules at the surface of the cluster. Most of the total energy of the cluster comes from the surface free energy if the clusters are small, since most of the molecules are at the surface in a small cluster. The total energy of the cluster is positive and the cluster soon dissipates. This is the situation at the freezing point of water.

Removing heat and lowering the temperature of water beneath the freezing point result in two changes.

1. The average size of the cluster increases. The increase in cluster size results in more water molecules in the interior of the cluster and less at the surface. Therefore, the total energy of the clusters becomes more negative from the increase in the negative volume free energy of the molecules in the center of the cluster.

2. The magnitude of the surface free energy decreases as the temperature decreases.

Eventually, the critical size is reached where the volume free energy is equal to the surface free energy. Any clusters that are now formed have an even or better chance of surviving long enough to become the nucleus of an ice crystal. At about −40°C, the average cluster size becomes equal to the critical size, and **homogeneous nucleation** (nucleation from only water molecules, compared to **heterogeneous nucleation** or nucleation from different types of molecules) has to occur. *At the temperature of homogeneous nucleation, a large number of clusters are present that can serve as nuclei for crystal formation. Therefore, ice crystals of very small size will be produced if the heat of fusion can be withdrawn quickly enough to keep the temperature from rising above the temperature of homogeneous nucleation.* If the water is cooled more slowly, large ice crystals will grow until solidification occurs. Ice crystals appearing earlier will have more time to grow and become larger than those that are formed later, resulting in a distribution of ice crystal sizes. *It is the formation of ice crystals of minimal size that is the goal of freezing for electron microscopy.*

Freezing of biological specimens. The speed at which cells are frozen determines the survival rate of the cells on thawing (Figs. 11-22 and 11-23). Cells will survive freezing if they are frozen very slowly (0.01°C/s) or very fast (1000°C/s). A moderate freezing speed of 1° to 10°C/s destroys the viability of the cells.

Very Slow Freezing. At very slow freezing rates (0.01°C/s), large ice crystals are formed in the aqueous medium outside the cells, attracting water from inside the cells to condense on the surface of the developing extracellular ice crystals. Thus the cells become dehydrated by the developing ice surrounding the cells. When the water inside the cells freezes, there is such a small amount of water available that small ice crystals are formed that do not significantly damage the cell. The cells are viable when they are thawed. While this method results in viable cells, it is of little use in scanning electron microscopy because there is considerable loss in volume and distortion of the specimen during the freezing.

Moderate Freezing Rate. Freezing at moderate rates of 1° to 10°C/s results in cells that have little change in volume or shape. However, the freezing rate is slow enough so that large needlelike ice crystals are formed inside the cell. These needlelike ice crystals pierce the membranes that compartmentalize the cell, destroying the viability of the cell. *Moderate freezing rates are too slow for preparing cells for scanning electron microscopy because the resulting large ice crystals in the protoplasm produce surfaces that are irregular (bumpy) and with holes (after the specimen has been dehydrated).*

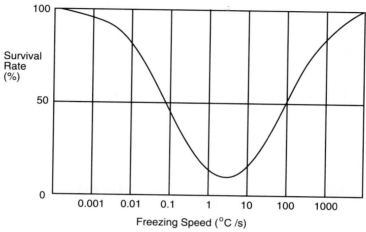

Figure 11-22   Survival rate of yeast cells at different temperatures. At slow freezing rates, large extracellular ice crystals dehydrate the cell, resulting in a high survival rate. At fast freezing rates, very small ice crystals are formed, resulting in a high survival rate. At moderate freezing rates, large ice crystals are formed inside the cell, piercing the membranes and resulting in a high mortality rate.

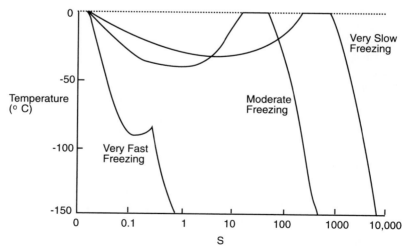

Figure 11-23   Curves illustrating how fast heat is removed from a specimen at different freezing rates.

Very Fast Freezing.   Very fast freezing rates of 1000°C/s are best for preparing cells for scanning electron microscopy. At these freezing rates, the specimen is cooled so fast that, after homogeneous nucleation of ice crystals has begun, the heat of fusion is removed so quickly that a very small ice crystal is formed around each nucleation point. The small ice crystals result in minimal damage to the compartments of the cell, so the cells are viable after thawing. Most important for scanning electron microscopy is that the size of the ice crystals is beneath the resolution of the microscope (usually 5 to 10 nm). In pure-water systems it is difficult to obtain ice crystals in the range of 10 nm since it requires freezing rates of about 10,000°C/s (1000°C/s to produce 1- to 2-μm ice crystals, while 100°C/s produces 4- to 5-μm ice crystals). However, the protoplasm of cells contains many dissolved substances that act as cryoprotectants during freezing, resulting in the formation of smaller ice crystals than would be produced if water alone were frozen. Cryoprotectants act by (1) increasing the number of ice crystals at a given temperature, leading to the formation of a larger number of smaller ice crystals, and (2) increasing the viscosity of the medium, thereby slowing the growth and size of the ice crystals. Using the common method of plunging a specimen into a cold liquid cryogen usually produces fast enough freezing rates so that small ice crystals are formed near the specimen surface that do not significantly alter the appearance of the specimen surface.

Depth of rapid freezing in the specimen.   Heat can only be removed from a specimen by placing the surface of the specimen in contact with a cooler medium. This results in rapid freezing (with the formation of small ice crystals) occurring only at the specimen surface (Figs. 11-24 and 11-25). Beneath the specimen surface, heat has to be removed through the previously formed ice. Ice is an extremely poor conductor of heat, resulting in significantly lowered freezing rates away from the specimen surface. At about 10 μm from the surface, the rate of cooling becomes too low to prevent the growth of large ice crystals. Here the heat of fusion released by the growing ice crystals further slows the rate of cooling of the specimen, resulting in the formation of large crystals in these areas. The importance of this is that *no matter how good the cooler medium is at removing heat from the cells, it is only near the surface of the specimen that small ice crystals are formed, producing minimum specimen distortion.* Research involving the production of small ice crystals in the interior of the specimen has moved away from attempting to produce better cooling media because of this fact and is focusing on other areas, such as high-pressure freezing, which has the potential to produce small ice crystals throughout small specimens. Fortunately, for the scanning electron microscopist, it is the surface that is of interest, the portion of the specimen that is best preserved by current cryotechniques.

Methods of rapid freezing.   Plunging the specimen into a cold liquid medium (**cryogen**) is the most commonly used method of freezing specimens for scanning electron microscopy. A high rate of heat transfer occurs since the speci-

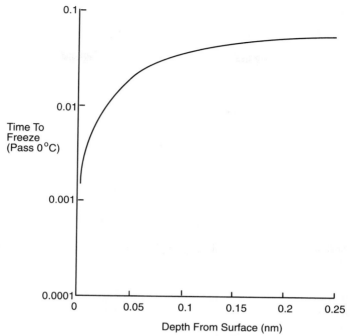

Figure 11-24   Graph showing how long it takes to pass 0°C at different places in a specimen plunged into a liquid halocarbon.

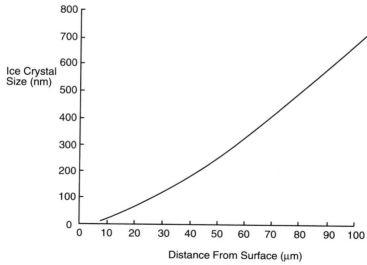

Figure 11-25   Sizes of ice crystals at increasing distances from the surface of a liver specimen plunged into liquid halocarbon.

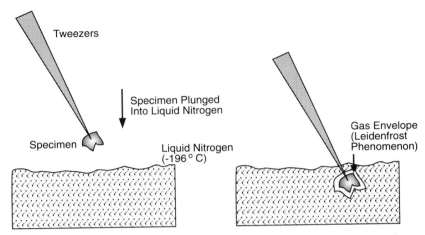

Figure 11-26 Leidenfrost phenomenon. A gas envelope is produced between the cryogen and specimen, retarding the transfer of heat from the specimen to the cryogen.

men is traveling through the cold cryogen and the specimen surface is continually being exposed to fresh cold cryogen, increasing the rate of removal of heat from the specimen.

A cryogen suitable for freezing specimens should have the following properties:

1. A very low freezing point (at least $-130°C$) so that the thermal gradient between the sample and the liquid cryogen is large, thus maximizing the heat removal from the specimen.

2. A large heat capacity and high thermal conductivity so that the liquid cryogen will act as an effective heat sink.

3. A boiling point that is well removed from the melting point to minimize the Leidenfrost phenomenon. The Leidenfrost phenomenon occurs in liquefied gases of low molecular weight, such as liquid nitrogen or helium (Fig. 11-26). These gases are very near their boiling point at 1 atm of pressure. Plunging a room temperature specimen into liquid nitrogen results in the nitrogen next to the warm specimen absorbing heat and changing into gas. The specimen becomes surrounded by a layer of gas (Leidenfrost phenomenon), which has a low thermal conductivity. Heat is removed slowly from the specimen, with the resulting slow freezing and formation of large ice crystals. *Therefore, cryogens should be used at their lowest liquid temperature to maximize the freezing rate of the specimen.*

4. It is preferable that the cryogen be safe to use and relatively inexpensive.

TABLE 11-5 CHARACTERISTICS OF LIQUID CRYOGENS

Liquid	Melting point °C (K)	Boiling point °C (K)	Specific heat $J\ g^{-1}\ K^{-1}$	Thermal conductivity $mJ\ m^{-1}\ s^{-1}\ K^{-1}$
Propane C_3H_8	−189.6 (84)	−42.1 (231)	1.92	219
Ethane CCH_3CH_3	−183.5 (90)	−88.8 (184)	2.27	240
Halocarbon 12 CCl_2F_2	−158.0 (115)	−29.8 (243)	0.85	138
Halocarbon 22 $CHClF_2$	−160.0 (113)	−40.8 (232)	1.08	152
Liquid nitrogen, N_2	−210 (63)	−195.8 (77)	2.0	153

The classes of compounds that have proved to be the best liquid cryogens are (1) liquid propane and ethane, (2) subcooled liquid nitrogen, and (3) the halocarbon refrigerants (Table 11-5).

Propane. Propane has a low melting point (−189.6°C), high boiling point (−42.1°C), good thermal conductivity, and a specific heat resulting in the fastest freezing rates of the commonly used cryogens, around 20,000°C/s (there is some variability in the reported absolute freezing rates of the different liquid cryogens, although ethane has been reported to give cooling rates twice that of propane). Propane has the advantage that it is easy to obtain, since it is the heating gas that is obtainable from local sources. A liquid propane solution is prepared for use by venting a propane cylinder into a liquid nitrogen-cooled container in which the liquid propane condenses. The major problem with propane is that it is highly flammable (it is combustible at concentrations as low as 2%), and it can be sub-cooled beyond the liquefaction point of oxygen, allowing oxygen to condense in the propane and producing a potentially explosive mixture. However, propane should cause no problem if it is used with common sense in a well-ventilated hood.

Subcooled Nitrogen. Liquid nitrogen (−196°C) is a poor liquid cryogen because it is near its boiling point, resulting in the formation of a gas envelope between the specimen and the liquid nitrogen. However, producing a vacuum over liquid nitrogen can bring nitrogen down to its triple point (−210°C, 1.01×10^{-2} mbar [13.5 kPa]), cooling the nitrogen from −196°C to −210°C (Fig. 11-27). Sub-cooled nitrogen is produced by placing 100 to 200 ml of liquid nitrogen into an insulated container (for example, a polystyrene cup) and placing the container with the liquid nitrogen into a desiccator with a wide-bore pumping exit. The desiccator is evacuated with a rotary pump. The nitrogen undergoes evaporative cooling, resulting in the formation of nitrogen ice mixed with liquid nitrogen (nitro-

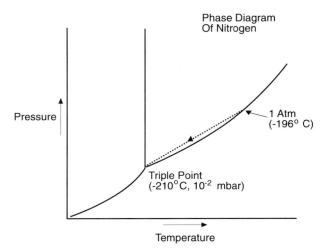

Figure 11-27 Phase diagram of nitrogen. Subcooling nitrogen results in the nitrogen being brought to the triple point at $-210°C$ and 10^{-2} mbar.

gen slush). The layer of nitrogen ice can form over the top of the liquid nitrogen, causing the liquid to explode through the surface ice and into the desiccator. This can be prevented by placing hollow drinking straws or antibump granules in the container. The nitrogen slush begins to warm up as soon as air is let into the desiccator. It is best to allow the nitrogen ice to change into liquid before freezing specimens because the liquid–ice mixture gives uneven freezing rates; the nitrogen ice is a poor conductor of heat. Subcooled nitrogen in a well-insulated container can be used for about 2 min after removal from a vacuum before the temperature begins to rise back up to $-196°C$. Plunging a specimen into subcooled liquid nitrogen does not result in the formation of a significant nitrogen gas envelope around the specimen because the nitrogen is relatively far from its boiling point. Good cooling rates of around $2000°C/s$ can be achieved with subcooled nitrogen. These freezing rates are not as fast as can be obtained with propane or the halocarbons. However, subcooled liquid nitrogen is safe and easy to produce.

Halocarbons. The commonly used halocarbons are 12 and 22 (Freon 12, Arcton 12, and others). The halocarbons were previously discussed in the section on critical point drying. The halocarbons have freezing rates of around $8000°C/s$, intermediate between propane and subcooled nitrogen. A liquid halocarbon bath can be prepared by venting a "dust-off" can of refrigerant onto the bottom of a container immersed in liquid nitrogen (Fig. 11-28). The halocarbon will solidify and has to be warmed up with a room temperature metal rod. The halocarbon should be allowed to begin to freeze again before plunging the specimen into the cryogen bath. Halocarbons are potentially damaging to the ozone layer of the atmosphere and may have limited availability in the future.

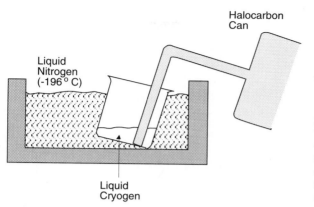

Liquid Nitrogen (-196° C)

Halocarbon Can

Liquid Cryogen

Figure 11-28 Method of producing a bath of liquid cryogen. The nozzle of the cryogen can is applied to the bottom of a liquid nitrogen-cooled container. The cryogen condenses on the bottom of the container as it is slowly vented from the can.

Freeze Drying

During freeze drying, water is removed from the specimen by **sublimation** (evaporation from the solid to the vapor state without any intervening liquid phase). In ice, molecules vibrate about their lattice positions, colliding with their neighbors, resulting in a distribution of the kinetic energies of the water molecules. A small portion of the surface molecules has enough kinetic energy to overcome the attractive forces of the solid and escape from the surface, entering the vapor phase above the solid. The commonest example of sublimation is the conversion of solid carbon dioxide (dry ice) into carbon dioxide gas without any intervening liquid phase. Sublimation of the water molecules from the specimen results in no surface tension forces and little distortion during the dehydration of the specimen, although there is some shrinkage. The amount of distortion and shrinkage, however, is the least of all the available techniques.

Recrystallization. Recrystallization of water molecules may occur during freeze drying if the process is carried out at too high a temperature. Frozen water molecules can recrystallize into larger ice crystals at specimen temperatures above $-80°C$. This recrystallization phenomenon means that freeze drying should be started at $-95°C$, a temperature at which there is no risk of recrystallization. At $-95°C$ and 2.7×10^{-5} mbar, 20 µm of ice will sublime in about 1 h. Portions of the specimen at depths greater than 20 µm are subjected to slow freezing rates during the initial freezing and contain large ice crystals anyway. Therefore, once the surface 20 µm has been freeze dried, there is little point in worrying about recrystallization of ice in the interior of the specimen. Accordingly, the specimen temperature can be raised to $-60°$ to $-70°C$ to more rapidly sublime away the ice in the interior of the specimen. Temperatures above this should be avoided because they may result in water vapor flows out of the specimen that are so rapid that there is the potential for damage of the specimen. This occurs from water that has been mixed with other molecules during freezing. As the specimen is frozen,

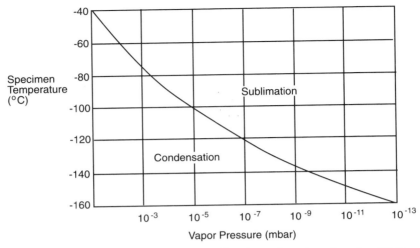

Figure 11-29   Curve representing the sublimation and condensation of water at different temperatures and vacuums.

water separates from any other molecules that are in solution, forming pure ice crystals. Eventually, the point is reached where the water mixed with other molecules no longer separates out, and the mixture of water and other molecules freezes. This freezing occurs at lower temperature than that of pure water. Therefore, during warming of the specimen, the mixture of water and other molecules melts before the pure water does. During freeze drying, this water exits rapidly from the specimen, producing fluxes that can damage specimen structure.

 Time Necessary to Freeze Dry a Specimen.   The total time necessary to sublime water off from a specimen depends on the specimen temperature and vacuum. Figure 11-29 gives the vapor pressure curve for water at different temperatures and pressure. There is total *sublimation* of water molecules from ice into the vapor phase at temperatures and pressures above the curve. There is total *condensation* of water vapor molecules onto the ice at temperatures and pressures beneath the curve. If the specimen is at $-100°C$ and the partial pressure of water is 1×10^{-5} mbar, an equal number of water molecules are condensing on the specimen as are subliming, and no freeze drying occurs. Either the temperature has to be raised or a better vacuum has to be attained to increase the sublimation rate. Usually, it is necessary to raise the temperature, since the configuration of the vacuum system is fixed and this is what dictates the vacuum and partial pressure of the water. Figure 11-30 shows the sublimation rate of water molecules from ice at a vacuum of 5×10^{-5} mbar. It is seen that sublimation rate for ice increases rapidly with increasing specimen temperature. At $-100°C$ it is only 1.5 nm/s, at $-80°C$ it is 50 nm/s, and at $-60°C$ it is 1000 nm (1 μm)/s.

 The actual time required to freeze dry a specimen can be significantly longer

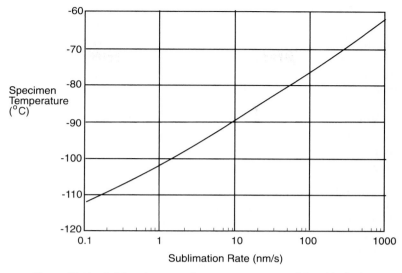

Figure 11-30 Sublimation rate of water at a vacuum of 5×10^{-5} mbar.

than the above theoretical calculation for pure ice. This is because the sublimation of water from frozen biological specimens leads to an increasing surface layer of dried material that further inhibits sublimation. The sublimation of water from a frozen specimen follows a general curve (Fig. 11-31). Initially, the sublimation rate is very high as the specimen, which is stored in liquid nitrogen, warms up to the freeze drying temperature ($-95°C$). The sublimation rate decreases as the drying front recedes from the surface to the interior of the specimen. A certain amount of water is still in the specimen at the end of the freeze drying period. This is the

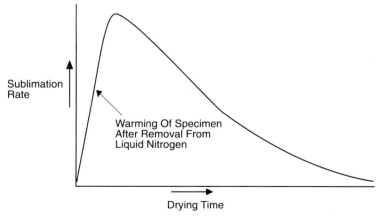

Figure 11-31 Sublimation rate of water during the period of freeze drying.

water that is bound to the macromolecules of the tissue and that can comprise 5% to 10% of the total water in the tissue. The frozen biological specimen has a mixture of pure water (ice) and the water bound to the biological molecules (bound water). The ice phase consists of unbound water. The bound water is linked to the hetero-polymers (proteins, fats, carbohydrates, and nucleic acids) that make up living matter. To remove most of this bound water, the specimen is warmed to a few degrees above room temperature before breaking the vacuum. This avoids distortion of the specimen structure by the evaporation of the bound water and maximizes drying. Most freeze dried specimens are very hygroscopic and should be placed in a desiccator for storage until the specimens are prepared for scanning electron microscopy.

Equipment for Freeze Drying.   Devices that produce high-quality freeze dried specimens are able to maintain a vacuum of 10^{-6} mbar. This means that the device is fitted with a cryopump, diffusion pump, or turbomolecular pump. Liquid nitrogen-cooled cold traps (shrouds) are commonly used next to the specimen to provide a better vacuum in the immediate area of the specimen. Freeze fracture devices are most commonly used for freeze drying since they already have a temperature-controlled specimen table, shrouds, and high-vacuum pumps.

Specimen Supports.   Freeze drying is often performed on specimens that are already mounted on a support that can be attached to a scanning electron microscope specimen stub after drying. Unfortunately, many adhesives crack or flake at liquid nitrogen temperatures, providing poor contact between the cold specimen table of the freeze drying device and the specimen. However, commercial silicone sealants provide an effective adhesive at liquid nitrogen temperatures. The specimen is attached to the support by the silicone sealant before freezing. After freeze drying, colloidal graphite can be painted up to a bulk specimen to provide electrical contact. If the specimens are suspensions of cells, they can be placed on a filter paper dot attached to the support by silicone sealant.

Specimen Coating

Almost all specimens that are examined in the scanning electron microscope are coated with a 10- to 20-nm layer of a metal or carbon. Better images are obtained from coated specimens for the following reasons:

1. Secondary electron and backscattered electron emissions are greater and form a smaller surface area of a coated specimen.
2. Thermal damage is reduced in coated specimens.
3. Specimens are more firmly attached together and to the support when they have a metal or carbon coat.
4. Specimen charging is reduced or eliminated.

Increased emission of electrons from a smaller surface area. Coating the specimen with a metal results in increased resolution of the image for two reasons:

1. Better image resolution is obtained by increasing the *total emission* of secondary and backscattered electrons from the specimen, thereby reducing the necessity of a large amplification of the signal through the photomultiplier, which is turn reduces the amount of noise in the image.
2. Better resolving power of the scanning electron microscope is obtained by *decreasing the area of the specimen surface* from which the backscattered and secondary electrons are emitted.

Increasing the Total Emission of Electrons. The greatest depth from which secondary electrons can escape in a metal, such as gold, is about 5 nm (Fig. 11-32). In an insulator, such as a biological specimen, the greatest escape depth is about 50 nm. The difference in the depth of the escape zone of secondary electrons is due to the number of conduction electrons in metals and insulators (see Chapter 7). After it is generated in the specimen, a secondary electron that has the energy to escape from the specimen surface collides primarily with conduction electrons. **Metals** have a large number of conduction electrons. Therefore, a secondary electron generated in a metal will undergo many collisions with conduction electrons per unit volume, reducing the chance that the secondary electron will escape from the specimen unless it is near the surface of the specimen. **Insulators** have few conduction electrons. Therefore, in an insulator, secondary electrons can travel

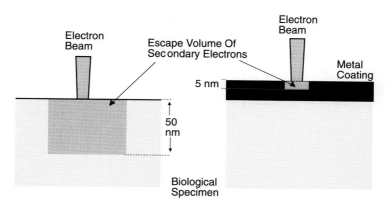

Figure 11-32 The escape volume of secondary electrons in a specimen that has a 10-nm coating of a metal is considerably smaller than the escape volume of secondary electrons in an uncoated specimen.

relatively great distances between collisions, increasing the chance that the secondary electron approaches the specimen surface with enough energy to escape.

Coating of a specimen usually involves the deposition of a 10- to 20-nm film of a metal such as gold. This means that the entire escape depth (about 5 nm) of the secondary electrons is in the metal coating. Most elements have about one secondary electron escaping from the specimen for each ten beam electrons of 20 keV striking the specimen. Thus, the **secondary electron coefficient** for most elements is about 0.1. Gold, however, has a secondary electron coefficient of about 0.2 at 20 kV. Therefore, twice as many secondary electrons are generated per beam electron from a specimen coated with gold.

The total number of backscattered electrons emitted from a coated specimen will also be higher, although the increase will not be as great as for secondary electrons. This is because backscattered electrons have much more energy and escape from about 100 times greater specimen depth than secondary electrons. Therefore, a 10- to 20-nm metal coating will only account for a small part of a 500-nm escape depth. However, the number of backscattered electrons produced per beam electron (**backscattered coefficient**) increases more dramatically as the atomic number of the specimen increases, when compared to the secondary electron coefficient (see Chapter 7). Gold has a backscattered coefficient of about 0.7, while carbon is only about 0.05. Therefore, more backscattered electrons are produced from a gold-coated specimen, even though the effect is reduced because most of the backscattered electrons escape from areas beneath the gold coating.

The increased emission of secondary and backscattered electrons from a metal-coated specimen results in increased signal to the detector. This results in less need to amplify the signal through the multiplier, a higher signal to noise ratio, and a better quality image on the cathode ray tube.

Emission of Electrons from a Smaller Surface Area of a Metal-coated Specimen.   The generation of secondary electrons by the electron beam occurs in a volume represented by the diameter of the electron beam, plus 0.5 nm for metals or plus 5 nm for insulators. Thus the diameter of the zone of secondary electron generation by beam electrons for a 5-nm beam diameter would be 5.5 nm in metals and 10.5 nm in insulators. Therefore, the secondary electrons generated by the primary electron beam emanate from the smaller surface area in metal-coated specimens, producing smaller pixel sizes on the specimen and images of better resolution (Fig. 11-32).

Backscattered electrons and secondary electrons produced by backscattered electrons interacting with specimen atoms also escape from a smaller surface area in a well-coated specimen. This is because the backscattered electron escape volume is smaller in high atomic number elements (for example, the metal coating of a specimen). However, this effect would not be as marked as the decrease in the secondary electron escape area, since a 10- to 20-nm metal coating represents only a small part of the escape zone of backscattered electrons and the secondary electrons that they generate.

Thermal damage of the specimen.   Interaction of the specimen with the electron beam can result in heating of the specimen, resulting in ultrastructural changes if the specimen is heat labile. Thermal damage can be minimized by coating the specimen with a metal. The metal coating carries the heat away to the relatively large mass of the specimen support, which represents a heat sink. The rate at which the heat is transferred from the specimen to the support depends on the thermal conductivity of the specimen and the metal coating. Carbon has a relatively low heat conductivity of 0.057 cal/cm^2/s, with biological specimens having even lower values. Metals have greater values; gold has a thermal conductivity of 0.71 cal/cm^2/s. Therefore, a gold coating over a biological specimen carries heat away from the specimen and helps to minimize thermal damage of the specimen. The metal coating is relatively thin, but it is over the surface of the specimen, and this is where most of the heat is generated during interaction of the specimen with the electron beam.

Mechanical integrity of the specimen.   Metal coatings are very strong and are able to hold particulate specimens together and attach the specimen to the support. Coatings are often of sufficient strength to hold the specimen onto the support without the use of adhesives, which are often nonconducting and lead to specimen charging.

Charging.   Charging is the buildup of an electrical charge on the specimen that results in alteration of the collecting field of the detection system as the electron beam travels from one pixel position to another on the specimen. This results in images that have bright and dark lines, or a shift in the position of part of the specimen image in the direction of the raster lines on the cathode ray tube (Fig. 11-33).

During viewing in the scanning electron microscope, the specimen is bom-

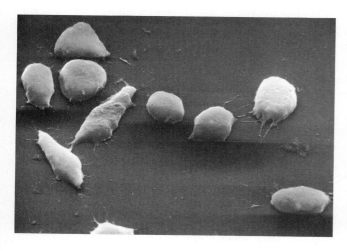

Figure 11-33 Charging is manifested as light and dark areas in the direction of the raster pattern.

barded by beam electrons. Each high-energy electron undergoes multiple collisions with specimen atoms until the energy of the beam electron is dissipated. The
collisions with the specimen atoms eject secondary and backscattered electrons
out of the specimen atoms. Some of the secondary and backscattered electrons
escape from the specimen surface and are collected to produce the image. The
number of secondary and backscattered electrons escaping from the specimen is
not the same as the number of beam electrons entering the specimen. This imbalance is corrected if the specimen conducts electrical current, because electrons
are drained to or supplied from the grounded specimen support. However, if the
specimen is nonconductive (an *insulator*), the specimen will become positively or
negatively charged because the imbalance in electrons leaving and entering the
specimen cannot be corrected from the grounded specimen support. *Biological
specimens in their dehydrated state are insulators and therefore generally charge
while being viewed in the scanning electron microscope.*

Three situations can arise while the specimen is bombarded by beam electrons (presuming the specimen is an insulator, resulting in little flow of electrons
to or from the grounded specimen support):

1. The specimen will charge *positively* if the number of secondary and backscattered electrons escaping from the specimen is *greater* than the number of
   electrons entering the specimen–beam interaction zone from the electron
   beam and grounded specimen support. Each electron has a charge of -1.602
   $\times 10^{-9}$ C (C = coulomb, a unit of electrical charge equal to the quantity of
   electricity transferred by a current of one ampere in one second), while each
   proton has a charge of $+1.602 \times 10^{-9}$ C. An excess of protons over electrons
   in the specimen results in a positively charged specimen.

2. The specimen will be *neutral* if the number of secondary and backscattered
   electrons leaving the specimen is the *same* as the number of electrons entering the specimen from the electron beam and specimen support. This is the
   ideal situation, since there is no charging of the specimen and the collecting
   field of the detection system is not distorted.

3. The specimen will be charged *negative* if the number of secondary and backscattered electrons leaving the specimen is *less* than the number of electrons
   entering the specimen from the electron beam and specimen support.

The number of secondary and backscattered electrons escaping from the
surface of a particular specimen depends on the accelerating voltage of the electron
beam. Figure 11-34 shows the relationship between the energy of the beam electrons and the number of secondary and backscattered electrons that escape from
the specimen surface per beam electron. There are two places, E_1 and E_2, where
the electrons escaping from the specimen surface are *equal* to the number of beam
electrons entering the specimen. Between E_1 and E_2, *more* electrons are escaping
from the specimen surface than are being replaced by beam electrons, resulting in

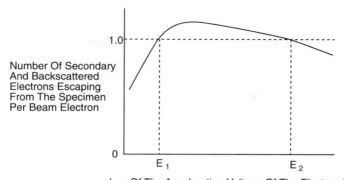

Figure 11-34 Relationship between the accelerating voltage and the number of secondary and backscattered electrons escaping from a specimen.

a *positively* charged specimen. At accelerating voltages less than E_1 and greater than E_2, *less* electrons are leaving the specimen than there are beam electrons entering the specimen, resulting in a *negatively* charged specimen.

Accelerating Voltages Less Than E_1. If the electron beam has an energy less than E_1 (usually about 200 V, depending on the specimen), the number of secondary and backscattered electrons escaping from the specimen is less than the number of beam electrons entering the specimen (Fig. 11-35). The negatively charged specimen slows the negative beam electrons, causing the beam electrons to lose energy before striking the specimen. The decrease in the energy of the beam electrons results in a further decrease in the number of secondary and backscattered electrons leaving the specimen, making the specimen more negatively charged. The process continues until the specimen acts as an electron mirror, completely repelling the electron beam.

Accelerating Voltages between E_1 and E_2. E_1 and E_2 will vary according to the type of specimen (Fig. 11-36). However, E_1 is usually around 200 V, while E_2 is in the range of 1 to 5 kV. Between E_1 and E_2 more secondary and backscattered electrons leave the specimen surface than are replaced by beam electrons, resulting in a positively charged specimen. Most of the secondary electrons escaping from the specimen surface have an energy of only a few electron volts and are not able to escape from the positively charged specimen. Many of these low-energy secondary electrons are not able to penetrate the positive voltage barrier set up by the charging and cannot escape from the specimen. The number of secondary electrons escaping from the specimen is reduced until the number of electrons escaping from the specimen is only slightly more than the number of electrons entering the specimen. The specimen is then maintained at a small positive voltage. Therefore, between E_1 and E_2 the specimen is very nearly neutral because an equilibrium is set up, with any positive bias to the specimen resulting in fewer secondary electrons escaping from the specimen, bringing the specimen back toward neutrality. Operating the scanning electron microscope at an accelerating voltage between E_1 and E_2 would eliminate most of the charging in an uncoated specimen. However, relatively few electrons are in the beam at low accelerating voltages (the brightness is low), resulting in a weak signal that requires a large

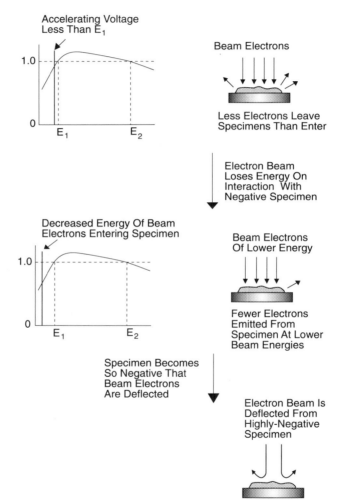

Figure 11-35 Operating the scanning electron microscope at an accelerating voltage of less than E_1 results in the specimen becoming negative since more beam electrons enter the specimen than leave as secondary and backscattered electrons. The negative specimen lowers the accelerating voltage of the beam electrons entering the specimen, resulting in even less electrons leaving the specimen per beam electron entering the specimen. Eventually, the specimen becomes so negative that the beam electrons are repelled from the specimen.

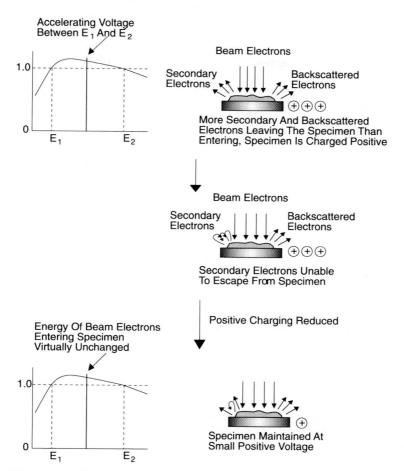

Figure 11-36 Operating the scanning electron microscope at an accelerating voltage between E_1 and E_2 results in less beam electrons entering the specimen than leave as secondary and backscattered electrons, resulting in a positively charged specimen. The positive bias to the specimen results in the lower-energy secondary electrons failing to escape from the specimen. This brings the specimen toward neutrality, with stabilization of the specimen potential occurring at a small positive voltage.

amount of amplification to obtain an image (resulting in a noisy image). Thus it is often not practical to examine specimens at these low voltages unless a field emission source is used to produce a large number of electrons in the beam at low accelerating voltages.

Accelerating Voltages above E_2. Above E_2, less secondary and backscattered electrons escape from the specimen than are entering from the electron beam (Fig. 11-37). This results in a negatively charged specimen, which decelerates the

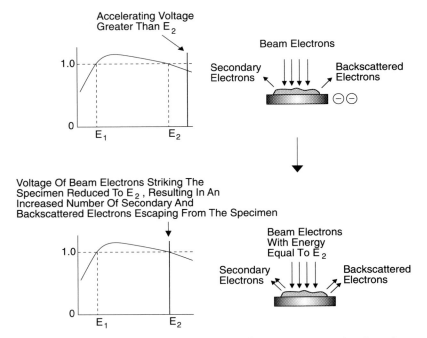

Figure 11-37   Operating the scanning electron microscope at an accelerating voltage above E_2 results in more beam electrons entering the specimen than leave as secondary and backscattered electrons. This results in the specimen charging negatively. The negatively charged specimen reduces the accelerating voltage of the beam electrons as they approach the specimen. This, in turn, increases the number of secondary and backscattered electrons leaving the specimen per beam electron entering the specimen. This continues until the voltage of the beam electrons striking the specimen is reduced to E_2.

beam electrons as they approach the specimen. The energy of the beam electrons is reduced until the number of secondary and backscattered electrons escaping from the specimen is the same as the number entering the specimen. An equilibrium is set up at E_2. Unfortunately, the actual E_2 voltage will vary over the surface of the specimen. At one place on the specimen, the surface potential will have a large negative value, resulting in a large number of electrons collected by the detector at those pixel positions and producing a bright image at those pixel positions on the cathode ray tube. At another position on the specimen, the surface potential will have a low negative value, resulting in few electrons collected by the detector at these pixel positions and a dark image at those pixel positions on the cathode ray tube.

Coating the specimen with a 10- to 20-nm layer of a metal results in the reduction or elimination of charging. The conductive metal coating drains excess electrons from the specimen surface to the grounded specimen support or provides electrons from the grounded specimen support to the specimen, depending on whether the specimen is charged positively or negatively.

Sputtering to coat specimens with a metal.   Sputtering is usually used to coat specimens for scanning electron microscopy. Sputtering is preferred to vacuum evaporation of a metal for the following reasons:

1. Sputtering produces a more uniform coating of metal over the specimen. Vacuum evaporation of metals is line of sight from the metal to the specimen, resulting in uneven coating of rough surfaces.

2. Sputter coaters are relatively inexpensive, primarily because a vacuum of only 10^{-2} mbar is required. A rotary pump can therefore be used to produce the vacuum. Vacuum evaporation of metals is performed at vacuums of 10^{-4} mbar, requiring more expensive vacuum systems with either diffusion, turbomolecular, or cryopumps.

3. Sputter coaters use relatively small amounts of the coating metal, which is usually expensive (gold or gold–palladium).

4. Sputter coaters are easy to use and require less time to coat specimens.

5. Sputter coaters produce less heating of the specimen because a magnetic field prevents most of the generated electrons from interacting with and heating the specimen.

6. Metal coatings produced by sputtering are stronger than those produced by vacuum evaporation.

Sputter Coater.   There are a number of designs for sputter coaters, but the most common is the diode (two electrodes, an anode and cathode) type.

A diode sputter coater consists of a small cylindrical vacuum chamber that is pumped by a rotary pump (Fig. 11-38). A Pirani gauge measures the vacuum. The removable top of the vacuum chamber has the cathode (**target**) that contains the material to be sputtered and a **magnet**. Although targets used for coating specimens for scanning electron microscopy usually contain gold or gold–palladium (60 : 40), targets composed of silver, palladium, platinum, copper, nickel, or chromium are commercially available. The cathode is usually flanked by a **cathode shield** that prevents the side and rear of the target from being sputtered. At the base of the vacuum chamber is the **specimen table**, which is at ground and serves as the anode. The distance between the cathode (target) and anode (specimen table) is typically on the order of 5 to 10 cm. The closer the specimen table is to the target, the more metal that condenses on the specimen, but also the more is the radiative heating of the specimen. There is often a ring around the specimen table that is kept slightly positive. Any electrons that approach the specimen table are attracted to the positive ring and do not interact with and heat the specimens. There is a valve for regulating the flow of argon gas into the vacuum chamber.

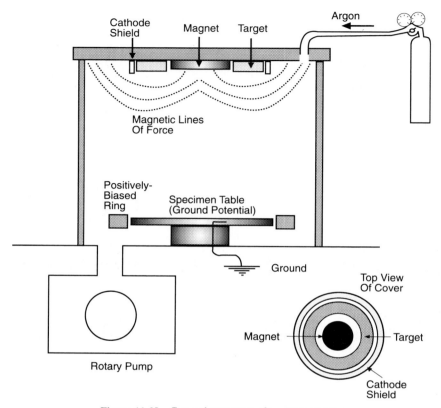

Figure 11-38   General structure of a sputter coater.

Process of Sputter Coating (Fig. 11-39).   The specimens to be sputtered should not contain volatiles (such as those from adhesives and colloidal silver or graphite) or it will not be possible to obtain sufficient vacuum in a reasonable time. Also, volatiles condense on the surface of the target, reducing the efficiency of sputtering. Specimens on supports are commonly placed in an oven at 40°C for 1 h before sputtering. The specimens are placed in the vacuum chamber and pumped to a vacuum in the range of 10^{-2} mbar. Argon is leaked into the vacuum chamber and pumped out to flush out residual gases and ensure that the atmosphere in the chamber is primarily argon. Argon is an economical inert gas that does not react with the specimen during sputtering. Nitrogen can also be used, but results in a sputtering rate that is half that of argon. This is because the mass of nitrogen (14) is much smaller than argon (40). Atoms of smaller mass sputter atoms out of the target more slowly, resulting in longer sputtering times to obtain the same thickness of metal coating over the specimen and increasing the possibility of specimen heating. The vacuum chamber is usually flushed a couple of times with argon to

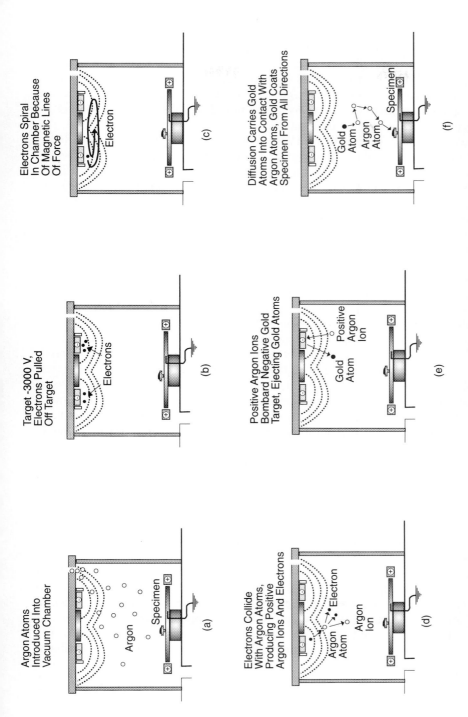

Figure 11-39 Process of sputter coating. (a) The specimen chamber is pumped by the rotary pump and argon gas is introduced into the vacuum chamber. (b) At a pressure of about 6×10^{-2} mbar, the target (commonly gold or gold–palladium) is made -3000 V. Electrons are pulled off impurities in the target by the potential. (c) The electrons are attracted toward the anode (specimen table), but encounter the lines of magnetic force that cause the electrons to spiral in the chamber. (d) An electron collides with an argon atom, ejecting an orbital electron out of the argon atom, creating a positive argon ion and a second electron. The electrons continue to spiral in the chamber, colliding with additional argon atoms. (e) A positive argon ion is propelled to the negative target. On colliding with the target, the positive argon ion ejects a metal (gold) atom from the target. (f) Because of the relatively poor vacuum (6×10^{-2} mbar), the mean free path is short, resulting in the gold atom colliding with a number of argon atoms before settling on the specimen. The gold atoms approach the specimen surface from all possible angles, resulting in an even coating of the specimen surface.

remove much of the water and carbon dioxide. These gases produce O_2 during sputtering, which reduces the rate of sputtering.

Sputtering is initiated at an argon pressure of about 6×10^{-2} mbar. The target is made 3000 V negative. A self-sustaining **glow discharge** or **plasma** (ionized gas) is set up between the cathode (target) and anode (specimen table) if sufficient argon atoms are present. Argon atoms are admitted into the vacuum chamber to keep the glow discharge at 15 to 20 mA.

The glow discharge results from the voltage field set up between the -3000-V target and the grounded specimen table. The discharge current is initiated when an electron is pulled from a speck on the cathode and is accelerated toward the anode. As the electron leaves the cathode, it encounters magnetic lines of force, which are perpendicular to the cathode–anode axis. The magnetic lines of force subject the electron to the right-hand rule of force (see Chapter 3). The magnetic lines of force push the electron at right angles to the magnetic lines of force and to the original direction of the electron. This causes the electron to spiral around in the upper portion of the vacuum chamber, where it eventually collides with an argon atom, imparting energy to the orbital electrons of the argon atom. Some of these electrons are energized to higher orbitals in that atom and then fall back to their original orbitals, resulting in the emission of the lost energy as light photons that produce the glow in the plasma. Some of the orbital electrons are ejected from the argon atoms, creating a positive argon ion and a negative electron. The electron can go on to make a collision with another argon atom. The positive argon ion is propelled toward the negative cathode target (the argon has such a large mass that it is minimally affected by the magnetic lines of force). The argon ion slams into the target, ejecting atoms out of the target (usually gold or gold–palladium) and a number of electrons. The electrons contribute to the glow discharge and make it self-sustaining (as long as the potential between the cathode and anode is maintained).

The mean free path of diffusion of the metal atoms ejected from the target is short at the sputtering vacuum of 6×10^{-2} mbar. The metal atoms travel 1 to 10 mm between collisions with argon atoms and ions, causing the metal atoms to be scattered widely. The metal atoms form a diffuse cloud and approach the specimen surface from all sides, resulting in a relatively even coating over the specimen. Even crevices in the specimen receive enough metal atoms to make the specimen surface conductive.

Heating of the specimen during sputtering can result from (1) heat radiation or (2) interaction of the electrons with the specimen. The specimen is usually 5 to 10 cm from the cathode, which minimizes any radiative heating. The magnetic lines of force restrict most of the electrons to the upper portion of the vacuum chamber. The positive ring around the specimen table attracts any electrons that manage to get to the area of the specimen. Thus, heating is usually not a problem during sputter coating.

Sputtering Yield.   The yield of sputtered atoms depends on (1) the negative

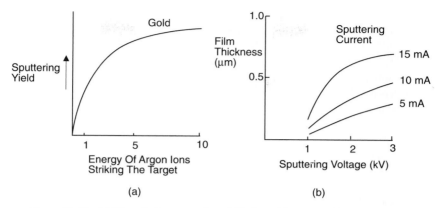

Figure 11-40   (a) The relative sputtering yield of a gold target at different energies of argon ions. (b) The thickness of the metal coating at different sputtering voltages and currents. (*Adapted from* J. I. Goldstein and others, *Scanning Electron Microscopy and X-ray Microanalysis,* Plenum Press, New York, 1981)

voltage of the cathode target, (2) on the number of electrons in the discharge current, and (3) the composition of the target.

1. The energy of the positive argon ions will increase as the cathode target is made more negative. The increased energy of the argon ions results in an increased number of metal atoms sputtered out of the target for each argon ion striking the target (Fig. 11-40). The argon ions of higher energy penetrate about two atomic layers into the target.

2. The current density or discharge current is the number of electrons producing the plasma. The current density is the most important factor in controlling the number of atoms sputtered out of the target (Fig. 11-40). The number of electrons in the discharge current is normally increased by admitting more argon atoms into the chamber. This results in more collisions of argon atoms with the spiraling electrons, resulting in more argon atoms striking the target, more sputtered atoms from the target, and more electrons contributed to the discharge current from the collisions of the argon ions with the target. The number of metal atoms sputtered out of the target will increase until the increased number of argon molecules results in an increase in the number of sputtered atoms returning to the target, resulting in a decrease in the sputtering rate. This begins to occur at a pressure of about 2×10^{-1} mbar (~20 Pa).

3. Targets composed of different substances have different numbers of atoms sputtered out of the target under the same sputtering conditions (Table 11-6). Gold has the highest sputtering yield, with gold-palladium (60:40) having somewhat less. Carbon forms a very hard target and is difficult to sputter.

TABLE 11-6   RELATIVE SPUTTERING YIELD OF SOME ELEMENTS

Element	Relative sputtering yield (atoms/600 eV)
Chromium	1.3
Copper	2.3
Gold	2.8
Iron	1.3
Palladium	2.4
Platinum	1.6
Silver	3.4
Tantalum	0.6
Titanium	0.6
Tungsten	0.6

Thickness of the Metal Coating.   The thickness of the metal coating should be at least 10 nm to be conductive. The metal atoms form islands as they are sputtered onto the surface of the specimen, with the islands becoming continuous only at a thickness of about 10 nm. The formation of a thin film on the specimen surface begins with the sputtered atoms arriving at the specimen surface. The sputtered atoms will move over the surface of the specimen until they lose enough energy to be bound to a position on the surface (Fig. 11-41). These atoms form nucleation sites. Other sputtered atoms arriving at the surface move along the surface until they reach a previously formed nucleation site. The newly arrived atom coalesces with the atom(s) at the nucleation site, and an island of sputtered atoms begins to grow. Eventually, the islands of sputtered atoms become so large that they grow together, forming a continuous layer. This occurs at a thickness of about 10 nm and makes the surface of the specimen conductive.

Gold or gold–palladium (60 : 40) are usually the metals of choice for sputtering. There is a certain amount of granularity in the thin film on the surface of the specimen (usually on the order of 5 to 10 nm). However, this granularity is so close to the resolution of the scanning electron microscope that is not seen.

Measurement of the Thickness of Sputtered Films.   Some sputter coaters have a device that provides an estimate of the film thickness by measuring the discharge current (number of electrons generated during sputtering) using the following formula:

thickness

$$= \begin{bmatrix} \text{plasma current} \\ \text{(mA)} \end{bmatrix} \times \begin{bmatrix} \text{potential between} \\ \text{anode and cathode} \\ \text{(kV)} \end{bmatrix} \times \begin{bmatrix} \text{time} \\ \text{(min)} \end{bmatrix} \times \begin{bmatrix} \text{gas constant} \\ (=5 \text{ for argon}) \end{bmatrix}$$

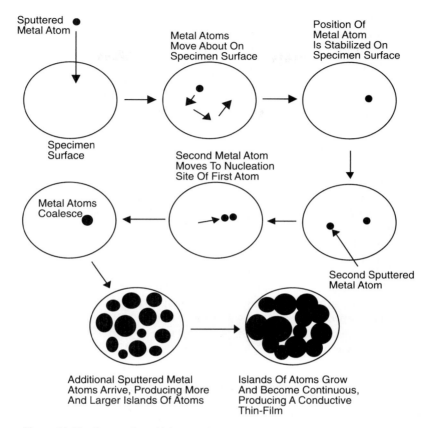

Figure 11-41 Process by which a continuous, electrically conducting thin film is produced on the surface of a specimen.

Carbon coatings produced by vacuum evaporation of a heated source. Sometimes specimens are coated with carbon instead of a metal. The most common application is when specimens are to be subjected to X-ray micro-analysis. The X-ray energies of carbon are so low that they do not coincide with the X-ray energies that are used to characterize most elements. A carbon coating is thus used to stabilize the surface of the specimen without interfering with the X-ray microanalysis. Carbon films are electrically conductive, thereby reducing charging, but carbon films do not have the high density of metal films, resulting in weaker signals emanating from the specimen. The images obtained from carbon-coated specimens are therefore usually poorer than those obtained from metal-coated specimens.

Carbon can be sputtered onto a specimen. However, sputter coating carbon takes considerably longer than sputtering metals because of the low sputtering rate off a carbon target. Therefore, carbon films are commonly produced by vacuum evaporation off heated graphite rods.

Carbon exists in two crystalline forms, diamond and graphite, and numerous, amorphous, less-ordered forms. The graphite form is used to produce specimen coatings for scanning electron microscopy. Graphite is electrically conductive, has relatively low density and softness (compared to diamond forms), and has a grayish-black appearance.

Vacuum Evaporation.   Solids and liquids always have a certain number of molecules escaping from their surface, producing vapors. The molecules are continually escaping from and condensing back onto the surface of the solid or liquid, producing a vapor pressure. The greater the number of molecules escaping and the lower the number of molecules condensing, the greater is the vapor pressure. The higher the temperature of the substance is, the more molecules that escape from the surface of the solid or liquid and the higher the vapor pressure. The **evaporation temperature** of a substance is the point at which evaporation of molecules from the solid or liquid phase becomes substantial. *The evaporation temperature is the temperature at which the vapor pressure of the substance reaches 1.33 $\times$ 10^{-2} mbar.* The evaporation temperature depends on the substance (Fig. 11-42). Some substances are liquid at their evaporation temperature, while other (carbon, chromium, and tungsten) are solids.

In vacuum evaporation, the substance is heated to its evaporation temperature in a vacuum of 10^{-4} mbar, resulting in appreciable evaporation of the sub-

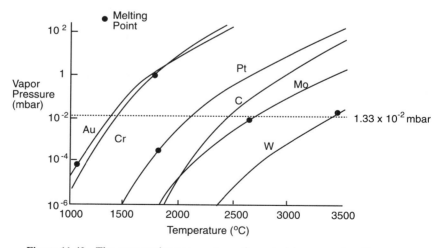

Figure 11-42   The evaporation temperature of a material occurs when the vapor pressure of the material reaches 1.33 $\times$ 10^{-2} mbar; at this point, evaporation becomes considerable. Molybdenum, gold, and platinum are liquids at their evaporation temperature while carbon, tungsten (W), and chromium are solids.

stance. Atoms evaporate from the substance mostly as individual atoms. The mean free path of diffusion at 10^{-4} mbar is so long that the evaporated atoms usually arrive at the specimen without colliding with any gas molecules. A thin film is built up on the cooler specimen from nucleation sites in the same manner that the thin film is built up during sputtering (see the previous section).

Carbon is evaporated onto specimens from graphite rods. At atmospheric pressure, carbon has no melting point. The triple point is 3850°C and 120 atm; at temperatures and pressures above this, liquid carbon can be formed. At pressures below 1 atm (for example, a vacuum), it is not possible to obtain carbon in a liquid form. This means that in a vacuum carbon in graphite form sublimes from solid to the vapor state. The evaporation temperature for carbon (where it has a vapor pressure of 1.33×10^{-2} mbar) is 2400°C. Therefore, bringing graphite to a temperature of 2400°C in vacuum will result in sufficient sublimation of carbon from a graphite rod, producing a coating of carbon over a specimen.

Apparatus for the Vacuum Evaporation of Carbon.   Carbon is usually evaporated using the Bradley process (Fig. 11-43). Two rods of spectrographic carbon (graphite) are used. One of the rods is fixed in position. This rod can be blunt or sharpened to a point. The second rod slides in an insulated bush and is pressed against the first by a spring. The second rod is sharpened to a point. A current of 20 to 30 A and 30 V is passed through the electrically conducting graphite rods. The points of the graphite rods have the greatest resistance to electrical current and heat the most. Significant carbon sublimation takes place when the tip reaches 2400°C, with most of the carbon atoms reaching the specimen unimpeded because of the vacuum of 10^{-4} mbar. As the tip evaporates, it is kept up against the other rod by the spring mechanism, ensuring continued evaporation of carbon. Small

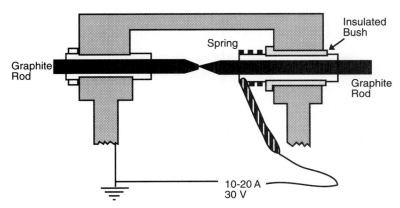

Figure 11-43   Carbon gun used for the evaporation of carbon from graphite rods using resistance heating.

glowing particles of graphite will also be emitted from the heated graphite rod, but these do not seem to adhere in any significant number to the specimen surface and cause problems. The specimen is often rotated and tilted during the deposition process to help obtain an even coating.

Characteristics of Carbon Films Produced by Vacuum Evaporation.   At thicknesses greater than 2 nm, the carbon coating over the specimen is continuous and conductive. The carbon film is essentially free from granularity and will offer a continuous surface when viewed in the scanning electron microscope. Carbon coatings produced by vacuum evaporation are of a relatively uniform thickness over irregular surfaces, even if the specimen is not rotated and tilted during evaporation. The reason for this is not known for sure, but it appears that the light carbon atoms bounce off the walls of the vacuum chamber, apparatus, and any gas molecules left in the chamber, approaching the specimen surface from a variety of angles.

12

X-ray
Microanalysis

HISTORY OF X-RAYS

X-rays were discovered in 1895 by Wilhelm Conrad Roentgen, professor of physics at the University of Wurzberg in Germany. While studying cathode rays (streams of electrons in vacuum tubes), Roentgen accidentally discovered that a screen covered with barium platinocyanide glowed brilliantly when held near the vacuum tube that contained the cathode rays. His further investigations showed that unknown rays (which he called X-rays) were able to penetrate the glass walls of the vacuum tube and excite the barium platinocyanide on the screen. He also observed that he could see the bones in his hand when he placed his hand between the cathode ray tube and the barium platinocyanide screen.

In 1912, Max von Laue showed that the wavelengths of X-rays were approximately the same order of magnitude as the spacing of atoms. In 1913, Henry G.-J. Moseley demonstrated that the wavelength of X-rays is directly related to the atomic number of the element that emits the X-rays. This discovery enabled investigators to qualitatively determine the elemental composition of an unknown specimen by measuring the wavelengths of the X-rays that were emitted from a specimen. The atomic events involved in the production of characteristic and continuum X-rays have been presented in Chapter 6.

In 1917, Sir William Henry Bragg and his son, Sir William Lawrence Bragg, received the Nobel prize for formulating the theory of diffraction of X-ray crystals. This discovery led to the development of the wavelength dispersive spectrometer, which is used to determine the elemental composition of specimens by measuring the wavelength of the X-rays emitted from the specimen.

X-ray microanalysis involves the excitation of X-rays within a microscopic volume of the specimen. Collection and analysis of the emitted X-rays enables the investigator to determine the elemental composition of that microscopic portion of the specimen. X-ray microanalysis was pioneered by Raymond Castaing (Fig. 12-1) in his 1948 doctoral thesis at the University of Paris titled *Application of Electron Beams to Methods of Local Chemical and Crystallographic Analysis*. Castaing modified an electron microscope to focus an electron beam with a diameter of less than 1 μm onto a specimen. The high-energy electrons resulted in the emission of X-rays from the specimen. Analysis of the X-rays enabled Castaing to determine the elements present in the X-ray escape volume of the specimen. Castaing's device was called an **electron probe microanalyzer**. The device had no electron-imaging capabilities (an electron image of the specimen could not be produced); the area of the specimen to be analyzed was moved under the beam using a light-optical microscope to orient the specimen. In 1956, V. E. Cosslett and P. Duncomb placed scanning coils in the electron column to produce a scanning electron microprobe. In this instrument it was possible to move the electron beam to illuminate different portions of the specimen. During these years, Sir Charles W. Oatley and his associates were developing the scanning electron microscope. The combination of the imaging capabilities of the scanning electron microscope and the X-ray microanalysis of the electron microprobe was the next logical advance. Today, many scanning electron microscopes are fitted with X-ray microanalytical devices. These enable the investigator to determine the elemental composition of the specimen while the specimen is being imaged. There are two basic types of X-ray microanalytical devices, **wavelength dispersive spectrometers**, which

Figure 12-1 Raimond Castaing. Born December 21, 1921. Education: Lycée de Monaco, Collège de Condom, 1930; Lycée de Toulouse, 1938; Ecole Normale Supérieure, 1940; Doctor in Physics, University of Paris, 1951. Career: Research Engineer, Office National d'Etudes et de Recherches Aeronautiques (ONERA) (1947–1951); Lecturer, University of Toulouse (1952–1955); Lecturer, University of Paris (1956–1959); Professor, University of Paris-Sud (1960–1987); Director General of ONERA (1968–1972); French Academy of Sciences (1977). (Photograph courtesy of Dr. Castaing)

measure the wavelengths of X-rays, and **energy dispersive spectrometers**, which measure the energy of X-rays.

WAVELENGTH DISPERSIVE SPECTROSCOPY

In wavelength dispersive spectroscopy, the wavelength of the X-rays is measured. The wavelength of X-ray photons is related to the energy according to Planck's law:

$$E = \frac{hc}{\lambda}$$

where E = energy of the X-ray photon

h = Planck's constant = 6.6262×10^{-34} J $\times$ S

λ = wavelength of the X-ray photon

c = speed of light = 3.0×10^8 m/s

Thus a Ca K_β X-ray with an energy of 4.012 eV would have a wavelength of 0.309 nm (3.090 Å) according to the following calculations:

$$E = \frac{hc}{\lambda}$$

$$\lambda = \frac{hc}{E} = \frac{1.2396}{E}$$

$$= \frac{1.2396}{4.012} = 0.309 \text{ nm}$$

There is an inverse relationship between energy and wavelength, the greater the energy, the shorter the wavelength (Fig. 12-2). The wavelengths of X-rays commonly measured using wavelength dispersive spectroscopy vary from 0.1 to 10.0 nm.

In wavelength dispersive spectroscopy, the wavelengths of the X-rays are commonly separated with a crystal that diffracts certain wavelengths onto a detector. A crystal has a basic pattern of atoms that is repeated in three dimensions (Fig. 12-3). A **unit cell** is a parallelogram in two dimensions that is defined by adjacent rows of atoms. In a crystal the unit cell is the same size no matter in which direction it is arranged. In 1912, in Germany, W. Friedrich, P. Knipping, and M. von Laue recognized that the crystal acts as a three-dimensional diffraction grating for X-rays. The amount of diffraction of X-rays is determined by the size and shape of the repeating unit of atoms in the crystal.

W. H. and W. L. Bragg recognized the similarity between reflection in a

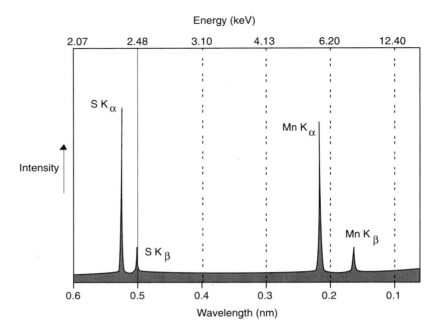

Figure 12-2 X-ray spectrum obtained using wavelength dispersive spectroscopy from a specimen containing sulfur and manganese. The longer X-ray wavelengths have less energy than the shorter wavelengths.

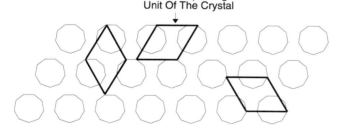

Unit Cell Defines
The Size And Shape
Of The Repeating
Unit Of The Crystal

Figure 12-3 A crystal has symmetry in a number of planes. A unit cell defines the repeating unit of the crystal. All unit cells have the same volume in the crystal, and the unit cell can be chosen in a variety of ways.

plane mirror and diffraction in a crystal. When X-rays enter the lattice of a crystal, they are scattered by the electrons in the crystal. In Fig. 12-4, two monochromatic (of the same wavelength) X-rays strike the same plane of a crystal in two different places, resulting in the diffraction of the X-rays. The two X-rays are diffracted at the same angle, resulting in reinforcement of the waves; the wavelengths of the diffracted X-rays remain synchronous. In a crystal, however, there are many lattice planes diffracting X-rays (Fig. 12-5). *The distance between the parallel*

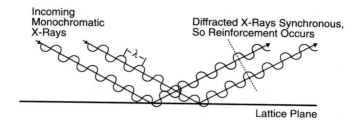

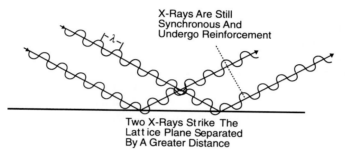

Figure 12-4 Two incoming monochromatic X-rays that are in phase will still be in phase after being diffracted off a crystal lattice, no matter what the distance is between the two X-rays.

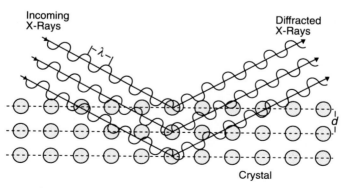

Figure 12-5 X-rays striking a crystal interact with a number of successive lattice planes in a crystal. The X-rays will be diffracted only if they strike the crystal at the correct angle.

*lattice planes and the angle at which the incident X-rays strike the crystal deter-
mines whether X-rays from different lattice planes will undergo general reinforcement.* General reinforcement occurs when Bragg's law is satisfied:

$$n\lambda = 2d \sin\theta$$

where λ = wavelength of the diffracted X-rays

d = interplanar spacing of the crystal

θ = angle between the crystal surface
and the incident and diffracted X-rays

n = integer (whole number)

Bragg's law states that it is necessary for $d \sin \theta$ to be a whole-number multiple
of the wavelength for general reinforcement of diffracted X-rays. If n is not a
whole number, destructive interference occurs and the X-rays are absorbed by
the surroundings. In Fig. 12-6a, two monochromatic X-rays strike each of two
successive lattice planes in a crystal. The lower X-ray travels an additional distance of AB + BC = $2d \sin \theta$ before leaving the crystal. General reinforcement
of the two X-rays occurs because the distance AB + BC = $2d \sin \theta$ = a whole
number of the wavelength. When this condition is met, the X-rays diffracted from
the crystal planes combine in phase to produce an intensity maximum due to

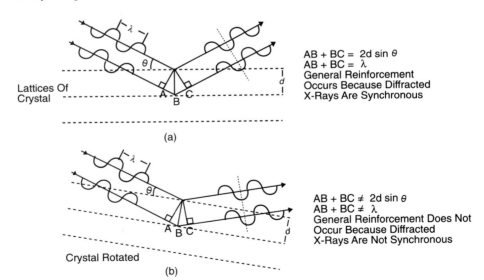

(a)

(b)

Figure 12-6 X-rays are diffracted from a crystal only if they meet the conditions
of Bragg's law, $n\lambda = 2d \sin \theta$. (a) The distance AB + BC equals λ and the X-rays
are diffracted. (b) The distance AB + BC does not equal λ and the X-rays are
absorbed.

TABLE 12-1 CRYSTALS COMMONLY USED IN WAVELENGTH DISPERSIVE SPECTROSCOPY

	2d (nm)	Lowest detectable atomic number	
		$K\alpha_1$	$L\alpha_1$
LiF (lithium fluoride)	0.40	19 (potassium)	49 (indium)
α Quartz	0.67	15 (phosphorus)	40 (zirconium)
PET (pentanerythritol)	0.87	13 (aluminium)	36 (krypton)
RAP (rubidium acid phthalate)	2.61	8 (oxygen)	33 (arsenic)
KAP (potassium acid phthalate)	2.66	8 (oxygen)	23 (vanadium)
PbSt	10.4	5 (boron)	

constructive interference (general reinforcement). If AB + BC does not equal a whole number of the wavelength, destructive interference occurs (Fig. 12-6b) and the X-rays are absorbed.

Therefore, general reinforcement occurs only if X-rays of a particular wavelength strike the crystal at a specific angle (θ) with respect to the interplanar spacings (d) of the crystal. Since each element emits X-rays of particular wavelengths that are characteristic for that element alone, the wavelength sorting of the crystal allows the operator to establish the presence of elements in specimens.

The smaller the spacings are of the atomic planes in the crystal (d), the higher is the angle of diffraction of the X-rays. A wavelength dispersive spectrometer cannot measure all the X-rays from 0.1 to 10.0 nm with a single crystal. Bragg's law establishes an upper limit of twice the atomic spacings of the crystal (2d), since sin θ cannot be greater than 1. Therefore, a crystal composed of α quartz with d equal to 0.334 nm cannot diffract X-ray wavelengths above 0.668 nm. The angle of incidence of the X-rays on the crystal (θ) has to be less than 90° or the detector would have to be at the X-ray source point, which is impossible. At θ = 90°, sin θ = 1. Crystals used in wavelength dispersive spectroscopy are able to diffract only a portion of the complete X-ray spectrum of wavelengths. Therefore, scanning electron microscopes with wavelength dispersive spectrometers usually have more than one crystal spectrometer so that a more complete spectrum of wavelengths can be obtained. Having more than one crystal spectrometer in the X-ray beam at one time allows the operator to analyze more than one element at a time. Table 12-1 lists some of the crystals commonly used in diffraction.

A wide range of X-ray wavelengths is generated from the specimen, but only those X-rays that strike the crystal at a certain angle (the angle defined by Bragg's law) reach the X-ray detector. The crystal is rotated to obtain a continuous spectrum of diffracted X-rays during wavelength dispersive spectroscopy. The geometrical relationship between the specimen, diffracting crystal, and X-ray detector has to be precise during acquisition of the X-ray spectrum to maintain calibration of the device and to maximize the number of X-rays striking the detector. To focus

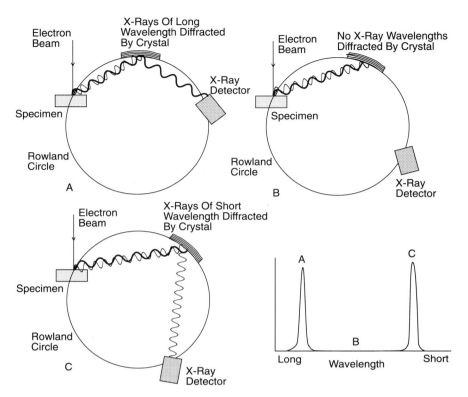

Figure 12-7   A wavelength dispersive spectrometer acquires a spectrum by movement of the diffracting crystal and X-ray detector along the Rowland circle. The points A, B, and C in the spectrum represent the three positions of the instrument in A, B, and C.

X-rays on the detector, the distance between the X-ray source (specimen) and diffracting crystal has to be the same as the distance between the diffracting crystal and the X-ray detector. To accomplish this, the X-ray source, diffracting crystal, and X-ray detector are all on the circumference of a circle of specified radius, called the **focusing circle** or **Rowland circle** (Fig. 12-7).

There are two basic types of wavelength dispersive spectrometers, the **Johann semifocusing** spectrometer and the **Johansson fully focusing** spectrometer (Fig. 12-8). The shape of the diffracting surface of the crystal is the main difference between the two types of spectrometers.

Johann Semifocusing Spectrometer.   In this type (Fig. 12-8) the radius of curvature of the diffracting surface of the crystal is twice that of the Rowland circle. This results in only a portion of the diffracting surface on the circumference of the Rowland circle. A crystal of this shape focuses only a portion of monochro-

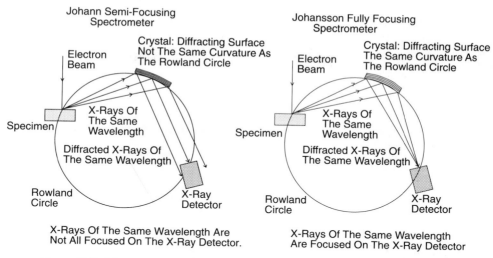

Figure 12-8 The two types of wavelength dispersive spectrometers. The diffracting face of the crystal of the Johann semifocusing spectrometer is not completely on the circumference of the Rowland circle, and only a portion of the X-rays are focused into the X-ray detector. The diffracting face of the crystal in the Johansson fully focusing spectrometer is on the Rowland circle, and all the diffracted X-rays strike the X-ray detector.

matic X-rays on the detector at one time, resulting in the counting of only a portion of the X-rays diffracted from the crystal. The only advantage of the Johann semifocusing spectrometer is that the positioning of the specimen on the Rowland circle is less critical since the X-rays are defocused by the system anyway. This system is usually not the system of choice since the wavelengths are not focused as well as they are in the Johansson fully focusing spectrometer.

Johansson Fully Focusing Spectrometer.   In this system (Fig. 12-8) the crystal is curved so that the diffracting face lies completely on the Rowland circle. This results in the focusing to the detector of all the X-rays of the same wavelength at one position of the crystal.

The X-ray detector is usually a gas proportional counter (Fig. 12-9) producing *a pulse of electrical charge that is proportional to the energy of the individual X-rays entering the counter*. The device consists of a gas-filled tube with a thin wire (commonly tungsten) in the center. The wire is held at a +1- to +3-kV potential. X-rays enter the tube through a thin window. The X-ray photons are absorbed by the gas in the tube by the **photoelectric effect** (the ejection of electrons from an atom by photons) in which an inner electron is ejected from a gas atom, producing a free electron and an ion (electron–ion pair). There follows the emission of either an Auger electron or an X-ray photon, which are absorbed by another atom, producing another photoelectron until the energy of the incident X-ray is dissi-

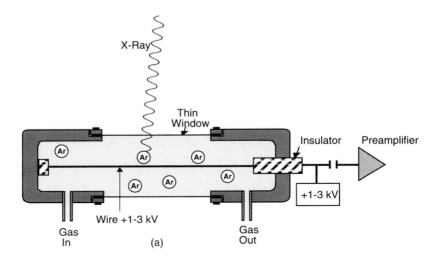

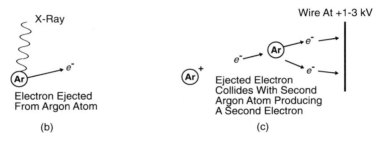

Figure 12-9 Gas proportional counter used as an X-ray detector. (a) Structure of the gas proportional counter. An X-ray enters through the thin window, striking an atom of the gas (argon) in the counter. (b) Interaction between the X-ray and the atom results in the ejection of an outer orbital electron from the argon atom. (c) The electron ejected in b interacts with a second argon atom, causing the ejection of a second electron. The electrons are attracted to the wire at +1 to +3 kV, producing a voltage pulse that passes to the preamplifier.

pated. About 28 eV is absorbed for each electron ejected from a gas mixture of 90% argon–10% methane. Thus a Ca K_α X-ray with a wavelength of 0.336 nm and an energy of 3.692 keV (3692 eV) would result in the production of 3692/28 or 152 electron–ion pairs for each Ca K_α X-ray entering the gas proportional counter. The greater the energy of the X-ray is, the more electron–ion pairs that are produced. The electrons are attracted to the central wire at +1 to +3 kV, colliding with other gas atoms and causing secondary ionizations. This results in the production of a charge pulse in the bias voltage of the wire. The charge pulse is proportional to the number of electron–ion pairs produced by the X-rays. The charge pulse passes to the electronics, which classifies the pulse as a Ca K_α X-ray.

The composition and thickness of the window determines the lowest atomic number that can be detected. The K_α of Al (atomic number = 13) has only 1.2% of the X-rays transmitted through a 34-μm Be foil, 55% through a 7.5-μm Be foil, 30% through a 1.5 μm Mylar foil, and 84% through a Formvar foil. X-rays from lower atomic number elements are transmitted even less, since their X-rays have less energy. Recent advances in using unsupported stretched polypropylene films have resulted in even better transmission of lower-energy X-rays, producing better detection of lower atomic number elements.

The charge pulse from the gas proportional counter is very small and difficult to detect without a low-temperature, low-noise preamplifier system (Fig. 12-10). The charge pulse passes to such a cryogenic preamplifier, which is located close to the gas proportional counter to minimize stray electronic noise. The pulse from the preamplifier is boosted sufficiently to allow transmission over several feet of coaxial cable to the main amplifier, where it is inverted, shaped, and amplified to produce a Gaussian (normal distribution) curve. The amplified signal is transmitted to a multichannel analyzer that determines the voltage of each X-ray pulse. The voltage of the X-ray pulse is used to assign the pulse to one of a series of memory locations representing voltage ranges. For example, if there are 1000 memory locations spanning 0 to 10,000 V, each memory location would have a voltage range of 10 V. Memory location 1 would have 0 to 10 V, memory location 2 would have 11 to 20 V, and so on. A Ca K_α X-ray of 0.336-nm wavelength and an energy of 3692 would be assigned to memory location 369 out of 1000 in a range of 0 to 10,000 V. Everytime a Ca K_α X-ray strikes the gas proportional counter, another pulse is assigned to channel 369. A cathode ray tube is used to display the number of pulses in each of the 1000 channels, producing an image of characteristic X-ray peaks on a background of continuum X-rays.

Gas proportional counters produce **escape peaks** that are characteristic of the type of gas in the counter. An escape peak is a peak in the X-ray spectrum that is

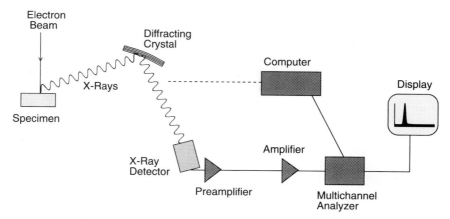

Figure 12-10 Parts of a wavelength dispersive spectrometer.

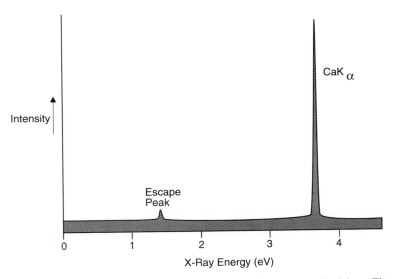

Figure 12-11 Escape peak in an X-ray spectrum of a sample of calcium. The spectrum was acquired by a wavelength dispersive spectrometer that used a gas proportional counter containing argon gas.

equal in energy to the difference between the characteristic X-ray peak and the energy of the K_α of the gas in the proportional counter. Argon is usually the main gas in the gas proportional counter. The K_α for argon is 2300 eV. Therefore, if Ca K_α of 3692 eV is the characteristic X-ray peak, the escape peak will be 3692 eV − 2300 eV, or 1392 eV and a small escape peak will appear at this position on the X-ray spectrum (Fig. 12-11). Escape peaks arise when an incoming X-ray or high-energy electron produced from a previous interaction is absorbed by an argon atom, resulting in the ejection of a K orbital electron (Fig. 12-12). This produces an argon K_α X-ray, with a subsequent reduction in the measured energy of the incoming X-ray from the specimen. Only 25% of Ar K_α X-rays are absorbed by 10 mm of argon gas. This means that Ar K_α X-rays have a good probability of escaping a gas proportional counter. An escape peak is produced only if the Ar K_α X-ray escapes. Usually, the escape peak is about 5% of the height of the main peak. X-rays of energy below 3.20 keV (the energy of the Ar K absorption edge) do not produce an escape peak.

Wavelength dispersive spectroscopy has the following advantages over energy dispersive spectroscopy:

1. *High resolution:* Resolution in X-ray microanalysis is commonly measured as the **full width, half-maximum** (FWHM) of the Mn K_α X-ray peak. To determine the FWHM, the vertical halfway point of the Mn K_α peak is determined and the width in electron volts is measured at this point (Fig. 12-13). Wavelength dispersive spectrometers usually have a FWHM for Mn K_α of

Production Of Ionizing Electron

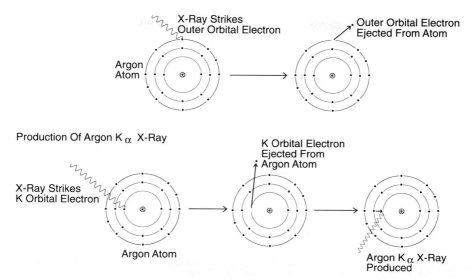

Figure 12-12 An X-ray entering into a gas proportional counter containing argon gas can interact with the gas in one of two ways. In the production of an ionizing electron, part or all of the energy of the X-ray is absorbed by the argon atom, resulting in an outer orbital electron being ejected as an Auger electron. The Auger electron is attracted to the positive wire in the counter. On the other hand, argon escape peaks are produced when the X-ray ejects a K orbital electron, resulting in the production of a K_α X-rays. An argon escape peak is produced if the Ar k_α X-ray escapes from the counter without interacting with other argon atoms. No Auger electron is ejected from the argon atom, resulting in a measured loss of energy from the counter that is equal to the difference between the energy of the X-ray and the energy of the argon K_α X-ray.

about 10 eV, while energy dispersive spectrometers commonly have an energy range of 150 eV (both can be compared to 2.3 eV for the energy range of Mn K_α X-rays emitted from the manganese in the specimen).

2. *Light element detection:* Low atomic number elements have characteristic X-ray peaks that are separated by small energy differences. For example, the boron (Z = 5) K_α is 0.183 eV, while carbon (Z = 6) K_α is 0.277 eV. The energy difference (94 eV) is beneath the resolution of energy dispersive spectroscopy (about 100 eV at these X-ray energies), so the peaks of the two elements cannot be separated using this method. However, because the energy of X-rays is inversely proportional to their wavelength, boron with a wavelength of 6.76 nm and carbon with a wavelength of 4.47 nm are easily distinguished in wavelength dispersive spectroscopy, providing the X-rays are able to traverse the window of the gas proportional counter.

3. *Good peak to background ratios:* The ratio between the peaks of the charac-

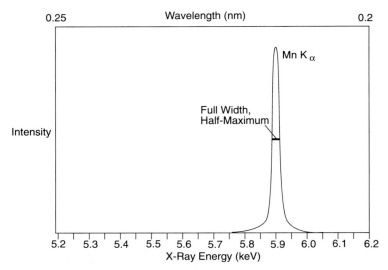

Figure 12-13 A full width half-maximum (FWHM) of the Mn K_α peak is determined by removing the background, determining half the height of the peak, and measuring the width of the peak in electron-volts at this point.

teristic X-rays and the background is 10 to 50 times better than in energy dispersive spectroscopy. This results in the ability to detect small quantities of an element in a specimen. The peak to background ratio is very good because the spectrometer accepts only a narrow band of wavelengths in the spectrum.

The disadvantages of wavelength dispersive spectroscopy are (1) the ability to measure only one element at a time, making the acquisition of a spectrum a long and tedious process, and (2) the low efficiency of X-ray collection. Typically, the solid angle of spectrometer acceptance is only 0.001 steradians (sr), compared to 0.01 to 0.1 sr for energy dispersive spectroscopy. In addition, only about 30% of the X-rays that enter the X-ray detector are counted in wavelength dispersive spectroscopy; many of the X-rays pass through the tube without ionizing gas atoms. In energy dispersive spectroscopy, nearly 100% of the X-rays that strike the detector are counted. The low efficiency of collection in wavelength dispersive spectroscopy means that higher electron beam currents on the specimen are needed (10^{-8} A), compared to energy dispersive spectroscopy (10^{-10} A). Such high electron beam currents in wavelength dispersive spectroscopy require relatively large beam diameters (about 200 nm), resulting in a large X-ray escape volume in the specimen, thus reducing spatial resolution in the specimen. In contrast, electron beam diameters in energy dispersive spectroscopy can be as low as 5 nm. The reduced number of X-rays counted in wavelength dispersive spectroscopy means that long counting times are required, typically 15 min for wavelength dispersive spectroscopy, compared to 1 min for energy dispersive spectroscopy. The long exposure in wavelength dispersive spectroscopy can cause damage to biological specimens, although metallurgical specimens are hardier and usually not harmed.

ENERGY DISPERSIVE SPECTROSCOPY

In energy dispersive spectroscopy, the *energy* of the X-rays is used to produce electron–hole pairs in a semiconductor crystal and the signal is used to produce a spectrum. In the mid-1960s, the semiconductor radiation detector crystal was developed at the Lawrence Berkeley Laboratory. In 1968, Raymond Fitzgerald, Klaus Keil, and Kurt Heinrich (Figs. 12-14 and 12-15) described the measurement of X-ray energies by means of a semiconductor radiation detector on an electron

Figure 12-14   Klaus Keil. Born November 15, 1934, in Hamburg, Germany. Keil received his primary and secondary school education in Jena, his M.S. from Fredrich Schiller University in Jena, and a Ph.D. in mineralogy and geochemistry from Johannes Gutenberg University in Mainz. From 1961 to 1963 he was a research associate in the Department of Geology at the University of California, San Diego; from 1963 to 1968 he was a staff research scientist with the Space Sciences Division, Ames Research Center, Moffett Field in California; from 1968 to 1990 he was in the department of Geology at the University of New Mexico, where he became professor of geology and director of the Institute of Meteoritics. In 1990 he moved to the University of Hawaii at Manoa, where he became head of the Planetary Geosciences Division and professor of geology and geophysics. (From *Meteoritics* 23: 316 [1986]). (*Photograph courtesy of Meteoritics*)

Figure 12-15   Kurt F. J. Heinrich. Born May 31, 1921, in Vienna, Austria. Heinrich received his Dr. Chem. from the University of Buenos Aires in 1948. He came to the United States, where he worked in private industry from 1948 to 1964. In 1964 he joined the National Bureau of Standards, where he became the section chief of the microanalytical section. His book, *Electron Beam Microanalysis,* is an authoritative work on the subject. (Photograph courtesy of Dr. Heinrich)

beam microprobe. Their original system had only a resolution of 500 eV and was barely able to resolve elements with adjacent atomic numbers. However, within a few years, resolution was improved to 150 eV. Today, almost all X-ray microanalytical units on scanning electron microscopes and many on electron beam microprobes use energy dispersive spectroscopy.

The main component of an energy dispersive spectrometer is the solid-state semiconductor detector. Solid-state detectors and some components of the amplification system in energy dispersive spectrometers are semiconductors. A short review of semiconductors therefore will be presented before discussing these parts of the energy dispersive spectrometer.

Semiconductors

Semiconductors are materials that have a conductivity between insulators (poor conductors), such as rubber, glass, and wood, and good conductors, such as the metals iron, aluminum, copper, and silver (Fig. 12-16). Two semiconductors that are widely used in the electronics industry are silicon and germanium. The intermediate conductivity of silicon and germanium can be attributed to their atomic structure of four valence electrons. Insulators have a large forbidden gap between

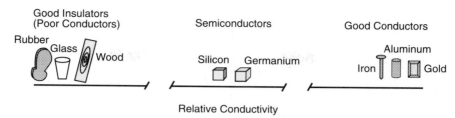

Relative Conductivity

Figure 12-16 Semiconductors, such as silicon and germanium, are neither good insulators nor conductors.

the valence band and conduction band, resulting in electrons that are strongly bound to the individual atoms (see Chapter 6). A good conductor has no forbidden gap, so the electrons are free to move from the valence band to the conduction band and carry electrical current. In pure form, semiconductors are actually good insulators. This is because of the four valence electrons in silicon and germanium. The outer shell of each atom is half-full and has four vacant positions. In the crystalline state, adjacent atoms contribute four electrons to fill the four vacancies in the outer orbital of each silicon or germanium atom. Therefore, in pure silicon or germanium crystals, all electrons are bound to their atoms (Fig. 12-17), and there are no electrons free to make electrical current. The crystals, however, can be made conductive by adding impurities to the crystal in the process called **doping.**

n-Type Semiconductors. A silicon crystal can be made conductive by adding a few phosphorus atoms, usually about one phosphorus atom for 2 million silicon atoms (the more phosphorus that is added, the more conductive the silicon crystal is). Phosphorus has five valence electrons occupying a space in the silicon lattice (Fig. 12-18). Four of the valence electrons bind to the silicon atoms, leaving one phosphorus electron free to wander through the crystal. This leaves the phosphorus atom as a positive site in the crystal, since the phosphorus nucleus has one more proton than the number of orbital electrons. This is an *n*-type crystal since there is a negative charge carrier in the crystal.

An *n*-type semiconductor crystal will conduct electrical current if one side of the crystal is made negative and the other side positive by attaching the terminals of a battery. The free electrons in the crystal are attracted to the positive

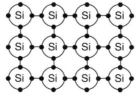

Silicon Semiconductor Crystal

Figure 12-17 A crystal of silicon has the four electrons of each silicon atom shared with adjacent atoms.

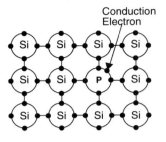
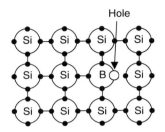

n - Type Silicon Semiconductor
Crystal Doped With Phosphorus

p - Type Silicon Semiconductor
Crystal Doped With Boron

Figure 12-18   Doped crystals of silicon. Phosphorus has five valence electrons, resulting in a valence electron that is not shared with adjacent silicon atoms. The creates an *n*-type (*n*egative) silicon semiconductor crystal. Boron has three valence electrons, resulting in a hole where a valence electron of an adjacent silicon atom is not shared. This creates a *p*-type (*p*ositive) silicon semiconductor crystal.

battery terminal, while electrons are pumped out of the negative battery terminal. Thus an *n*-type semiconductor crystal will conduct electrical current.

   p-Type Semiconductors.   A second type of semiconductor is produced by doping a semiconductor crystal with a small amount of boron, gallium, or indium, all with three valence electrons in their outer shell. Inclusion of atoms with three valence electrons in the silicon or germanium lattice results in one of the four valence electrons of the semiconductor atom not being paired (Fig. 12-18). There is a **hole** in the adjacent valence 3 atom. This hole can be filled by an electron from an adjacent silicon atom, leaving the boron, gallium, or indium as a negative site in the crystal (the four electrons surrounding the three protons in the nucleus). The hole is able to migrate across the crystal by successive filling of the hole by electrons, the hole moving in one direction and the electrons moving in the other. Silicon or germanium doped with a small amount of a valence 3 atom such as boron, gallium, or indium is a *p*-type semiconductor since the charge carriers are *p*ositive holes.

   A *p*-type semiconductor will conduct current when attached to a battery. The semiconductor crystal pumps an electron into the positive battery terminal, leaving a positive hole behind. The new hole repels an old hole all the way to the other end of the crystal, where the negative battery terminal is attached. Here a negative electron from the battery enters the crystal to fill the positive hole, and migration of electrons occurs across the crystal to the positive battery terminal. Thus the electrons move through the crystal from the negative to the positive battery terminal, while the positive holes move in the opposite direction.

   pn-Junction Diodes.   A *pn*-junction diode (diode means having two electrodes) is produced when a *p*-type and an *n*-type semiconductor are joined. *Diodes*

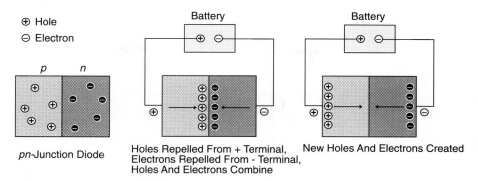

⊕ Hole
⊖ Electron

pn-Junction Diode

Holes Repelled From + Terminal,
Electrons Repelled From - Terminal,
Holes And Electrons Combine

New Holes And Electrons Created

Figure 12-19   A *pn*-junction diode is created when *p*-type and *n*-type semiconductors are joined. The diode is conductive when the positive terminal of a battery is attached to the *p*-type semiconductor and the negative terminal of the battery is attached to the *n*-type semiconductor.

are structures that allow electrical current to flow in only one direction. In *pn*-junction diodes, electrical current flows only when the positive terminal of a battery is attached to the *p*-type side and the negative terminal is attached to the *n*-type side (Fig. 12-19). The positive holes in the *p*-type side are repelled from the positive terminal and attracted to the negative terminal. The negative electrons in the *n*-type side are repelled from the negative terminal and attracted to the positive terminal. At the junction between the *p*-type side and the *n*-type side, the electrons meet holes and fill them in. At the same time, electrons are being pulled from the *p*-type side by the positive terminal, creating new holes, and electrons are being provided to the *n*-type side from the negative terminal. The result is a continuous flow of electrical current in the *pn*-junction diode connected in this manner.

Reversing the positions of the battery terminals changes the *pn*-junction diode from a conductor to an insulator (Fig. 12-20). The electrons in the *n*-type side are attracted to the positive terminal, and the holes in the *p*-type side are attracted to the negative terminal. A **depletion zone** with no electrons or holes is produced in the center of the junction. This results in no crossover of electrons and holes at the junction of the *p*-type and *n*-type semiconductor and no flow of current, making the diode an insulator.

Solid-state X-ray Detectors

A solid-state X-ray detector is a *pn*-junction diode set up as an insulator. The negative–positive bias placed on opposite sides of the semiconductor crystal segregates the holes and electrons, preventing the flow of electrical current. X-rays entering the detector create electron–hole pairs, producing current that is used to measure the energy of the X-rays.

A *minimal amount* of **residual** or **base line conductivity** occurs in a semiconductor crystal used as a solid-state detector in an energy dispersive spectrometer.

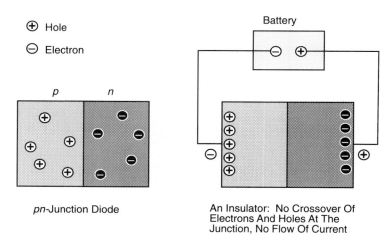

pn-Junction Diode

An Insulator: No Crossover Of
Electrons And Holes At The
Junction, No Flow Of Current

Figure 12-20   A *pn*-junction diode is an insulator (no conduction) when the negative terminal of a battery is attached to the *p*-type semiconductor and the positive terminal of the battery is attached to the *n*-type semiconductor. In this configuration the positive holes are attracted to the negative terminal of the battery, and the negative electrons are attracted to the positive terminal of the battery. This results in no crossover of electrons and holes at the junction and no flow of electrical current.

Even a perfectly formed semiconductor crystal has some residual conductivity due to random thermal excitation of electrons across the gap between valence and conduction bands. The valence electrons in a pure silicon or germanium crystal occasionally absorb enough heat energy to push a valence electron into the conduction band, where it is free to roam the crystal as electrical current (**leakage current**). Such residual conductivity enters into the energy dispersive spectrometer as electrical noise, on top of which are the pulses created by the X-rays. The greater the electrical noise is, the more difficult it becomes to distinguish the X-ray peaks. To minimize the thermally induced residual current, detectors are operated at low temperatures, usually by incorporating a liquid nitrogen-cooled device called a **cryostat**.

Since the best semiconductor crystals available are less than perfect, the residual conductivity created by the imperfections and impurities in the crystalline structure results in an unacceptable leakage current and electronic noise. Silicon crystals cannot be made pure enough to maintain the required bias necessary to use the crystal as a solid-state detector. The impurities introduce additional levels in the forbidden band between the valence and conduction band. To correct for this, silicon crystals have the impurities and imperfections compensated by drifting lithium atoms into a *p*-type silicon or germanium semiconductor crystal containing B^- (boron minus) sites. Lithium is used because its small radius (0.06 nm) enables it to easily diffuse into silicon. A zone that has a very high resistance (very low

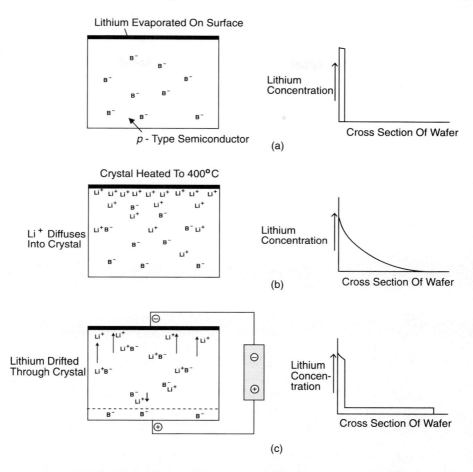

Figure 12-21 The production of a lithium-drifted silicon or germanium crystal. First, lithium is evaporated on the surface of a *p*-type silicon or germanium crystal doped with boron. The crystal has an excess of electrons on the boron sites. Next the crystal is heated to 400°C so that the lithium diffuses into the crystal. Finally, the lithium is drifted through the crystal by applying a potential across the crystal, as shown. The lithium is not completely drifted across the crystal, thereby leaving a small volume that still contains only B⁻ in a *p*-type layer.

conductivity) is created when each B^- atom has been compensated by a Li^+ atom. After compensation by lithium drifting, the creation of electron–hole pairs in the intrinsic zone occurs only by the action of an external event (for example, the absorption of an X-ray).

The process of lithium compensation (Fig. 12-21) begins by the evaporation of lithium in a vacuum or inert gas onto one side of a wafer of *p*-type silicon or germanium (less than 10-mm diameter) containing B^- sites. The wafer is heated to

400°C for a few minutes, causing lithium to diffuse into the crystal. A concentration gradient of lithium is established in the wafer (Fig. 12-21) with three zones: (1) an *n*-type zone rich in lithium, (2) a thin **intrinsic zone** in the center in which the *acceptor levels* of boron are compensated for by the *donors* of lithium, and (3) a *p*-type zone poor in lithium. An *inverse* voltage is applied to the wafer by placing a negative charge on the side of a silicon wafer at 150°C. The negative lead is attached to the side of the wafer that has the evaporated lithium, attracting the uncompensated lithium ions back to this side of the wafer. This increases the width of the intrinsic zone. The very high internal resistivity of the compensated region results in the migration of some lithium ions in the other direction toward the *p*-type zone. The drifting of lithium ions continues until the intrinsic zone reaches the desired width (2 to 3 mm). The *p*-zone is largely removed to allow access to the intrinsic zone, the portion of the detector that produces the electron–hole pairs. A small layer of the *p*-zone (the **dead layer**, about 100 nm thick) is left without lithium ions having drifted into it, and containing only B^- sites. Deep, circular grooves or guard rings are cut into the intrinsic region to ensure uniformity of the field and to avoid edge effects (Fig. 12-23).

A problem with lithium-drifted crystals is that lithium can drift back out of the crystals. Lithium is relatively mobile at room temperature in germanium crystals. In silicon crystals, lithium is relatively immobile at room temperature but can be made mobile by raising the temperature. Lithium-drifted crystals are therefore usually kept at liquid nitrogen temperatures to immobilize the lithium in the crystal. The Si(Li) crystals are more widely used than Ge(Li) in energy dispersive spectrometers partly because the lithium does not drift out of the crystal as easily, but also because of greater analytical sensitivity. As long as the bias voltage is off, there is usually no damage done to Si(Li) crystals by warming to room temperature.

The lithium-drifted crystal functions as a *pn*-junction diode (Fig. 12-22). The layer from which the lithium has drifted functions as an *n*-type region because it has an excess of electron carriers. The Li^+ ions are lodged in the crystal, with their free electrons roaming the volume as electron carriers. The dead layer on the opposite side into which lithium has not drifted remains as a *p*-type region with an excess of holes. In the middle is the depletion zone (intrinsic zone), where Li^+ ions have exactly compensated for the B^- ions. There are therefore no charge carriers in the depletion zone. Attaching the negative side of a bias supply to the *p*-type side and the positive lead to the *n*-type side (with lithium) results in the crystal becoming an *insulator*. The holes in the *p*-type side are attracted to the negative lead, while the electrons in the *n*-type side are attracted to the positive lead. The holes and electrons are not able to cross over at the junction, and there is no electrical current flowing across the semiconductor crystal. In the depletion zone in the center of the crystal, there are no holes or electrons. Any holes or electrons that are produced thermally by residual conductivity are immediately swept to the respective poles, maintaining the depletion zone.

A Si(Li) or Ge(Li) crystal, approximately 3 mm wide and 7 mm in diameter,

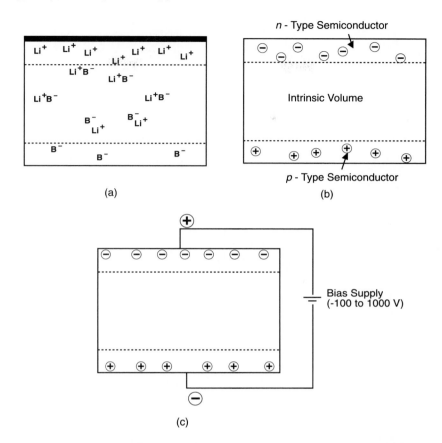

Figure 12-22 The lithium-drifted detector crystal. (a, b) The crystal contains three regions. An *n*-type layer occurs where there is an excess of Li$^+$ ions in the crystal. A *p*-type layer occurs where there is an excess of B$^-$ ions in the crystal. In between the *n*-type and *p*-type layers there is an intrinsic or depletion layer where the B$^-$ sites have been complexed with Li$^+$. (c) When a reverse bias is placed on the crystal, the free electrons in the *n*-type semiconductor are attracted to the positive terminal, while the positive holes are attracted to the negative terminal. No flow of current occurs, and the lithium-drifted crystal is an insulator.

has approximately 20 nm of gold evaporated on the front surface and 200 nm on the back surface of the crystal. The gold surfaces serve as electrical contacts, so potentials of -100 to -1000 V can be applied to the front surface of the crystal while the rear surface is relatively positive at ground (Fig. 12-23). The Si(Li) or Ge(Li) crystal has valence band states filled, while the conduction band states are empty. An X-ray strikes the crystal, producing photoelectrons, Auger electrons, and secondary X-rays (Fig. 12-24), as occurs in a gas proportional counter. The X-ray photon is first absorbed by a silicon or germanium atom, pushing an electron

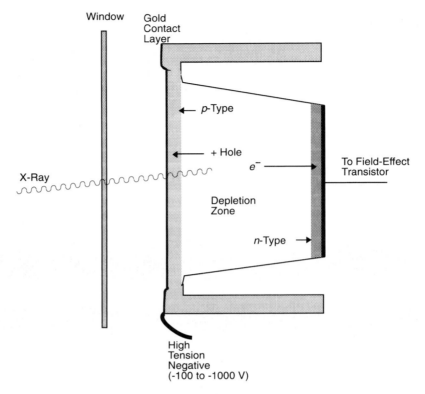

Figure 12-23 The lithium-drifted detector crystal has a 20-nm layer of gold on the front of the crystal and a 200-nm layer of gold on the back of the crystal. The front of the crystal is made −100 to −1000 V negative, while the rear of the crystal is at ground. An incident X-ray travels through the window and into the crystal, producing electron–hole pairs in the depletion zone. The positive hole is swept to the negative front of the crystal, while the negative electron moves to the rear of the crystal. This creates a voltage pulse that travels to the field-effect transistor.

(photoelectron) into the conduction band and leaving a hole in the valence band. The photoelectron usually has sufficient energy to create additional electron–hole pairs as the photoelectron scatters inelastically in the Si(Li) or Ge(Li) crystal. Approximately 3.8 eV is lost everytime an electron–hole pair is created in a Si(Li) crystal. The photoelectron continues through the crystal until its energy is dissipated. Meanwhile, the silicon or germanium atom that absorbed the original X-ray photon is left in an energized state because of the ejection of the photoelectron. This energy is released in the form of a silicon X-ray or as the ejection of an outer orbital electron (Auger electron). The Auger electron scatters inelastically through the crystal, creating electron–hole pairs. The silicon X-ray can be reabsorbed (starting the sequence again), can scattered inelastically, or can escape the crystal

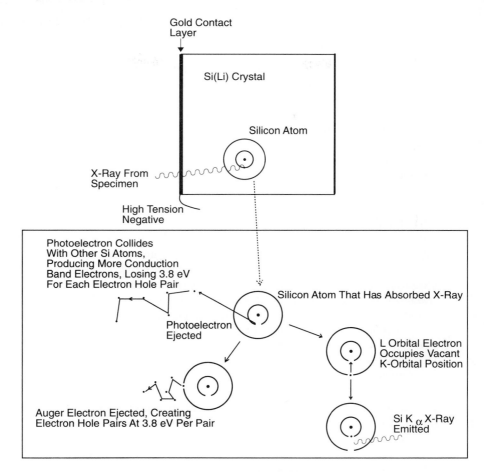

Figure 12-24 The process by which an incident X-ray creates photoelectrons, Auger electrons, and Si K$_\alpha$ X-rays in the lithium-drifted crystal.

(resulting in the formation of a silicon escape peak). Thus, all the energy of the X-ray photon is used in the production of electron–hole pairs in the crystal (unless the silicon X-ray escapes the crystal).

The number of electron–hole pairs created by each incident X-ray photon in a Si(Li) crystal is equal to the energy of the X-ray divided by 3.8 eV. A Ni K$_\alpha$ X-ray (7471 eV) incident on a Si(Li) crystal will produce 7471/3.8 or 1966 electron–hole pairs. As soon as the electron–hole pairs are produced, they are swept to their opposite charge poles, creating a charge pulse on the opposite side of the Si(Li) crystal. The collection of charge is rapid (about 100 ns) compared to the other processes in the energy dispersive spectrometer, which occur in the microsecond range. *The key result is that the voltage pulses produced by the Si(Li) or Ge(Li) detectors are on the average proportional to the energy of the incident X-ray photon.*

The 3.8 eV of energy used in the production of an electron–hole pair is much lower than the equivalent energy (28 eV) needed in the gas proportional counter. The greater number of electron–hole pairs created per unit energy in energy dispersive spectroscopy results in better statistics in the production of X-ray pulses. This, in turn, results in better resolution of individual X-ray peaks in the X-ray spectrum.

Preamplifier (Head Amplifier)

The amount of charge produced by the Si(Li) crystal from X-ray interaction is very small; a 4-keV X-ray will only produce 1050 electron–hole pairs, which is equivalent to a charge of 1.6×10^{-10} C. This signal has to be amplified about 10^{10} to be useful. At the same time, electronic noise has to be minimized during the amplification process. The initial amplification of the signal occurs in the preamplifier. The first and most important part of the preamplifier is the **field-effect transistor** (FET), which is positioned immediately behind the detector to minimize electronic noise caused by a long attachment cable.

A field-effect transistor amplifies the weak signal from the solid-state X-ray detector. A field-effect transistor is a single *pn*-junction with three connections (Fig. 12-25). The weak signal from the lithium-drifted detector is fed into the *p*-type semiconductor that sits in the middle of the *n*-type semiconductor. The positioning of the *p*-type semiconductor is such that it separates the *n*-type semiconductor into two halves joined by a narrow channel. A working current is fed into the *n*-type semiconductor at the source; it passes through the channel and exits at the drain. There is a steady flow of leakage current that feeds into the

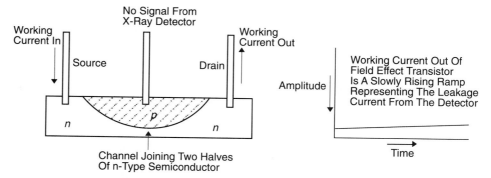

Figure 12-25   Field-effect transistor. The middle electrode receives the current from the lithium-drifted detector. Only the thermal leakage current travels to this electrode if no current pulses, representing X-rays, are being received. In this situation the working current out is a slowly rising ramp.

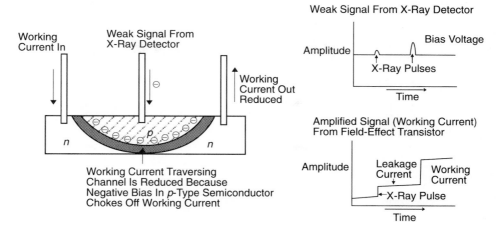

Figure 12-26   Field-effect transistor when current pulses representing X-rays are being received from the lithium-drifted detector. The current into the middle electrode produces a negative bias in the channel of the *n*-type semiconductor, reducing the working current out of the field-effect transistor. The signal from the detector into the field-effect transistor is represented as pulses. The signal out of the field-effect transistor is a ramped staircase. Each step represents the amplified X-ray pulses.

central electrode of the field-effect transistor when no X-rays are striking the detector. This makes the *p*-type semiconductor volume progressively more negative, setting up a negative field in the adjacent volume of the *n*-type semiconductor that slowly chokes off the working current passing through the channel. The result is a steadily decreasing amount of working current leaving the field-effect transistor. Current pulses, representing incident X-rays on the detector, cause steps in the output of the working current from the field-effect transistor (Fig. 12-26). The height of the steps is proportional to the energy of the incident X-rays. Thus the greater the signal from the X-ray detector, the more negative is the channel of the *p*-type semiconductor, the more the working current is choked off in the channel of the *n*-type semiconductor, and the greater is the reduction in the working current leaving the *n*-type semiconductor. The key to the action of the field-effect transistor is that *a small change in the incoming signal (from the X-ray detector) produces a large change in the working current,* which subsequently passes to the main amplifier.

Cryostat

The Si(Li) or Ge(Li) detector and the field-effect transistor are kept at liquid nitrogen temperature to minimize thermally induced electronic noise and to minimize lithium migration in the detector crystal. The detector and field-effect transistor assembly are mounted on a solid copper rod (**cold finger**), the other end of

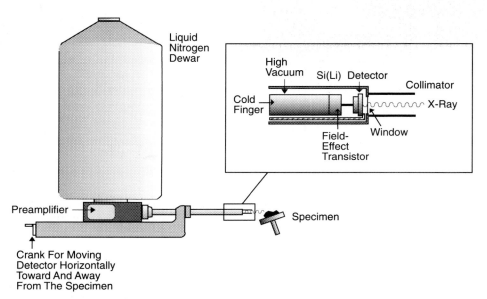

Figure 12-27   Physical layout of the detector, cryostat, and preamplifier.

which is immersed in liquid nitrogen at $-196°C$ (Figs. 12-27 and 12-35). The liquid nitrogen dewar usually contains 5 l and needs to be refilled every three days.

The very cold detector, field-effect transistor, and cold finger have to be kept at a high vacuum (10^{-6} mbar) or atmospheric gases will condense onto these components, shorting them out and making them ineffective. Also, minor amounts of contamination on the detector will absorb incident X-rays, reducing the efficiency of the system. A metal tube, with a window at the detector end, is placed around the detector, field-effect transistor, and cold finger to hold the vacuum. The vacuum insulates the tube from the cold components and prevents condensation of moisture on the housing.

Windows.   The window over the detector has to be strong enough to withstand pressures from 1 atm (when the specimen chamber is vented) to 10^{-6} mbar (when the specimen chamber is at operating pressure). At the same time, the window cannot be so thick that it absorbs a significant proportion of the low-energy X-rays, preventing them from reaching the detector. The amount of absorption of X-rays depends on the composition and thickness of the window. This, in conjunction with the resolution of the system, determines the lowest atomic number element that can be detected by the energy dispersive spectrometer. This is because the lower the atomic number of the element is, the lower the energy of the particular series (K, L, M) of the X-rays. X-ray detectors are usually offered in one of three types of window configurations: (1) beryllium, (2) thin organic polymer, or (3) windowless.

Beryllium. Beryllium is rolled to a thin sheet. The thinner the sheet is, the more X-rays that are transmitted through it, but the weaker the sheet is. A beryllium window 25 μm (1 mil) thick transmits only 12% of 1-keV X-rays, while a beryllium window 8 μm (0.3 ml) thick transmits 50% of 1-keV X-rays. Beryllium windows in energy dispersive spectrometers are usually 7.5 to 8 μm thick. Generally, an energy dispersive spectrometer with a 8-μm beryllium window is not able to effectively detect K series X-rays emitted from atoms beneath an atomic number of sodium (Z = 11, K_α X-ray = 1.041 keV). About 60% of the Na K_α X-rays pass through an 8-μm-thick beryllium window. The same window will only pass 1% of oxygen K_α X-rays (0.52 keV).

Thin Organic Polymers. A large number of thin polymer sheets can be used as windows. Mylar (polyethylene terephthalate, $C_{10}H_8O_4$), 2 to 6 μm thick, can be used to make a strong window. Thin organic polymers transmit lower-energy X-rays. Some manufacturers claim that boron X-rays (Z = 5, K_α = 0.183 keV) can be transmitted and differentiated from carbon X-rays (Z = 6, K_α = 0.277 keV) using their energy dispersive spectrometers fitted with thin organic polymer windows.

Windowless Detectors. Some energy dispersive spectrometers have a device that allows the operator to rotate the beryllium window out of the way when high vacuum has been attained in the specimen chamber. This windowless mode allows the X-rays to reach the detector without being impeded by a window. The main problem with the windowless setup is that the crystal detector is close to liquid nitrogen temperature, so any gases remaining in the specimen chamber or created on interaction of the electron beam with the specimen are absorbed on the detector. This contaminating layer on the detector absorbs some of the incident X-rays. The advantage of a windowless setup is that all the X-rays reach the detector, and the lowest atomic number element that can be detected is determined by the resolution of the spectrometer (usually around 100 eV when low atomic number elements such as carbon and boron are being measured).

Interception of Electrons Directed Toward the Detector. Electrons from the interaction of the electron beam with the specimen and specimen chamber cannot be allowed to reach the detector, create electron–hole pairs, and contribute to the background continuum. About 1000 backscattered electrons are created for each X-ray on interaction of the specimen with the electron beam. Since the backscattered electrons are of high energy, they are the ones that cause most of the background if they enter the detector. Beryllium windows absorb almost all the incident backscattered electrons below 25 keV. In addition, detectors with beryllium windows usually have collimators that absorb all the electrons and X-rays that do not come directly from the specimen. A collimator is a hollow metal tube that sits outside the detector (Fig. 12-27). The collimator is directed at the specimen so that only those X-rays and electrons generated from the specimen pass up the center of the collimator. Those detectors that are windowless or have a thin organic polymer window have a magnetic electron trap outside the detector (Fig. 12-28).

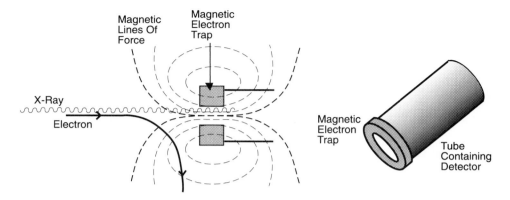

Figure 12-28 A magnetic electron trap is used for detectors with thin organic polymer windows and windowless setups. The magnetic field of the electron trap repels electrons and prevents them from reaching the lithium-drifted detector.

The magnetic field of the electron trap repels the electrons and prevents them from entering the detector chamber.

Resetting the Ramp Voltage from the Field-effect Transistor.   When no X-ray pulses are coming from the Si(Li) detector, the signal from the field-effect transistor is a slowly increasing ramp due to the leakage current from the detector (Fig. 12-25). X-ray voltage pulses are added on top of the leakage ramp to produce a stepped ramp (Fig. 12-26). Eventually, the voltage of the stepped ramp reaches a maximum value and has to be reset in order to keep receiving charge pulses from the detector. This is usually accomplished by means of a pulsed optical feedback system (POF) (Fig. 12-29). When the voltage leaving the field-effect transistor reaches a predetermined value, the voltage pulse is shunted through a light-emitting diode, resulting in a flash of light. The light falls on a photodiode that is in the line between the detector and field-effect transistor. Light striking the photodiode causes the photodiode to become conductive, draining off the signal current from the detector. In this way the voltage is reset to the bottom of the ramp (Fig. 12-30).

Considerable noise is generated when the pulsed optical feedback system is turned on. It is therefore necessary to turn off the main amplifier during this time. This introduces the concept of **deadtime**, or time at which the energy dispersive spectrometer is not accumulating X-rays and is dead. Deadtime occurs here and at two other places in the signal-processing chain (in the pulse pileup rejector and the analog to digital converter). The deadtime caused by the pulsed optical feedback system depends on the energy and number of X-rays incident on the detector. For example, a specimen emitting five hundred 10-keV X-rays per second causes roughly twice as much current to flow to the field-effect transistor as does a specimen that emits five hundred 5-keV X-rays per second. The field-effect circuit resets itself twice as often for the 10-keV X-rays as for the 5-keV X-rays.

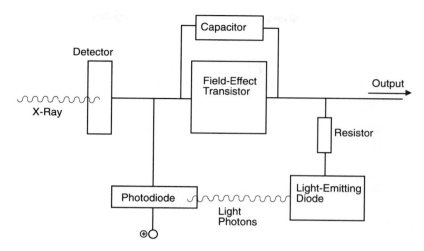

Figure 12-29   Pulsed optical feedback system used to reset the voltage of the field-effect transistor. When the voltage of the field-effect transistor reaches a maximum value, the current is shunted through a light-emitting diode and is used to emit light. The light strikes a photodiode, making the photodiode conductive.

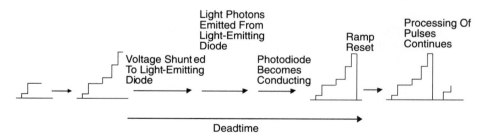

Figure 12-30   Resetting of the ramped voltage from the field-effect transistor. The voltage out of the field-effect transistor is a staircase ramp. When the ramp reaches a maximum value, the voltage is shunted to a light-emitting diode and used to produce light photons. The photons strike a photodiode, making it conductive. In this way the ramp is reset to a base line voltage and the field-effect transistor begins a new voltage ramp.

Main Amplifier

The main amplifier converts the ramped, staircase output from the preamplifier (in the millivolt range) to individual positive pulses that are linearly amplified up to 10 V in amplitude. In this process, the proportionality of the energy in the incident X-rays to the height of the voltage pulses is maintained. The conversion to individual pulses is necessary for acceptance of the signal by the next step, the analog to digital converter.

The main amplifier produces bell-shaped voltage pulses. The longer the main amplifier takes to produce each voltage pulse, the less sensitive is the process to electronic noise and the more accurate is the height of the voltage pulse. Ideally, it would appear that relatively long times (time constants) would be desirable in producing each voltage pulse. However, X-ray voltage pulses are continually entering into the main amplifier, and it is necessary for the main amplifier to return to base line voltage before another pulse can be accurately processed. Figure 12-31 shows two voltage pulses processed for the three different times of 1, 5, and 10 μs. At a 1-μs time constant, there are two 3-V pulses. At a 5-μs time constant, the first pulse is tailing off into the peak of the second pulse, creating a second pulse of 3.5 V, instead of the accurate value of 3 V. At a 10-μs time constant, the contribution of the tail of the first voltage pulse raises the peak of the second voltage pulse to a value of 4.5 V. The high value for the second pulse for the 5- and 10-μs time constants causes the second pulse to be assigned to an incorrect channel in the multichannel analyzer, resulting in the assignment of the X-ray pulse to an incorrect X-ray energy in the multichannel analyzer.

A **pulse pileup rejection circuit** in the main amplifier is responsible for the rejection of X-rays that arrive at the main amplifier before the previous voltage pulse has returned to base line voltage (Fig. 12-32). In pulse pileup rejection, the signal from the field-effect transistor is sent to the main amplifier and to a parallel fast amplifier. The main amplifier processes the pulse over the specified time constant (for example, 5 μs plus 15 μs to return to base line voltage). The parallel fast amplifier processes the same signal very rapidly (for example, 1 μs plus 5 μs

Main Amplifier Voltage Pulse Shapes For Different Time Constants

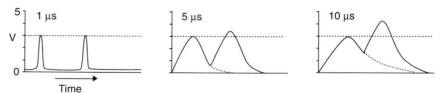

Figure 12-31 Output voltage from the main amplifier at three different time constants. Two distinct 3-V pulses occur at time constants of 1 μs. The second pulse has a value of 3.5 V at time constants of 5 μs. The second pulse has a value of 4.5 V at time constants of 10 μs.

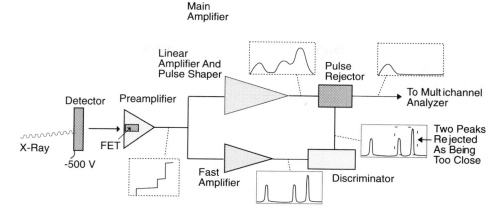

Figure 12-32   Pulse pileup rejector circuit in the main amplifier. The signal from the preamplifier is processed in parallel through a fast and a slow circuit. Pulses that arrive in the fast circuit, before the previous pulse has been processed by the slow circuit, cause rejection of the same pulses in the slow circuit.

to return to base line voltage) and is therefore able to sense the arrival of another pulse during the period that the main amplifier is still processing the first pulse. If the second pulse arrives in the fast discriminator circuit before the main amplifier has finished processing the first pulse, a pulse rejector is set to reject both pulses, or only the second pulse, from reaching the multichannel analyzer. The time blanked by the pulse processor during rejection of the pulse(s) is added to the deadtime already established from resetting the ramp voltage of the field-effect transistor.

Multichannel Analyzer

The output from the amplifier is an analog signal that can be measured only by an analog device such as a voltmeter. For further processing, it is necessary to digitize (transform into numbers) the analog signal. In the first stage of the multichannel analyzer (MCA), the signal is digitized by an analog to digital converter (ADC) (Fig. 12-33). The analog signal into the ADC charges a capacitor. Once the analog signal into the capacitor has peaked, there is an input gate that prevents the signal from entering into the capacitor, and the capacitor discharges at a constant rate. The time that the capacitor takes to discharge to a predetermined voltage level is measured by a rapidly oscillating crystal clock. The large voltage pulses (representing large X-ray energies) result in long capacitor discharge times, while the small voltage pulses result in short capacitor discharge times. The total number of oscillations of the crystal clock during the discharge is recorded by a scaler as a number. It is this number that is the digital readout from the ADC. The digitization of an analog pulse takes about 5 μs when a 100-MHz clock is used. The discharge time

Multichannel Analyzer

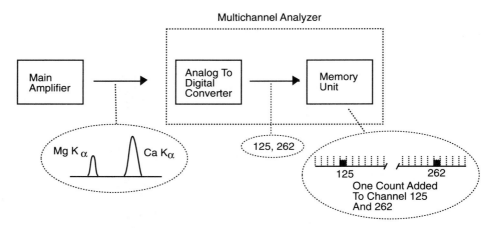

Figure 12-33 The multichannel analyzer has an analog to digital converter that converts the voltage pulses from the main amplifier into numbers. These numerical values are sent to the memory unit, which places the values into one of 1024 channels. In this example, each channel has a value of 10 eV.

contributes to deadtime, since the capacitor will not accept a second voltage pulse during the discharge time. However, there is little danger that a second X-ray pulse will be received during this period of time, since the main amplifier has already rejected X-ray pulses received within this time frame.

The digitized X-ray pulses from the analog to digital converter are specific numbers that are passed to the memory unit, the second stage of the multichannel analzyer (Fig. 12-33). The memory unit is usually part of the microcomputer of the microanalytical unit. The memory unit typically has 1024 channels, with successive channels storing increasing digital values. The X-ray energy range of each channel can be varied by the operator. The most common X-ray energy range for the whole X-ray spectrum is 0 to 10, 240 eV, resulting in each channel covering 10 eV (for example, channel 1 ranges from 0 to 10 eV, channel 2 from 11 to 20 eV, and so on). As the digitized X-ray pulses come out of the ADC, they are stored in one of the channels of the multichannel analyzer on an add-one basis.

For example, Fig. 12-33 shows an X-ray pulse from a Mg K_α (1254 eV) and a Cl K_α (2622 eV) entering the analog to digital converter. The ADC analyzes the height of the voltage pulses and converts them to the numbers 125 and 262. These numbers are passed to a memory unit that has 1024 channels, with each channel having an energy range of 10 eV. The digitized Mg K_α number (125) results in one count added to channel 125. The digitized Cl K_α number (262) results in one count added to channel 262. If the memory unit is set so that each channel covers 20 eV (spectrum range from 0 to 20,480 eV), the digitized Mg K_α number (125) would be added to channel 62, and the digitized Cl K_α number (262) would be added to

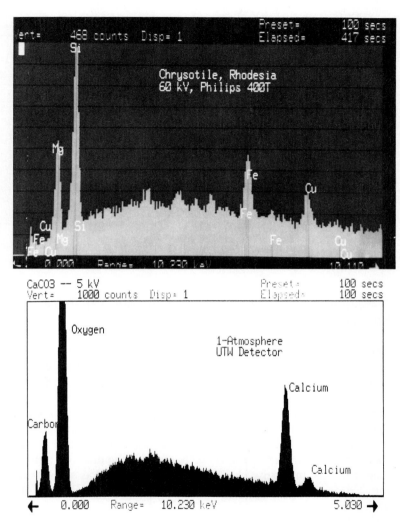

Figure 12-34 Hard copies from an energy dispersive spectrometer. (a) Video monitor display of a spectrum from chrysotile type of asbestos. (b) Plotter representation of a spectrum of $CaCO_3$.

channel 131. The results of the accumulated X-rays are displayed as a spectrum on a video monitor with the X-ray energy range on the horizontal axis and the number of accumulated X-rays in each channel on the vertical axis. Hard copies are obtained by use of a plotter with a pen recorder or a photographic device fitted to a high-resolution cathode ray tube (Fig. 12-34).

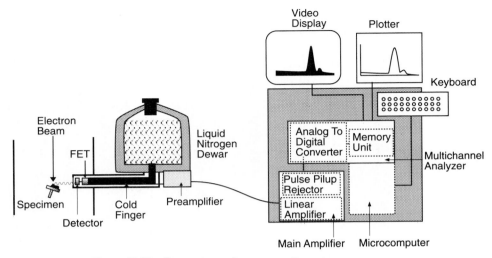

Figure 12-35   Components of an energy dispersive spectrometer.

Summary of the Components of an Energy Dispersive Spectrometer

The X-rays emanating from the specimen enter into the Si(Li) or Ge(Li) crystal, producing charges proportional to the energy of the incident X-rays (Figs. 12-35 and 12-36). These charge pulses pass to the field-effect transistor of the preamplifier, whose output is a staircase ramp, with each step of the ramp proportional to the energy of the original X-rays. The signal is next processed by the main amplifier, which linearly amplifies the signal. Up to now the signal has been in analog form. In the analog to digital converter, the analog signal is digitized to numbers that are placed in the appropriate channels of the memory unit of the multichannel analyzer. The number of X-ray counts is displayed as a spectrum in a video monitor. Hard copies from the multichannel analyzer are made photographically or with a plotter.

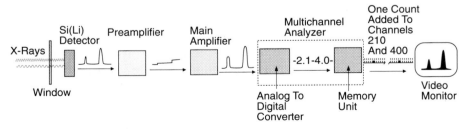

Figure 12-36   Shape of the signal at different places in an energy dispersive spectrometer.

Deadtime

The **deadtime** in the system is the difference between **real time** (actual elapsed time during an analysis run) and **live time** (the real time minus the time lost when the energy dispersive spectrometer is not recording incident X-rays). The deadtime is time lost at three locations in the energy dispersive spectrometer: (1) resetting the voltage of the field-effect transistor of the preamplifier, (2) rejection of pulses by the pulse pileup discriminator when a pulse arrives in the main amplifier before the previous pulse has been processed, and (3) in the ADC converter of the multi-channel analyzer when a second voltage pulse arrives before the previous one has been digitized.

Deadtime results when nearly coincident X-rays are rejected. However, an increase in the incident X-rays entering the detector does not necessarily result in an increase in the rate that X-rays are accepted and processed. This is because the higher the input rate of X-rays is, the greater the number of rejected pulses. As a general rule, maximum throughput of X-rays occurs when deadtime is about 60%.

Artifacts Produced by the Energy Dispersive Spectrometer

X-rays produced by translocation of orbital electrons in an atom (for example, a L orbital electron moving to a vacant K orbital) have a 2-eV range of energies when measured at half the maximum of the peak intensity (FWHM) (see section on wavelength dispersive spectrometers for a discussion of FWHM). The distribution of X-ray energies within this energy range results in a Gaussian (normal or bell-shaped) curve. Any variation in the output of the energy dispersive spectrometer from this Gaussian peak of the 2-eV energy range constitutes an artifact (modification by an individual of a natural event) of the spectrometer. The output from the energy dispersive spectrometer, in fact, has X-ray peaks that are not Gaussian and have a much greater energy range than 2 eV. These output peaks, therefore, contain a number of artifacts produced by the energy dispersive spectrometer. These artifacts are produced in three different places within the energy dispersive spectrometer: (1) in the detector, (2) in the detector environment, and (3) in the main amplifier. Some artifacts produce serious distortions of the X-ray peaks, while some are insignificant.

Artifacts produced by the detector. The artifacts produced in the detector include broadening and distortion of the X-ray peaks, the production of silicon X-ray escape peaks, and absorption discontinuities produced by silicon and gold atoms in the detector.

Broadening and Distortion of X-ray Peaks. The broadening of X-rays peaks is the most serious artifact produced by the energy dispersive spectrometer. In broadening, the width of X-ray energies of the natural X-ray peak is expanded

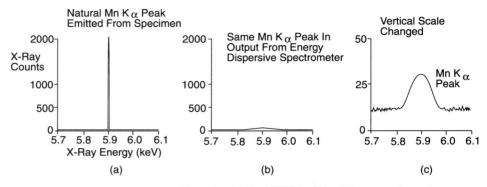

Figure 12-37 Manganese K_α peaks. (a) The FWHM width of X-ray energies emitted from manganese atoms is 2.3 eV. (b) The FWHM width of the Mn K_α peak is about 150 eV in the output from the energy dispersive spectrometer. The Mn K_α peak is only 2.5% of height of the peak in (a) because the X-rays are spread over a greater energy range. (c) Changing the vertical scale of X-ray counts in the output of the energy dispersive spectrometer results in a Mn K_α peak that is easier to see, but it also results in a magnification of the background continuum.

approximately 75 times. X-ray spectrum resolution is measured as the FWHM of Mn K_α (5.898 keV), which has an energy range of 2.3 eV (Fig. 12-37a), when generated, but which is broadened by the detector into a peak with a FWHM of about 150 eV (Fig. 12-37b). Broadening of the X-ray peaks also results in a great decrease in the height of the peak, since the X-ray energies are spread over more energy channels in the spectrum. The output Mn K_α peak from the detector is 2.5% of its natural height. This decrease in height is caused by the broadening of the peak from 2.3 eV to 150 eV. The decrease in the height of the X-ray peak results in a decrease in the peak to background ratio. This makes it more difficult to distinguish a small peak from the background noise of the continuum. For example, a natural peak of Mn K_α with a 2.3-eV energy range (FWHM) may be 2000 counts high if measured accurately by the energy dispersive spectrometer (Fig. 12-37a). However, the output of the energy dispersive spectrometer produces a peak with a FWHM of 150 eV and 30 X-ray counts (Fig. 12-37b). For the X-ray spectrum to be useful, the operator has to change the vertical scale of X-ray counts to magnify the size of the X-ray peak. In doing so, the background continuum is increased, also (Fig. 12-37c). The poorer the resolution of the system (greater than 150 eV), the greater is the difficulty in resolving small peaks from the background.

The peak broadening in the Si(Li) detector is due to the statistical nature of electron–hole pair production in the detector. On the average, 3.8 eV of the energy of an X-ray is used in the production of an electron–hole pair in the Si(Li) crystal. However, 3.8 eV is only the average energy, and there is a broad range on each side of this energy that can be consumed in the production of electron–hole pairs. Thus, two X-rays of the same energy can result in the production of a different

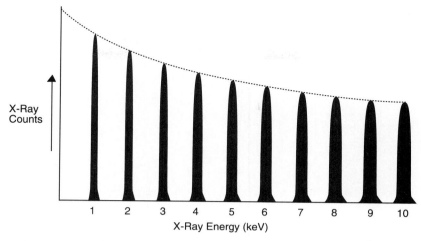

X-Ray
Counts

X-Ray Energy (keV)

Figure 12-38   The greater the energy of X-rays entering the detector, the more
the peaks are broadened by the detector. The energy widths of the X-rays peaks
are exaggerated in the illustration.

number of electron–hole pairs in the Si(Li) crystal. This results in the broadening
of the X-ray energies within a single X-ray peak. Added to this is the noise produced by fluctuations in the leakage current in the detector and thermal noise in
the field-effect transistor, which also result in peak broadening.

Peak broadening increases with increasing energy of the X-rays entering the
detector. The FWHM for the K series of low atomic number elements (and therefore low-energy X-rays), such as carbon, is around 100 eV. High atomic number
elements have FWHM around 200 eV (Fig. 12-38).

Distortion from the ideal Gaussian X-ray peak occurs when inefficiencies in
the collection of holes and electrons occur in the detector. This results in the
formation of low-energy tails off the X-ray peaks. These are, however, relatively
insignificant compared to the broadening of the peaks.

Silicon X-ray Escape Peaks.   An incident X-ray entering the Si(Li) detector
causes the ejection of an orbital electron as a photoelectron from the silicon atom.
This is followed by emission of an Auger electron or Si K_α X-ray from the silicon
atom (if a K orbital electron is ejected from the atom followed by an L orbital
electron falling to the vacant K orbital). The Auger electron travels only a fraction
of a micrometer before being reabsorbed in the crystal and releasing its energy in
the formation of electron–hole pairs. The Si K_α X-ray, however, will travel much
farther; a Si K_α X-ray has a 10% chance of traveling 30 μm in the detector. This
means that some of the Si K_α X-rays generated near the front or rear of the detector
crystal can escape without producing electron–hole pairs. The escape of the Si K_α
X-rays from the detector robs the output X-ray peaks from the specimen of an

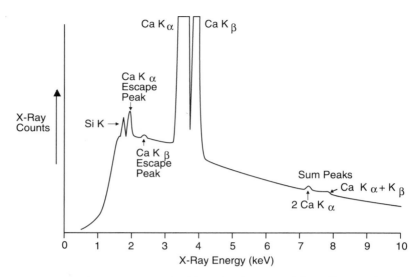

Figure 12-39 X-ray spectrum of calcium showing some of the artifacts produced by the energy dispersive spectrometer.

amount of energy equal to the energy of the X-ray minus the energy of the Si K_α X-ray. This is 1.740 keV for the Si K_α X-ray and 1.832 keV for the Si K_β X-ray. An artifact peak, called the **silicon escape peak**, is created at an energy equal to the energy of the X-ray entering into the detector, minus the energy of the silicon X-ray. Both Si K_α and Si K_β escape peaks are produced. However, the Si K_β is only 2% of the Si K_α, so only one Si K escape peak is observed in the spectrum. Figure 12-39 is an X-ray spectrum of calcium showing the Ca K_α (3.692 eV) and Ca K_β (4.012 keV) peaks. The silicon escape peaks occur at these energies minus the energy of the Si K_α X-ray. Thus the silicon escape peak for Ca K_α is at 1.952 keV(3.692 − 1.740), and for the Ca K_β it is at 2.272 keV (4.012 − 1.740).

The probability that a silicon K X-ray will escape depends on how deeply the silicon K X-rays are generated in the detector crystal. The deeper the generation of the silicon X-rays, the less likely the silicon K X-rays are apt to escape from the front of the detector crystal and the smaller is the silicon escape peak. The angle and energy of the incident X-ray entering the detector determines the height of the silicon escape peak. Those incident X-rays that enter perpendicular to the detector surface will penetrate deepest and have the least probability of escaping the detector. Conversely, the X-rays that enter the detector at a small angle, produce Si K X-rays near the front of the detector, resulting in the escape of more Si K X-rays from the detector. The greater the energy of the X-rays entering the detector, the deeper are the Si K X-rays generated in the detector and the fewer Si K X-rays that escape from the detector to contribute to the silicon escape peaks. Figure 12-40 shows the escape probability of a Si K_α X-ray as a function of the energy of

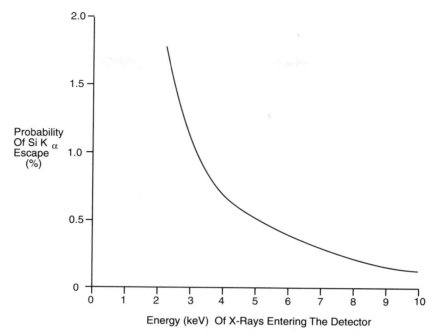

Figure 12-40 The energy of X-rays entering a Si(Li) detector determines the probability of escape of Si K_α X-rays from the detector.

the X-ray incident on the detector. Certain generalizations can be made about silicon escape peaks. First, incident X-rays with an energy less than the critical excitation energy of Si K X-rays (1.840 keV) cannot generate silicon X-rays and therefore do not have silicon escape peaks. Thus, no silicon escape peaks are generated from elements below atomic number 15. Second, low atomic number elements produce relatively large silicon escape peaks (about 1.5% to 2% of the parent peak) if the incident X-rays are above the critical excitation energy. Third, metallurgical specimens are usually composed of high atomic number elements that produce high-energy X-rays that have very small escape peaks (about 0.1% to 0.2% of the parent peak).

Silicon Peaks. A very small silicon K peak usually occurs in the X-ray spectrum (Figs. 12-39 and 12-42). The silicon peak is mostly due to the interaction of incident X-rays and electrons passing through the window, with silicon atoms in the dead layer (20 to 200 nm thick) of the detector crystal. Lithium has not been drifted into this dead layer, so electron–hole pairs are not produced here. Interaction of the incident X-ray or electron with a silicon atom in the dead layer can result in the emission of a silicon X-ray. Electron–hole pairs can be produced if the silicon X-ray passes into the intrinsic zone of the detector crystal, resulting in a charge pulse that is registered as a Si K X-ray. Another source of X-rays that contributes to this silicon X-ray peak is the silicon-based vacuum greases, oils, and sealants that are used to maintain the vacuum of the scanning electron microscope. The window of the detector is a couple of degrees cooler than the rest of the specimen chamber, so contaminant gases condense on it. Any such silicon compounds on the window result in the formation of Si K X-rays when they are struck by incoming electrons and X-rays.

Silicon and Gold Absorption Edges.   The Si(Li) detector crystal has a front surface electrode of gold about 20 nm thick and an inactive dead layer of silicon through which the incident X-rays have to travel to reach the active intrinsic zone of the crystal. The gold and silicon layers absorb X-rays if the X-rays have enough energy to eject electrons from the atoms. The **critical energy** or **critical excitation energy** is the minimum energy necessary to overcome the binding energy of the atomic nucleus and to remove an orbital electron (see Appendixes II and III for critical excitation energies of the elements). As the energy of the incident X-ray increases, the probability of ejecting an orbital electron from a gold or silicon atom decreases until the X-ray energy equals the critical energy, at which time the probability of ejecting an orbital electron increases greatly. This discontinuity in the probability is called an **absorption edge** (Fig. 12-41) (see Appendix VI).

The output spectrum from an energy dispersive spectrometer contains discontinuities in the spectrum just above the absorption edge of the silicon K line (1.849 eV) and, to a lesser extent, the gold M line (2.223 keV) (Fig. 12-42). At

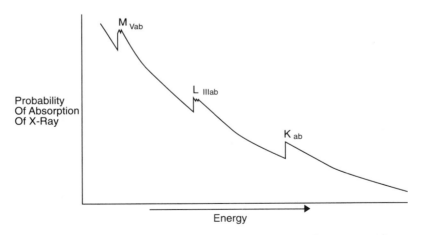

Figure 12-41   There is an increase in the absorption of X-rays by an atom at those energies required to eject orbital electrons from the atom. These discontinuities are the absorption edges for that particular atom.

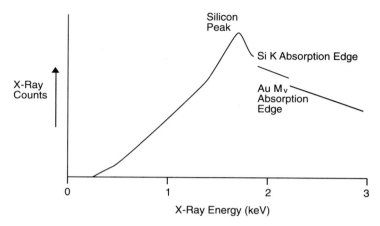

Figure 12-42   X-ray spectrum showing a silicon peak and discontinuities in the spectrum due to the absorption of X-rays at the Si K absorption edge and the Au M_v absorption edge.

these energies there is a drop in the measured X-ray counts. The discontinuities are due to a sharp increase in the absorption of X-ray energies at the absorption edges of the silicon in the dead layer and the gold in the contact layer. X-rays absorbed in these layers do not produce electron–hole pairs in the detector, so no contribution to the output signal from the detector is made for the X-ray energies, resulting in a discontinuity in the output spectrum. The absorption of X-rays by the silicon dead layer and the gold contact layer is relatively insignificant in the spectrum because these layers are thin.

Artifacts of the detector environment.   Artifacts produced by the detector environment include the absorption of low-energy X-rays by the window and background noise created by microphony.

Absorption of Low-energy X-rays by the Detector Window.   Low-energy X-rays that are absorbed by the detector window do not reach the detector, resulting in an output X-ray spectrum that slopes downward from the low energies toward zero (Fig. 12-39). The amount of loss depends on the type of window and its thickness. A 7.6-μm-thick beryllium window absorbs almost all of the X-rays beneath 600 eV, while transmitting virtually all the X-rays above 2 keV. Between these X-ray energies, the absorption of X-rays increases with decreasing X-ray energy. X-rays with an energy of 1.5 keV are about 70% transmitted, while at 1.0 keV about 45% of the X-rays are transmitted. Windows made of thin organic polymers transmit more lower-energy X-rays, resulting in less tailing toward zero of the spectrum (Fig. 12-43).

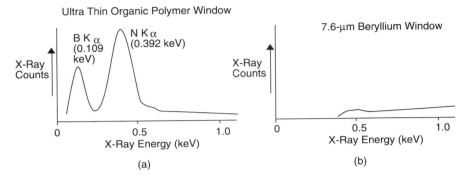

Figure 12-43   An energy dispersive spectrometer that has an ultrathin organic polymer window over the detector will transmit a large number of low-energy X-rays. (a) Spectrum of a specimen containing boron and nitrogen produced from a detector containing an ultrathin window. (b) X-ray spectrum from the same sample made with a 7.6-μm beryllium window over the detector. The beryllium window absorbs the low-energy X-rays.

Microphony.   Microphony is the transformation of sound waves into electrical currents. The detector and electronic circuitry of the energy dispersive spectrometer are very sensitive to vibrations that cause microphony. Vibrations due to motors or even conversation can produce added noise in the X-ray spectrum. Accumulation of ice at the bottom of the liquid nitrogen Dewar of the cryostat causes vibrations as the ice "dances" during boiling of the liquid nitrogen. Ice at the bottom of the liquid nitrogen Dewar can also isolate the cold finger from the liquid nitrogen, resulting in elevated temperature of the detector, possibly above that of the operating range of the detector.

Artifacts produced by the main amplifier (pulse processor).   The only significant artifact produced by the main amplifier is **sum peaks**. The pulse pileup rejector of the main amplifier recognizes an X-ray pulse that arrives before the previous X-ray pulse has been processed. In such a case, the pulse-pileup rejector rejects both pulses or only the second pulse. The pulse pileup rejector is not perfect and sometimes does not reject nearly coincident X-ray pulses. This occurs particularly at high count rates (above 3000 counts per second). This results in two X-ray pulses being counted as one X-ray, with an energy equal to almost the sum of the two incoming X-rays. The resulting output spectrum from the energy dispersive spectrometer has sum peaks that are nearly equal in energy to two valid peaks in the spectrum. Figure 12-39 shows a spectrum of calcium. The spectrum has two small sum peaks, one nearly equal to the energy of two Ca K_α X-rays and a second smaller peak nearly equal in energy to one Ca K_α plus one Ca K_β peak. Theoretically, there is also a sum peak nearly equal to the energy of two Ca K_β X-rays. However, this peak is so small that it cannot be seen in the background

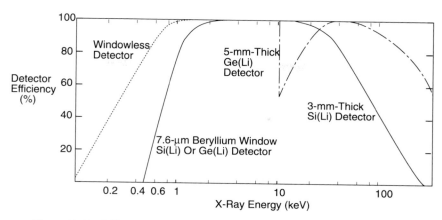

Figure 12-44   Efficiency of a Si(Li) detector and a Ge(Li) detector at different X-ray energies. The efficiency of both detectors at low X-ray energies is determined by the window over the detector. A Ge(Li) detector is more efficient at higher X-ray energies, although it has an absorption edge aberration from 11.1 to about 30 keV.

of the spectrum. As can be seen in Fig. 12-39, all the sum peaks are relatively insignificant.

Detector Efficiency

The efficiency of the detector refers to the percentage of different X-ray energies striking the window that are converted into current pulses by the detector. A Si(Li) detector is nearly 100% efficient from 2 to 20 keV (Fig. 12-41). Below 2 keV, the incident X-rays from the specimen can be absorbed by the window over the detector, the gold contact layer, and the silicon dead layer. The lower the energy of the X-rays, the more the X-rays are absorbed by the window. A windowless detector results in the greatest detection efficiency for low-energy rays, followed by thin organic polymer windows and beryllium windows. Above X-ray energies of 20 keV, the efficiency of the detector falls off due to the passage of X-rays completely through the intrinsic layer without production of electron–hole pairs.

A Ge(Li) detector has greater efficiency in the detection of X-rays above 20 keV (Fig. 12-44). This is because the higher atomic number of germanium ($Z = 32$, compared to $Z = 14$ for silicon) makes it a denser detector that more efficiently absorbs higher-energy X-rays. A disadvantage of a Ge(Li) detector is the lowered detection efficiency between the Ge K absorption edge of 11.1 and 30 kV. Ge(Li) detectors are usually used to analyze energies above 40 keV. Ge(Li) detectors are not commonly used on scanning electron microscopes since these instruments rarely use accelerating voltages above 30 keV and therefore cannot produce high-energy X-rays.

Advantages of Energy Dispersive Spectroscopy

The advantages of using an energy dispersive spectrometer, instead of a wavelength dispersive spectrometer, include (1) acquisition of the whole X-ray energy spectrum at one time, (2) very quick analysis, (3) greater collection of X-rays from the specimen, and (4) the possibility of using a smaller electron beam diameter to acquire the X-ray spectrum.

Acquisition of Complete X-ray Spectrum at One Time.   In energy dispersive spectroscopy, the complete X-ray spectrum that the investigator is interested in can be acquired at one time. To set the energy range, the investigator sets the amount of X-ray energy in electron volts that is placed in each of the 1024 channels. For example, each channel receives 5 eV if the investigator is interested in the energy range from 0 to 5.120 keV. Acquisition of the entire range is begun when the investigator begins acquiring the X-rays. In comparison, the wavelength dispersive spectrometer will only accept a very narrow range of X-rays at one time.

Speed of Analysis.   An energy dispersive spectrometer acquires a complete spectrum of X-rays much faster because it accepts a complete X-ray spectrum at one time. A wavelength dispersive spectrometer has to mechanically scan a few electron volts of the spectrum one at a time, requiring some time to acquire a complete spectrum of X-ray energies.

Greater Collection of X-rays from the Specimen.   The detector in an energy dispersive spectrometer can be moved very close to the specimen. This results in a large solid angle of collection of X-rays (0.01 to 0.1 sr), resulting in more of the X-rays collected and high count rates. A wavelength dispersive spectrometer has fewer X-rays collected by the detector and a smaller solid angle of collection (0.001 sr).

Possibility of Using a Small Probe (Electron Beam) Diameter to Produce X-rays in a Specimen.   Electron beam diameters as small as 5 nm provide enough high-energy electrons to generate sufficient X-rays for measurement by an energy dispersive spectrometer. The large number of X-rays collected by an energy dispersive spectrometer result in the need to generate less X-rays in the specimen. Small electron beam diameters result in slightly better spatial resolution in bulk specimens since the escape volume of the X-rays is slightly smaller with smaller electron beam diameters. The decrease in X-ray escape volume with decreasing electron beam diameter is much more dramatic in thin sections and foils, where electron beam scattering (producing X-rays) is very restricted.

Disadvantages of Energy Dispersive Spectroscopy

The disadvantages of energy dispersive spectroscopy include decreased spectral resolution, artifacts produced in the detection process, and lower maximum count rates when compared to wavelength dispersive spectroscopy.

Decreased Resolution.   Resolution of X-ray peaks (FWHM) in an energy dispersive spectrometer varies from about 100 to 200 eV depending on the energy of the X-rays. A wavelength dispersive spectrometer has a resolution in the 5- to 10-eV range. The better resolution of a wavelength dispersive spectrometer results in better peak to background ratios, little overlap of X-ray peaks that are close in energy, and better minimum detectable limits of elements. The decrease of X-ray resolution is the most serious disadvantage of the energy dispersive spectrometer.

Artifacts in the Detection Process.   An energy dispersive spectrometer produces a number of additional peaks, such as escape peaks, sum peaks, and silicon peaks, as well as gold and silicon discontinuities in the spectrum. The true X-ray peaks are also distorted. With the exception of escape peaks, these aberrations do not occur in a wavelength dispersive spectrometer.

Lower Maximum Count Rate.   An energy dispersive spectrometer has an optimal count rate of 2000 to 3000 counts per second. Above this, the time resets involved in the counting process add to excessive deadtime and the process becomes inefficient. A wavelength dispersive spectrometer, however, can have count rates of 100,000 counts per second without loss in energy resolution.

Operation of the Scanning Electron Microscope during Energy Dispersive Spectroscopy

X-ray microanalysis using scanning electron microscopes is almost always performed with energy dispersive spectrometers. Maximum utilization of the energy dispersive spectrometer requires the correct setting of some parameters of the scanning electron microscope. The most important parameters are the accelerating voltage, the diameter of the electron beam, and the geometrical relationship of the specimen to the X-ray detector.

Accelerating voltage.   The accelerating voltage of the electron beam of the scanning electron microscope has to be greater than the critical excitation voltage of the X-ray lines that the investigator is interested in. The most efficient generation of an X-ray line (K, L, or M) occurs at accelerating voltages of 2.5 to 3 times the voltage of the X-ray line. For example, Fig. 12-45 shows X-ray spectra of pure copper taken at accelerating voltages of 10 and 20kV. At 10 kV, only the Cu L lines are efficiently excited. At 20 kV, both the Cu K and L lines are efficiently excited as can be seen in the relative heights of the Cu K lines in the two spectra. Note that the background continuum is also higher in the spectrum made with an accelerating voltage of 20 kV.

Usually, X-ray microanalysis involves looking at the K lines of lower atomic number elements and the L or M lines of higher atomic number elements. All these lines are in the 0- to 10-keV energy range, and this is the X-ray energy range that is usually used by investigators. However, sometimes the X-ray lines of elements in a specimen are close together, resulting in overlap of the X-ray peaks, making

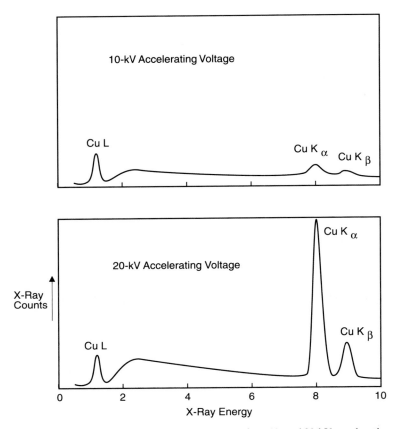

Figure 12-45 X-ray spectra of a pure copper sample at 10- and 20-kV accelerating voltage. At 10 kV, only the Cu L is optimally excited, while at 20 kV both the Cu L and Cu K lines are optimally excited.

it difficult to distinguish the peaks of each element. For example, sulfur K lines overlap molybdenum L lines very closely at 2.3 keV. In this case, it is advantageous to look for molybdenum K lines just above 17 keV, where there is no interference from sulfur. However, almost all X-ray microanalysis can be performed by looking at lines in the 0- to 10-keV range. All the X-rays in this range can be excited by the accelerating voltages available in scanning electron microscopes (20 to 30 kV).

Utilizing the scanning electron microscope at higher accelerating voltages than necessary to excite applicable X-ray lines has two disadvantages. The first is that high accelerating voltages result in deeper penetration of the electron beam into the specimen. This results in greater electron scattering and a larger X-ray escape volume in the specimen. The second disadvantage is that higher accelerating voltages result in a greater level of background continuum without an increase in characteristic X-rays peaks (Fig. 12-45). This makes it more difficult to distinguish small X-ray peaks from the background continuum.

Thus, X-ray microanalysis should be performed at an accelerating voltage 2.5 to 3 times that of the highest-energy X-ray line the investigator is interested in. For example, if a specimen contains Al (K_α = 1.486 keV), Ca (K_α = 3.312 keV), and Ag (L_α = 2.984 keV), the optimal accelerating voltage would be 2.5 to 3 times the highest energy line (Ca K_α = 3.312 keV), or about 8 to 10 kV.

Electron beam diameter. The larger the diameter of the electron beam, the greater is the number of electrons in the electron beam and the more X-rays that are generated from the specimen. Low atomic number specimens (such as biological specimens) generate relatively few X-rays per incident electron, so it is often necessary to increase the diameter (spot size) of the electron beam to produce the necessary X-rays. Larger beam spot sizes, however, result in larger X-ray escape volumes in the specimen and poorer spatial resolution.

Geometrical relationship of the specimen and detector. The geometrical relationship between the specimen and the detector will determine the number and energy of the X-rays that are received by the detector. The operator can change the working distance and tilt of the specimen and, usually, the distance between the detector and specimen. Changing these parameters affects the line of sight between the specimen and the detector, the solid angle of collection, and the take-off angle of the detector, which in turn affects the number and energy of X-rays collected by the detector.

Line of Sight. The operator should remember that the X-rays are emanating from the portion of the specimen that is being illuminated by the electron beam. It is possible to illuminate the specimen with the electron beam by using an area scan, a line scan, or a stationary electron spot. Whatever the mode of illumination, the X-rays travel in a line of sight to the detector. If a raised portion of the specimen is between the area being illuminated and the detector, the X-rays will not reach the detector (Fig. 12-46). If a part of the X-rays directed toward the detector is obstructed, then an X-ray shadow will occur, and only those X-rays that reach the detector will be counted.

Solid Angle of Collection. X-rays generated in the specimen and collected by the circular, flat surface of the detector define a cone-shaped volume (Fig. 12-47). The three-dimensional angle created by this cone is the solid angle of collection of X-rays (often noted as omega [Ω]) and is measured in steradians (see Appendix I for an explanation of a steradian). The number of X-rays striking the detector increases as the solid angle of collection increases. Usually, the detector of an energy dispersive spectrometer can be moved close to the specimen during X-ray microanalysis by the operator (Fig. 12-47). When the energy dispersive spectrome-

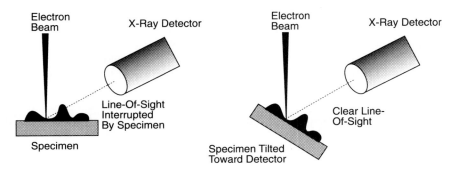

Figure 12-46   X-ray detection is a line of sight process. No X-rays will be detected
if a portion of the specimen is between the detector and the area where the X-rays
are generated. Tilting the specimen toward the detector often helps reduce the
chance that X-rays will be intercepted by parts of the specimen.

ter is not being used, it is backed away from the specimen to ensure that it is not
damaged. The number of X-rays received by the detector increases proportionally
as the inverse square of the distance from the specimen to the detector. It is
therefore advantageous to increase the solid angle of collection of X-rays during
microanalysis by moving the detector as close as possible to the specimen. At the
same time, the operator must be careful not to move the detector so close that the
detector window is damaged.

   Take-off Angle.   The take-off angle (commonly written as psi [Ψ]) is the
angle between the specimen surface and the line of sight to the center of the
detector (Fig. 12-48). The take-off angle determines the absorption of X-rays by
the specimen. The smaller the take-off angle, the longer is the path the X-rays

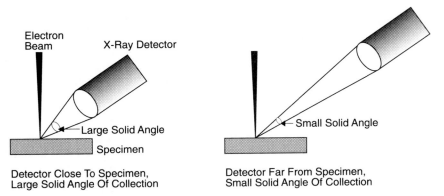

Figure 12-47   A large solid angle of collection can be created by moving the X-
ray detector close to the specimen. Increasing the distance between the specimen
and the detector results in a smaller solid angle of collection.

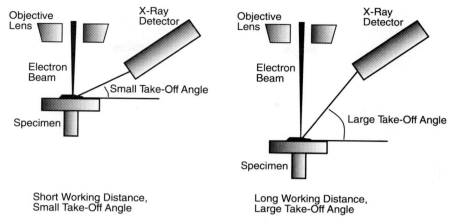

Short Working Distance,
Small Take-Off Angle

Long Working Distance,
Large Take-Off Angle

Figure 12-48 A short working distance results in a smaller take-off angle, while a long working distance produces a larger take-off angle.

have to travel to escape to the surface of the specimen and the fewer X-rays that escape from the specimen (Fig. 12-49). The energies of the resulting X-ray spectrum are shifted since more of the lower-energy X-rays are absorbed by the specimen.

Minimal absorption of X-rays in the specimen occurs when the detector is 90° to the specimen surface. The detector cannot, of course, be placed at 90° because this is the position of the electron beam. In actual fact, take-off angles greater than 30° are usually sufficient to ensure adequate collection of X-rays. The

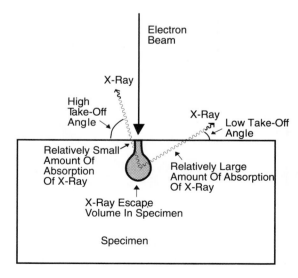

Figure 12-49 A low take-off angle results in X-rays traveling a greater distance in the specimen, with a greater chance that the X-rays are absorbed. A high take-off angle results in the X-rays traveling a shorter distance through the specimen, with a lesser chance of absorption.

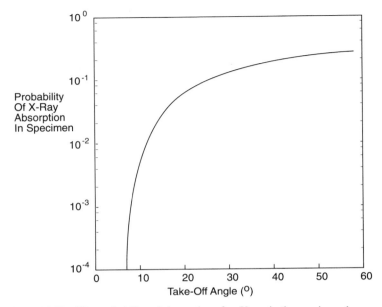

Figure 12-50   The probability of absorption of an X-ray in the specimen decreases as the take-off angle increases.

number of X-rays that escape from the specimen surface increases greatly between take-off angles of 10° to 20° and then plateaus at 30° (Fig. 12-50). The take-off angle can be increased in two ways by the operator. The first method is to tilt the specimen toward the detector to minimize the distance escaping X-rays have to travel through the specimen to reach the detector (Fig. 12-51). A second method is to increase the working distance (Fig. 12-48). However, increasing the working distance also decreases the solid angle of collection, which decreases the number of X-rays collected.

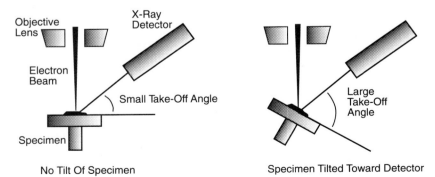

Figure 12-51   Tilting the specimen toward the detector results in a larger take-off angle.

QUALITATIVE ANALYSIS

In qualitative analysis, the elements present in the specimen are identified. Qualitative analysis is based on Moseley's law:

$$E = C_1(Z - C_2)$$

where E = the emission line for a given X-ray series
(for example, the K series)

Z = atomic number of the element

C_1 and C_2 are constants

Since C_1 and C_2 are constants, Moseley's law states that X-ray energies are proportional to atomic number. Moseley's law therefore enables the determination of the atomic number of the elements present in the specimen if the investigator knows the energy of the K, L, or M X-rays given off from the specimen. The determination of the energy of the X-rays basically involves the identification of the peaks present in the X-ray spectrum. The energy of an X-ray series (K, L, or M) increases progressively as the atomic number of the elements increases (Fig. 12-52; see Appendixes III through VI). This fact somewhat simplifies the identification of elements. In modern X-ray microanalytical units, three methods are available for identifying X-ray peaks in the spectrum.

Software Identification of X-ray Peaks. In this first method the microcomputer of the X-ray microanalytical unit uses a software routine to identify the X-ray peaks. The operator types in the correct command, and the microcomputer

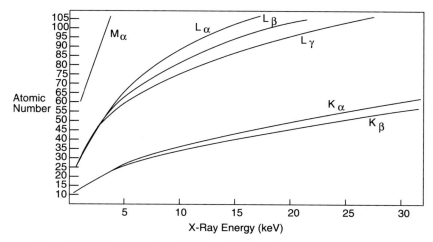

Figure 12-52 X-ray energies of K, L, and M X-ray lines from the elements in the atomic chart.

compares the energy of the X-ray peaks in the spectrum with known X-ray peaks for each element in the memory of the microcomputer. At the same time, the microcomputer checks for inconsistencies in the X-ray peaks, such as the relative height of the K_α peak to the K_β peak, which might indicate another element buried in one of the K peaks. The microcomputer displays the elements that have the closest X-ray energy fits. Unfortunately, many K X-ray peaks are similar in energy to the L or M X-ray peaks of other elements, making it very difficult for the microcomputer to narrow each peak to less than two or three elements. For example, an X-ray energy peak of 2.30 keV could be from sulfur (K_α = 2.308 keV), molybdenum (L_α = 2.293 keV), or lead (M_α = 2.346 keV). Fortunately, the operator usually has a reasonable idea of the elements present in the specimen and instructs the computer to ignore certain elements in the atomic chart during its search.

Operator Identifies the X-ray Peaks Using Internal Standards. In this method the operator displays internal X-ray energy standards (stored in the memory of the microcomputer) on the X-ray spectrum displayed on the video monitor. The operator overlays the standard X-ray lines on the X-ray spectrum, one element at a time, and chooses the best fit with the X-ray peaks in the spectrum. The K, L, and M lines for each element are displayed at one time. This usually enables the operator to narrow each X-ray peak to a couple of possible elements. The operators knowledge of the elements present in the specimen usually allows the selection of the correct element for the X-ray peak.

Operator Uses a Table or Slide Rule of X-ray Energies. In this method the operator calls a cursor (vertical line) onto the X-ray spectrum displayed on the video monitor. The cursor indicates the position of a certain X-ray energy in the X-ray spectrum. The operator places the cursor in the center of the X-ray peak of interest. The operator reads the X-ray energy position of the cursor and refers to an X-ray energy table (see Appendixes III through VI) or slide rule to find an element that produces X-rays of equivalent energy. The slide rule is a device that displays an element plus its X-ray lines.

Calibration of the X-ray Microanalytical Unit

Identification of elements in a spectrum using the X-ray microanalytical unit presupposes that the displayed X-ray energies are correctly calibrated. X-ray microanalytical units usually have a software routine that calibrates the unit by accepting low and high-energy X-ray peaks and identifying the two X-ray peaks with specific X-ray energies. Calibrating low- and high-energy X-ray peaks results in calibration of the whole spectrum. This is commonly done with the Al K_α peak (1.487 keV) and the Cu K_α peak (8.048 keV) (Fig. 12-53). An aluminum–copper specimen for this purpose is easily prepared by placing a piece of copper tape on an aluminum specimen support.

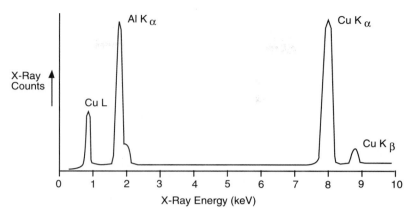

Figure 12-53 X-ray spectrum of a specimen that consists of a piece of copper tape on an aluminum specimen support. This is the type of specimen that is commonly used to calibrate an X-ray microanalytical unit. The Al K_α X-ray is used to calibrate the low end of the spectrum, while the Cu K_α is used to calibrate the high end of the spectrum.

Distinguishing X-ray Peaks

Qualitative microanalysis is not difficult once the X-ray peaks have been distinguished as valid entities. There are, however, three basic problems in distinguishing X-ray peaks: (1) elements present in small amounts produce X-ray peaks that are difficult to resolve from the background continuum, (2) the extraneous silicon peaks, escape peaks, and sum peaks complicate the identification of peaks, and (3) X-ray peaks that are close in energy are difficult to distinguish from each other.

Clarification of Small Peaks from the Background Continuum. Elements present in small amounts produce small peaks in the spectrum that are difficult to distinguish from the background continuum. The smallest peak that can be resolved from the background is called the **minimum detectable level** of the element. Minimum detectable level is essentially a statistical term, and there are different certainty degrees for the minimum detectable level. Basically, the minimum detectable level of a peak becomes better as more X-ray counts are used to produce the spectrum. The higher the X-ray counts used to produce the spectrum, the smoother is the background and the easier it is to distinguish a small X-ray peak. Statistically, there is nearly a 98% certainty that a peak is real if the height of the peak exceeds the height of the background by $2\sqrt{\text{background}}$. If the X-ray counts in a peak exceed the background by $3\sqrt{\text{background}}$, there is nearly a 99.9% certainty that the peak exists. Figure 12-54 shows a small X-ray peak in a spectrum that was acquired for 0.1, 0.5, and 2.5 min. The longer the acquisition, the smoother is the background and the better the minimum detectable level, as can be seen from the following three examples based on Fig. 12-54.

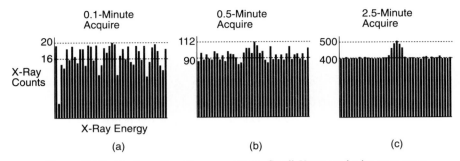

Figure 12-54 Portion of an X-ray spectrum. Small X-ray peaks become more statistically certain as the number of counts used to produce the spectrum increases.

Example 1 (Fig. 12-54a)   0.1-min count time

Average background count = 16 counts
Peak height = 20 counts

$$\text{MDL} = \text{background} + 2\sqrt{\text{background}} \text{ for 98\% certainty}$$
$$= 16 + 2\sqrt{16}$$
$$= 16 + 2(4)$$
$$= 24$$

Therefore, the X-ray peak has to have 24 counts to fulfill the minimum detectable level of a valid peak at 98% probability. The peak in Fig. 12-54a is only 20 counts and therefore cannot be counted as a real peak, because it is beneath the minimum detectable level at a certainty of 98%.

Example 2 (Fig. 12-54b)   0.5-min count time

Average background count = 90 counts
Peak height = 112 counts

$$\text{MDL} = \text{background} + 2\sqrt{\text{background}} \text{ for 98\% certainty}$$
$$= 90 + 2\sqrt{90}$$
$$= 90 + 2(9.49)$$
$$= 109.98$$

Therefore, the X-ray peak has to have 109.98 counts to fulfill the minimum detectable level of a valid peak at 98% certainty. The peak in Fig. 12-54b is 112 counts and can be considered a valid peak with a 98% certainty.

Example 3 (Fig. 12-54c)   2.5-min count time

Average background count = 400 counts
Peak height = 500 counts

$$\text{MDL} = \text{background} + 3\sqrt{\text{background}} \text{ for } 99.9\% \text{ certainty}$$

$$= 400 + 3\sqrt{400}$$

$$= 400 + 3(20)$$

$$= 460$$

Therefore, the X-ray peak has to have 460 counts to fulfill the minimum detectable limit of a valid peak at 99.9% certainty. The peak in Fig. 12-54c is 500 counts and can be considered a valid peak at a certainty of 99.9%.

Thus, the minimum detectable level is a function of acquired X-ray counts, which increase with acquisition time. Obviously, there is a practical limit on the amount of time spent in acquiring an X-ray spectrum. However, minimum detectable levels as low as 0.01% of the total number of atoms present in the specimen are feasible under conditions where there is no interference from another peak. The **mass limit**, or the absolute amount of an element that can be detected, can be as low as 10^{-15} g.

Artifactual Peaks.   X-ray peaks that are artifacts of the energy dispersive spectrometer have already been discussed. These peaks include silicon escape peaks, silicon peaks, and sum peaks. An operator has to be careful if an X-ray peak is present at the X-ray energies of any of these peaks.

Modern X-ray microanalytical units have a software program that calculates the *silicon escape peak* of each element and removes the peak from the spectrum. If a software program is not available, the operator should closely examine small peaks that occur at the energy of a major peak minus 1.74 keV, as the peak may be a silicon escape peak.

A small *silicon K peak* will occur in every spectrum acquired with a Si(Li) detector. To determine if silicon is truly present in the specimen, the operator needs to acquire a second blank spectrum and compare the two spectra.

Sum peaks occur at energies nearly twice the energy of major peaks in the X-ray spectrum. Any small peaks in the spectrum at these X-ray energies are probably sum peaks.

Spectral Resolution.   The spectral resolution of an energy dispersive spectrometer is around 150 eV at the 5.89-keV Mn K_α X-ray line. Lower X-ray energies have better resolution while higher X-ray energies have poorer resolution. The spectral resolution determines the ability of the operator to distinguish two X-ray peaks that are very close in energy. If two X-ray peaks are less than 50 eV apart, it is virtually impossible to distinguish them as two separate peaks. For example, the Pb M_α is 2.346 keV, while the S K_α is 2.308 keV. The two peaks are separated by only 38 eV and cannot be distinguished using energy dispersive spectroscopy. Therefore, an investigator looking for sulfur in biological sections stained with lead would never be able to see a small sulfur peak.

Further difficulties arise when a minor constituent of the specimen produces an X-ray peak that is close in energy to the X-ray peak of a major constituent. In such a case, it may not be possible to resolve the peak of the minor constituent, even if it is separated by as much as 200 eV from the major peak. However, such a minor element can be detected by one of two methods. (1) If the element is of a high atomic number, it will produce K, L, and possibly M lines. It is probable that only one of the X-ray series lines (K, L, or M) is buried in another peak, and the presence of the element can be deduced from the other lines. (2) The relative heights of X-ray peaks within a series are known, and any deviation from these knowns probably indicates the presence of another peak buried within the second peak. In the K series the ratio is K_α (1.0) to K_β (0.1), In the L series the ratio is L_α (1.0), L_{β_1} (0.7), L_{β_2} (0.2), while in the M series it is M_α (1.0), M_β (0.6), M_γ (0.05). Figure 12-55 shows an X-ray spectrum for a sample containing chlorine and palladium. An initial X-ray spectrum at 0 to 10 keV shows only the Cl K_α (2.621 keV)

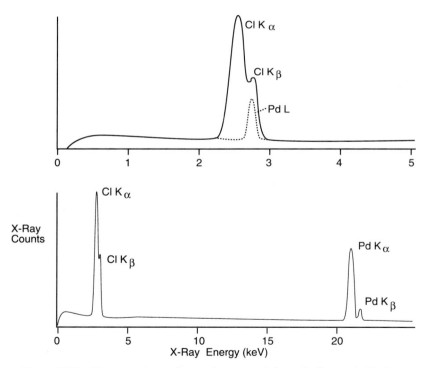

Figure 12-55 X-ray spectrum of a specimen containing palladium and chlorine. When the X-ray spectrum is acquired at a low accelerating voltage (for example, 15 kV), the only palladium peak is the Pd L peak, which is buried in the Cl K_β peak. This makes it very difficult to determine if palladium is present. However, acquiring the spectrum at 30 kV (lower spectrum) excites the Pd K peaks, which have no interference from chlorine X-ray peaks. The higher accelerating voltage allows the operator to clearly see that palladium is present.

TABLE 12-2 ELEMENTS THAT ARE USED AS BUFFERS, FIXATIVES, COATINGS, AND MOUNTS IN BIOLOGICAL SPECIMENS PRODUCE X-RAY LINES THAT INTERFERE WITH X-RAY LINES OF OTHER ELEMENTS

Element associated with specimen	Interfering X-ray line	Interferes with:	Interfered X-ray line
Os (fixative)	M	Al, P, S, Cl,	K_α
		Sr	L_α
As (buffer)	L	Na, Mg, Al	K_α
Au (coating)	M	P, S,	
		Zr, Mo	L_α
Pd (coating)	L	Cl, Ar	K_α
Ag (mounting)	L	Ar	K_α
Cu (mounting)	L	Na	K_α
Al (mounting)	K	Br, Kr	L_α

Adapted from G. I. Goldstein and others, *Scanning Electron Microscopy and X-ray Microanalysis*, Plenum Press, New York, 1981.

and Cl K_β (2.815 keV). However, instead of the usual K_α to K_β ratio of 10 : 1, the Cl K_α to Cl K_β in this figure is 2 : 1, indicating that another element is present producing an X-ray of approximately the same energy as the Cl K_β. A second X-ray spectrum at 0 to 40 keV shows the presence of a Pd K_α (21.121 keV) and Pd K_β (23.807 keV), indicating that the Cl K_β peak also contains the Pd L peaks (L_α = 2.838 keV, L_β = 2.990 keV). Thus, looking for additional series lines and checking relative peak heights in a series can often enable the investigator to detect elements that are not initially obvious in the spectrum.

Elements that are commonly present in biological and material science specimens with X-ray lines similar in energy are presented in Tables 12-2 and 12-3.

TABLE 12-3 ELEMENTS THAT OCCUR IN MATERIAL SCIENCE SPECIMENS AND THEIR X-RAY LINES THAT INTERFERE WITH OTHER X-RAY LINES OF OTHER ELEMENTS

Element	Interfering X-ray line	Interferes with:	Interfered X-ray line
Ti	K_β	V	K_α
V	K_β	Cr	K_α
Cr	K_β	Mn	K_α
Mn	K_β	Fe	K_α
Fe	K_β	Co	K_α
Pb	M_α	S	K_α
		Mo	L_α
Si	K_α	Ta	M_α
Ba	L_α	Ti	K_α

Adapted from G. I. Goldstein and others, *Scanning Electron Microscopy and X-ray Microanalysis*, Plenum Press, New York, 1981.

X-ray Mapping

In X-ray mapping, the output from the X-ray detector is used to produce an X-ray image on the cathode ray tube of the scanning electron microscope. The image is produced basically the same way as the secondary electron and backscattered electron image, except that, instead of the signal coming from the Everhart–Thornley detector, the signal now comes from the X-ray detector. Like the signal from the Everhart–Thornley detector, the signal from the X-ray detector can be displayed as either a line scan or an area scan (X-ray dot map). X-ray mapping allows the operator to spatially locate the position of elements in the specimen quickly.

Line Scan.   In a line scan, one of the 1000 lines on the cathode ray tube is continually scanned. The height of the line scan is controlled by the signal from the X-ray detector. In this Y-modulated mode, the greater the signal from the detector, the higher are the peaks of the line scan. All the X-ray energies striking the detector can be used to produce the Y-modulated line scan. Usually, however, the investigator is interested in the X-ray energies representing certain elements. In this case the investigator selects the X-ray energies that cover the major peak of the element in question. This is usually done by first making an area scan of a relatively large portion of the specimen to obtain an X-ray spectrum indicating the elements that are present. A window is painted over the X-ray energies that cover the major peak of the element in question (Fig. 12-56). In creating a window over the X-ray energies of the major peak, the operator is instructing the X-ray unit to send only those X-rays within that energy range to the cathode ray tube to produce the line scan. The peaks in the line scan represent the position of the particular element in the specimen. For example, Fig. 12-56a shows an X-ray spectrum from an integrated circuit that contains silicon and aluminum. A line scan, showing the concentration of silicon, is produced by first painting a window over the Si K peak (Fig. 12-56b). The cathode ray tube now only accepts X-rays in that energy range to modulate the line scan in the vertical direction. The photographic visualization of the location of silicon in the line scan over the integrated circuit requires three exposures of the same film. First, an area scan using the signal from the Everhart–Thornley detector is used to produce a secondary electron and backscattered electron image of the specimen on the film. Next, the cathode ray tube is programmed to produce a line scan over the specimen, and the line scan is recorded on the film as the second exposure. In the third exposure the cathode ray tube is switched to receive the energies in the Si K X-ray range from the X-ray detector, which modulates the intensity of the X-ray signal in the vertical direction. The third exposure of the film is made. These multiple exposures produce the photograph in Fig. 12-56c, which shows the relative portion of silicon in a line scan over the specimen.

Area Maps (X-ray Dot Maps).   An X-ray dot map shows the operator where one or more elements occur in a specimen. The specimen is scanned in a series of lines (for example, 1000 lines) as it would be to produce an image of a specimen

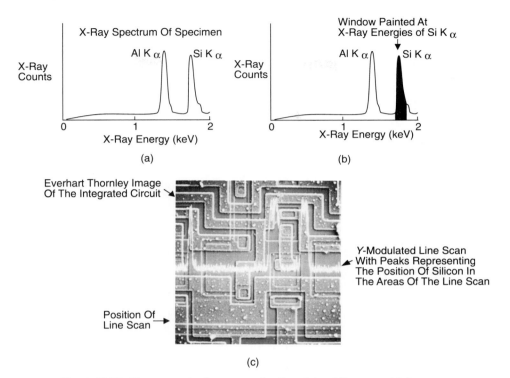

Figure 12-56 X-ray mapping by means of a Y-modulated line scan. (a) An area scan of the specimen is made to determine the elements present in the specimen. (b) The X-ray energies in the Si K are painted, thereby instructing the cathode ray tube to receive only these X-ray energies when producing the Y-modulated line scan. (c) A photograph is produced by making three exposures on the film. First, an Everhart–Thornley image is produced with an area scan. Second, a base line is made showing the part of the specimen that is covered by the electron beam. Third, the Y-modulated line scan representing the concentration of X-rays in the energy range of Si K X-rays is produced. The height of the peaks in the line scan represents the concentration of Si in the line scan. (Photograph courtesy of Kevex Instruments, Inc.)

using an Everhart–Thornley detector. However, instead the signal now comes from the X-ray detector, and this signal is used to determine the intensity of the pixels on the cathode ray tube. Usually, each pixel is either black (when no X-ray is received by the detector) or white (when an X-ray is received from the detector). The resulting image has white pixels in the areas where X-rays have been emitted from the specimen and black pixels in the areas where no X-rays have been emitted from the specimen. In such a setup, all energies of X-rays produce white pixels on the cathode ray tube. However, the operator is usually interested in the location of one or a few elements representing specific X-ray energies. In this case, a window is painted over the X-ray energies of a major peak of the element of

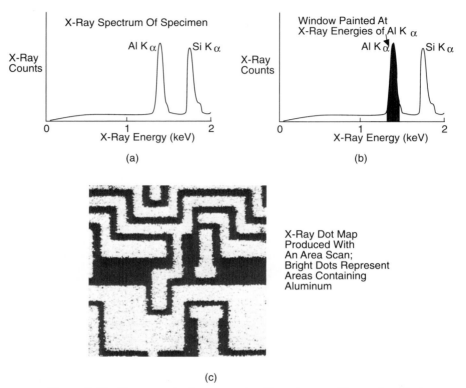

(a)   (b)

(c)

Figure 12-57   X-ray mapping by means of an X-ray dot map. The specimen is the same as in Fig. 12-56. (a) An area scan of the specimen is made to determine the elements that are present. (b) The X-ray energies in the Al K peak are painted to instruct the cathode ray tube to accept only these X-ray energies in producing the X-ray dot map. (c) An X-ray dot map of the specimen is made by making an area scan of the specimen and driving the pixel value in the cathode ray tube with the output from the X-ray detector. The areas with white pixels are the areas containing aluminum. (*Photograph courtesy of Kevex Instruments, Inc.*)

interest, and the cathode ray tube is instructed to receive only X-rays in that energy range. For example, an X-ray dot map of the integrated circuit containing aluminum and silicon in Fig. 12-56 is produced in Fig. 12-57. To determine the position of aluminum in the specimen, a window is painted in the X-ray energies of the AlK$_\alpha$ peak (Fig. 12-57b). The resulting X-ray dot map (Fig. 12-57c) has white pixels in the area scan when X-rays in the energy range of Al peaks are emitted from the specimen. Modern X-ray microanalytical devices allow the operator to acquire four or five elements at the same time. This is done by painting four or five windows over X-ray energies of the peaks of the elements in question. The dots for each element are displayed as different colors on a color video monitor.

Characteristics of X-ray mapping that limit data.   X-ray maps have certain characteristics that the investigator has to be aware of in interpreting the results. These include depth of analysis, spatial resolution in the specimen, potential magnification of the specimen, diameter of the electron beam, peak to background ratios, and X-ray shadows.

Depth of Analysis.   Like any X-ray technique, the X-ray mapping represents only those elements in the volume of the specimen from the bottom of the X-ray escape zone (a depth of 1 to 10 μm) to the specimen surface.

Spatial Resolution.   It is difficult to produce an X-ray escape volume of less than 1 μm^3 in the specimen because of the relatively long path that the X-rays travel in the specimen between collisions.

Potential Magnification.   The relatively large escape volume of the X-rays limits the highest magnification of X-ray dot maps to 3000×. Above this, hollow magnification occurs, with the information from a pixel position in the specimen being placed into so many pixels in the cathode ray tube that blurring of the image occurs.

Electron Beam Diameter.   Reducing the diameter of the electron beam below 100 nm will not significantly increase the resolution of the image because the minimal escape volume is around 1 μm^3, even with the smallest spot size. However, small spot sizes result in a decreased number of electrons in the beam and therefore a decreased number of X-rays. A relatively large number of X-rays (20,000 counts) are needed to produce one area scan to form an X-ray dot map, so spot sizes of 50 to 100 μm are needed to produce the required number of X-rays. The spatial resolution is not seriously decreased because of the large escape volume of the X-rays in the specimen at even the smallest beam spot sizes.

X-ray Shadows.   X-ray detection is a line of sight process. An X-ray shadow results if a portion of the specimen intercepts X-rays directed toward the detector. Such shadows will appear to be areas where there are no X-ray dots representing the element in question, when the element may actually be present. An X-ray map of the whole spectrum of X-ray energies will reveal shadow areas as areas of no X-ray dots on the cathode ray tube.

QUANTITATIVE MICROANALYSIS

In quantitative microanalysis, the quantities of the elements present in the specimen are determined. The results are usually expressed as the percentage of each element in the total detectable sample concentration. Quantitative microanalysis involves three basic steps: (1) **acquisition** of an X-ray spectrum from unknown or standard specimens under a controlled set of operating conditions, (2) **processing** of the X-ray spectrum to remove escape peaks, identify the peaks that are present, and model and remove background, and (3) **analysis** of the X-ray spectrum by extracting net peak intensities and using a quantitative analysis software routine to compute the concentration of elements in the specimen.

Acquisition of an X-ray Spectrum for Quantitative Microanalysis

An X-ray spectrum has to be acquired under a controlled set of parameters in order to obtain accurate quantification of the elements in the spectrum. The software routines require very specific data for accurate quantification of the elements in the specimen. The controlled parameters include certain specimen specifications, accelerating voltage of the electron beam, specimen–detector geometry, specimen–beam geometry and the number of electrons in the beam.

Specimen specifications. Almost all specimens that are examined in the scanning electron microscope are bulk specimens. Such specimens must be thicker than the volume of penetration of the electron beam into the specimen, the specimen volume analyzed for X-rays has to be homogeneous, and the specimen has to have a flat surface (Fig. 12-58).

Specimen Thickness. The specimen must be thick enough to contain the entire interaction volume of the electron beam in the specimen. The penetration of the electron beam depends on the density of the specimen (in g/cm^3) and the accelerating voltage of the electron beam. The nomogram in Fig. 12-59 shows this relationship. For example, steel containing chrome, iron, and nickel has an average density of 8 g/cm^3. With an electron beam of 25 kV, the specimen would have to be at least 1.1 μm thick for the electron beam interaction zone to be contained within the specimen. The absorption factors used in the quantitative analysis soft-

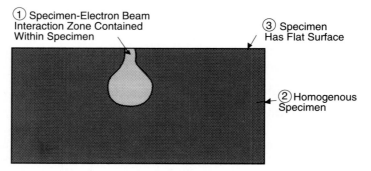

Figure 12-58 Quantitative analysis of X-ray spectra can only be performed on specimens that fulfill three criteria: (1) the specimen–electron beam interaction zone has to be completely in the specimen, (2) the volume from which the X-rays are collected has to be homogeneous, and (3) the specimen has to have a flat surface.

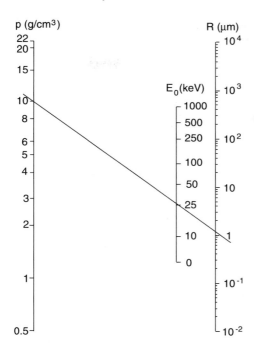

p (g/cm³)

E₀(keV)

R (μm)

Figure 12-59 Nomogram used for computing electron penetration depths for given accelerating voltages and sample densities. In this example, an electron beam with an accelerating voltage of 25 kV will penetrate a little over 1 μm into a steel sample with a density of 8 g/cm³.

ware would not be accurate if the specimen were thinner than this. Fortunately, almost all bulk specimens examined in the scanning electron microscope are thicker than the electron beam interaction volume in the specimen.

Homogeneity. The escape volume of the X-rays should be homogeneous in composition. This is important for the accuracy of the absorption and fluorescence factors used in the quantitative analysis software. These factors will not be accurate if the X-rays pass through different compositions of the specimen on the way toward the X-ray detector. This means that layered materials such as semiconductors, coated materials, and powdered materials cannot be measured accurately.

Flat Specimen Surface. The surface of a bulk specimen has to be flat for accurate quantitative analysis. Roughness of the surface results in X-rays traveling different distances from their point of generation. This results in differences in the path length of X-rays in the specimen, which produces variation in the absorption of X-rays. Metallurgical specimens have to be polished to submicrometer levels, making sure that any abrasives (such as diamond grit) are removed from the surface before X-ray microanalysis. Bulk biological specimens cannot be polished to a flat surface, so almost all X-ray microanalysis with biological specimens is performed on irregular surfaces. X-ray microanalysis on such rough biological surfaces cannot be truly quantitative and is usually called semiquantitative, indicating that the results contain a greater degree of error.

Accelerating voltage.   The accelerating voltage should be at least twice the critical excitation energy of the highest-energy X-ray line the investigator is interested in.

Specimen–detector geometry.   The geometry between the specimen and detector can be changed by tilting and changing the Z-axis (working distance) of the specimen and by moving the detector closer or farther from the specimen. These parameters, and other parameters dependent on these, produce changes in absorption and fluorescence that affect matrix corrections. These parameters are established at the beginning of an X-ray acquisition and placed in the microcomputer of the X-ray microanalytical system. These parameters include the following (Fig. 12-60):

Specimen tilt

Horizontal distance: distance from the detector to the optic axis of the column (area of specimen bombarded by the electron beam)

Solid angle of collection: the closer the detector is to the specimen, the larger the solid angle of collection

Take-off angle

Working distance

Specimen–electron beam geometry.   A specimen at 0° tilt will receive an electron beam with an incidence angle of 90°. The incidence angle will increase as the specimen is tilted, with the X-ray escape volume in the specimen becoming shallower.

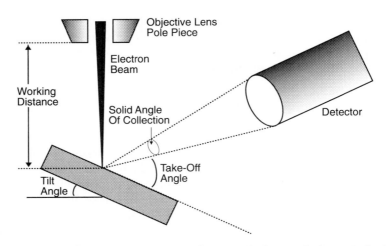

Figure 12-60   Some parameters that are important in the quantitative analysis of X-ray spectra.

Number of electrons in the electron beam.   The greater the number of electrons in the beam, the greater is the number of X-rays produced in the specimen. The number of X-rays reaching the detector should be adjusted to between 1000 to 2000 counts per second so that the deadtime is between 20% and 30%. Higher deadtimes result in sum peaks, unreasonably long acquisition times, and specimen damage or contamination. The number of electrons in the beam can be adjusted by changing the size of the electron beam diameter or by decreasing or increasing the electrons leaving the cathode by adjusting cathode saturation.

Processing of the X-ray Spectrum for Quantitative Microanalysis

Processing the X-ray spectrum involves removing escape peaks, identifying the elements that have produced the X-ray peaks, and removing the background continuum.

Removing Silicon Escape Peaks.   The microcomputer of the X-ray microanalytical unit has a program that determines the position of the silicon escape peaks by subtracting 1.74 keV from the energy of the real peaks in the spectrum. This program calculates the height of the silicon escape peaks and removes them from the specimen.

Identifying the Elements in the X-ray Spectrum.   The quantitative analysis software program requires the operator to identify the elements in the spectrum. The methods outlined in the section on qualitative analysis are used to identify the elements in the X-ray spectrum.

Removing the Background Continuum.   The software routine used to quantitate the X-ray spectrum requires the removal of the background before the spectrum can be quantified. There are three ways that the background can be removed.

1. *Recalling a stored background curve:* In this method (Fig. 12-61) the operator recalls a curve that represents the background, places the curve on the spectrum, and instructs the microcomputer to remove all X-ray counts under the curve. This method works well when similar specimens are being analyzed under the same conditions since all the spectra have backgrounds with similar curves.

2. *Painting a background curve:* In this method (Fig. 12-62) the operator paints a background curve over the X-ray spectrum displayed on the video monitor. This is done by selecting a couple of X-ray energy channels at one end of the spectrum that contain only background counts and painting the channels by running a cursor over them. Next, the operator moves the cursor about 2 keV and paints another couple of channels that contain only background. The microcomputer joins the two painted portions of the X-ray spectrum with a straight line. The process is repeated until the whole spectrum is modeled. The microcomputer then removes the X-ray counts under the modeled background.

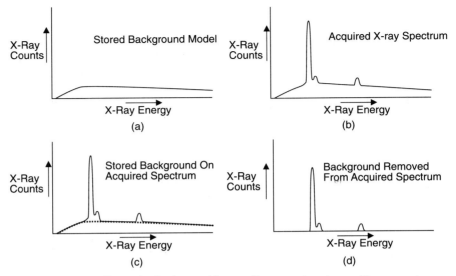

Figure 12-61   Removing background from an X-ray spectrum by recalling a stored modeled background from the computer memory.

3. *Use of algorithms to model the background:* This method uses algorithms to calculate a model of the background, including absorption edges.

Analysis of the X-ray Spectrum

Quantitative analysis of the processed X-ray spectrum would be simple if the electrons entering the specimen resulted in characteristic X-rays reaching the detector that were proportional to the elements in the specimen. In other words, a specimen containing 60% Al : 40% Cu would result in X-rays escaping the specimen that had a 60% Al K_α : 40% Cu K_α X-ray peak ratio. If this were the case, the investigator would simply have to determine the relative number of X-ray counts under each peak to determine the relative proportion of elements in the specimen. Unfortunately, this is not the case and a number of corrections have to be made before an X-ray spectrum can be quantitatively analyzed. These corrections involve three basic operations: (1) the deconvolution (separation) of overlapping X-ray peaks, (2) the calculation of *k* ratios, and (3) the application of ZAF correction factors.

Deconvolution of overlapping peaks.   The X-ray counts under the characteristic X-ray peaks are the data used to produce the quantitative results of the X-ray spectrum. Two methods are used to produce the relative X-ray counts under

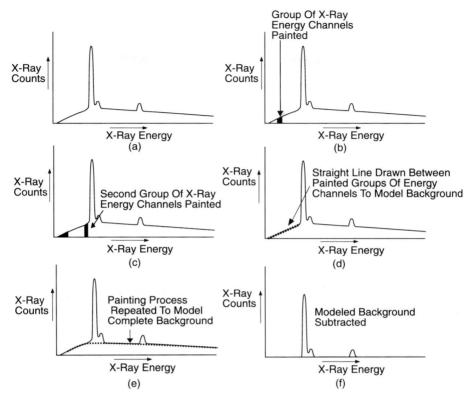

Figure 12-62 Modeling the background continuum by painting channels that represent the background.

the characteristic X-ray peaks: (1) the height of the peaks, or (2) estimation of the number of X-ray counts under each peak.

Peak Heights. In this method the height of the X-ray energy channel with the grestest number of X-ray counts in a peak (that has had background subtracted) is used to calculate the amount of that particular element in the specimen. Thus, in Fig. 12-63 the highest channel of the Cu K_α is twice as high as the Fe K_α. A 2 : 1 ratio of copper to iron is fed to the ZAF corrections to calculate the relative proportions of copper and iron. Errors can be introduced in this method if the peaks in a spectrum have different shapes (Fig. 12-64). A thin, sharp peak will contain less X-ray counts than a wide, blunt peak of the same height.

Number of X-ray Counts in Each Peak (Integral Method). The integral is the sum of the X-ray counts under a peak (after the background X-ray counts have

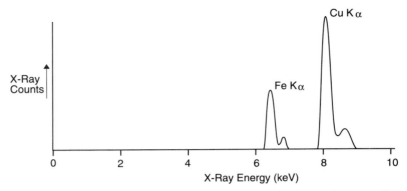

Figure 12-63 X-ray spectrum from a specimen containing iron and copper. The peak intensity of the Cu K_α is twice that of the Fe K_α. However, this does not mean that the copper concentration in the specimen is twice that of the iron.

been subtracted). The total number of X-ray counts under each peak theoretically produces an accurate count of the X-rays for that particular element. There are, however, potential problems associated with calculating the total number of X-ray counts under each peak in the X-ray spectrum:

1. The background has to be accurately subtracted or it is difficult to determine the number of X-ray counts under each peak. This is especially true at the peak edges where the peak X-ray counts blend into the background X-ray counts.

2. X-ray spectra frequently contain peaks that overlap. The resolution of the energy dispersive spectrometer varies from 100 to 200 eV depending on the energy of the X-rays. This results in the overlap of peaks from adjacent elements in the periodic chart. Overlapping also occurs when K, L, and/or M lines that are close in energy occur in the same spectrum. Overlapping X-ray peaks have to be separated and reconstructed by deconvolution techniques before the X-ray spectrum can be quantitatively analyzed.

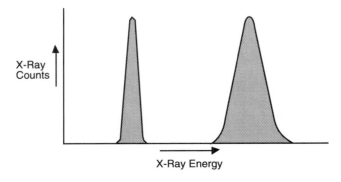

Figure 12-64 Using peak intensities to calculate the number of X-rays in a peak will be inaccurate if the peaks in the same specimen have different shapes.

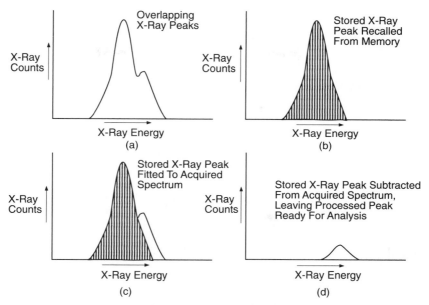

Figure 12-65 Overlapping X-ray peaks can be separated by recalling the X-ray peak for one of the elements from the memory. The counts in each X-ray energy channel are subtracted from the overlapping peaks to produce the processed peak of the second element, ready for analysis.

Deconvolution is therefore used to derive the relative contributions of constituent X-ray peaks to a single unresolved X-ray peak. There are two basic ways to perform deconvolution of overlapping peaks, with or without standards, although there are a number of variations of each method.

Deconvolution using standards requires the acquisition and storage of an X-ray spectrum for each element in the X-ray spectrum that is quantitatively analyzed. The stored X-ray spectra used as standards cannot have overlapping X-ray peaks. The microcomputer of the X-ray microanalytical unit recalls the stored spectrum of one of the overlapping elements. The standard spectrum has been acquired under the same conditions and time as the unknown spectrum. The microcomputer subtracts the X-ray counts in the standard X-ray peak from the overlapped, unresolved peak (Fig. 12-65). The resulting "stripped" peak is used to produce the number of X-ray counts of the element in the peak using the integration method. The deconvolution process is repeated to produce the modeled peak of the second element in the overlapping peak.

The method of deconvolution that does not use standards presumes that the X-ray peaks are Gaussian (bell-shaped or normal) curves. The process of deconvolution is basically the same as with standard spectra, except that a Gaussian curve of the correct energy is used to strip away the overlapping peak.

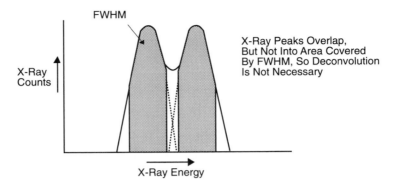

Figure 12-66 Deconvolution of overlapping X-ray peaks is often not necessary when partial integrals are used to determine the number of counts in each X-ray peak. This is true if the area of the peak used to measure the partial integral (FWHM in this case) does not cover the area of overlap.

Deconvolution of overlapping peaks requires accurate calibration of the X-ray microanalytical unit. A minor calibration error of 5 eV can result in significant integral errors of the total X-ray counts under each peak.

Methods that use partial integrals can be used to determine the relative X-ray counts in different X-ray peaks. In this method, only the X-ray counts in those energy channels with the most X-ray counts are used. Thus the low-intensity edges of an X-ray peak are excluded. This is an advantage since it is difficult to distinguish the X-ray counts of the peak from the background in these edges. A second advantage is the decreased probability of peaks overlapping, since a smaller portion of an X-ray peak is counted (Fig. 12-66).

There are a number of different ways of applying partial integrals. The methods vary in the portion of the X-ray peak that is used to produce the X-ray counts. The full width, half-maximum (FWHM) method (Fig. 12-67) measures half the

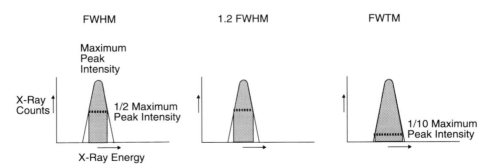

Figure 12-67 Three different methods of measuring partial integrals of X-ray peaks.

TABLE 12-4 RELATIVE ACCURACY OF
DIFFERENT METHODS OF MEASURING Cu
K_α X-RAY COUNTS WITHIN A PEAK

Peak intensity	1.6%
Integral	0.53%
Partial integral	
FWHM	0.59%
1.2 FWHM	0.56%
FWTM	0.53%

height of the X-ray peak and takes the X-ray counts in these energy channels. Uniformity of analysis is obtained by measuring the FWHM of every peak in the X-ray spectrum. Another partial integral is 1.2 full width, half-maximum, where the number of counted X-ray energy channels is increased by 20% over the FWHM method (Fig. 12-67). Full width, tenth maximum (FWTM) counts those X-ray energy channels that are one-tenth the height of the maximum height channel (Fig. 12-67). The minimum relative counting errors for Cu K_α peaks for the different methods are presented in Table 12-4.

K ratios. A k ratio is a comparison between the size of an X-ray peak from a pure element and the size of the same peak from an X-ray spectrum of a multielement specimen:

$$k \text{ ratio} = \frac{\text{peak in a multielement spectrum}}{\text{peak from a pure element}}$$

Both spectra have to be acquired under identical conditions. For example, Fig. 12-68 shows a spectrum of a steel sample containing iron and chromium. The Fe K_α peak in the spectrum contains 9000 counts. The k ratio of iron is determined by acquiring an X-ray spectrum from a specimen of pure iron under the same analytical conditions. The X-ray spectrum from pure iron has an Fe K_α peak that contains 10,000 counts. The k ratio for iron in the iron–chromium specimen is 90%. The Cr K_α peak in the iron–chromium specimen is 1100 counts. A spectrum acquired from a pure chromium specimen under the same conditions contains 9900 counts. Therefore, the k ratio for chromium in the iron–chromium specimen is 10%.

The k ratio is used as a first approximation of the concentration of each element in the specimen. The k ratio is not an accurate measurement of the elements in a multielement specimen. Figure 12-69 shows the relationship between the k ratios of chromium and iron in a specimen and the actual concentrations of the elements. The plot of the k ratios is nonlinear, meaning that the k ratios do not accurately reflect the true concentration of each element in the specimen. In an iron–chromium specimen the k ratio will indicate a concentration of chromium that is too high and a concentration of iron that is too low. One reason for this is

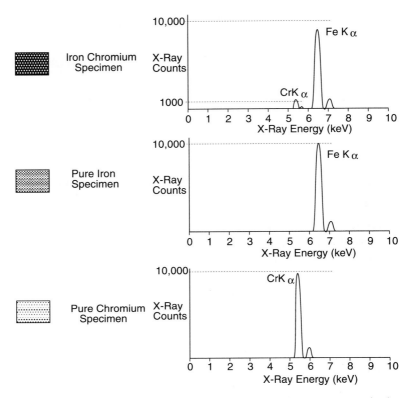

Figure 12-68 Calculating *k* ratios involves comparing the X-ray counts in the peaks in the multielement specimen with pure standards of each element in the multielement specimen. In this case the *k* ratio of iron is 90%, while that of chromium is 10%.

secondary emission or fluorescence, where a Fe K_α X-ray (6.404 keV) is absorbed by a chromium atom in the specimen. The 6.404 keV is greater than the critical excitation energy of the Cr K_α (5.988 keV), so a Cr K_α X-ray (5.415 keV) can be emitted. Thus a greater number of Cr K_α X-rays and a lesser number of Fe K_α X-rays are emitted from the specimen than are generated.

Even though the *k* ratios are not accurate representations of the quantity of elements in the specimen, the *k* ratios are used as a first approximation to calculate the true quantities of elements. This is done by multiplying the *k* ratios of each element by the ZAF correction factors.

ZAF corrections.   ZAF correction factors have to be taken into consideration before X-ray data can be quantified. There are three different types of ZAF correction factors: (1) the effects of the atomic number (**Z**) of the specimen; (2)

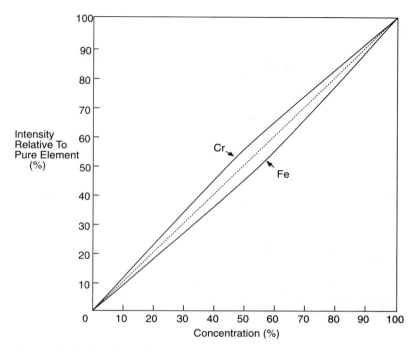

Figure 12-69 *K* ratios of iron and chromium (curves) compared to the actual concentration of iron and chromium (straight line) in specimens containing iron and chromium.

the absorption **(A)** of X-rays in the sample and the detector, and (3) the production of secondary X-rays by fluorescence **(F)**.

ZAF Correction Factors.   The number of X-rays produced in a specimen on interaction with the electron beam depends on atomic number **(Z)** of the elements in the specimen. The two most important Z factors are the variation in X-rays caused by variation in the *backscattered coefficient* and the *stopping power* of different elements in the specimen. The number of backscattered electrons produced per incident electron **(backscattered coefficient)** increases as the *atomic number of the elements in the specimen increases* (Fig. 12-70). The backscattered electrons are mostly high-energy beam electrons that are backscattered out of the specimen. Therefore, the more backscattered electrons that are produced per beam electron, the less high-energy beam electrons that penetrate into the specimen to produce X-rays. This means that specimens with low atomic number elements have more beam electrons penetrating into the specimen with the potential to produce X-rays. The **stopping power** (energy loss effect) indicates the likelihood of a collision between a high-energy electron and a specimen atom (which is neces-

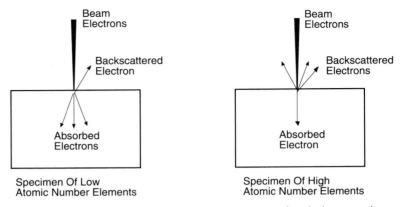

Figure 12-70 Many more backscattered electrons are produced when a specimen containing high atomic number elements is bombarded by an electron beam.

sary for the production of an X-ray). The stopping power *increases as the atomic number of the elements in the specimen increases* (Fig. 12-71). This means that specimens containing high atomic number elements are more likely to have collisions between high-energy electrons and specimen atoms and produce large numbers of X-rays.

Thus the atomic number of the elements in the specimen has an opposite effect on the backscattered coefficient and the stopping power. In fact, the effect of the backscattered coefficient and the stopping power tend to cancel each other out, with the result that the overall Z correction factor is relatively small.

A Correction Factor.   The A correction factor **(absorption coefficient)** is used to adjust for differences in absorption of X-rays by the specimen. X-rays are generated throughout the volume of penetration of the electron beam in the specimen. A significant number of the X-rays are reabsorbed by specimen atoms and

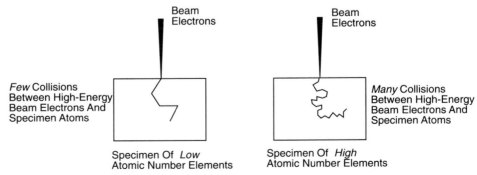

Figure 12-71 More collisions occur between high-energy electrons and specimen atoms in a specimen containing high atomic number elements.

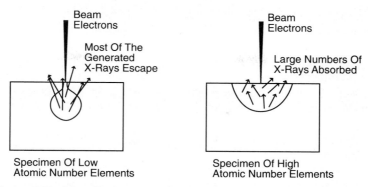

Figure 12-72 Most X-rays escape from a low atomic number specimen bombarded by an electron beam of low accelerating-voltage. Fewer X-rays escape from a high atomic number specimen bombarded by an electron beam of high accelerating voltage.

do not exit the specimen. The amount of reabsorption of X-rays is determined largely by the accelerating voltage of the electron beam and by the density of the specimen. These determine the shape and size of the volume in the specimen where the X-rays are generated. Absorption of X-rays is small at low accelerating voltages and in specimens containing low atomic number elements. Under these conditions, the size of the volume of X-ray generation is small, and the low density of the specimen allows most of the X-rays to escape the surface of the specimen (Fig. 12-72). Specimens composed of high atomic number elements that are illuminated with an electron beam of a high accelerating voltage have a greater absorption of X-rays (Fig. 12-72).

The take-off angle and the incidence angle of the electron beam on the specimen are important in calculating the absorption coefficient since these parameters affect the size and shape of the volume of primary excitation of X-rays in the specimen.

The absorption of X-rays in specimens containing more than one element is calculated by adding the absorption coefficient of its element multiplied by its mass coefficient. The specimen surface has to be flat in order to accurately calculate the absorption coefficient. An irregular surface will result in such a variety of X-ray escape paths that the calculation can only be an approximation of the element quantities in the specimen.

F Correction Factors. The F correction factor is used to correct for aberrations caused by **fluorescence (secondary emission)** (production of X-rays by interaction of atoms with higher-energy X-rays). This occurs when a high-energy X-ray emitted from one atom strikes a second atom with enough energy to cause the emission of an X-ray of lower energy from the second atom. An erroneous result will occur if the second X-ray is from a different element, since the X-ray that is received by the X-ray detector is of an energy characteristic of the second element, not of the first element. For example, electron beam interaction with a copper–iron specimen causes the emission of Cu K_α X-rays (8.04 keV), which are above the critical excitation energy of Fe K_α X-rays (7.11 keV). The interaction of a Cu K_α X-ray with an iron atom can produce an Fe K_α X-ray. As a result, the measured iron concentration is enhanced, while the intensity of copper is suppressed in the resulting X-ray spectrum. Continuum X-rays of sufficient energy can also result in the production of characteristic X-rays by fluorescence.

Computer Programs Used to Quantify X-ray Spectra

The quantification of an X-ray spectrum basically involves the multiplication of the k ratio for each element in the acquired X-ray spectrum by the ZAF correction factors for that element:

$$\text{quantity of an element} = (k \text{ ratio}) \times (\text{ZAF correction factors})$$

MAGIC is probably the most widely used computer program in X-ray quantitative microanalysis. MAGIC is a FORTRAN-based program for ZAF correction that was developed by J. W. Colby at Bell Laboratories in 1967. The current version of MAGIC is indicated by a Roman numeral (for example, MAGIC V).

BENCE-ALBEE is a program designed for ZAF correction of geological materials. It is aimed at analysis of common geological specimens containing phosphates, carbonates, silicates, sulfates, and oxides.

FRAME was developed at the National Bureau of Standards and is used in some software programs for X-ray microanalysis.

All these software programs theoretically require X-ray spectra of pure elements as standards to produce the k ratios. However, variations of these programs have been developed that allow standardless analysis of specimens. The standardless programs calculate theoretical k ratios by taking the peaks in the acquired X-ray spectrum and calculating the size of each peak in an X-ray spectrum of a pure element acquired under the same operating conditions as the multielement spectrum.

Appendixes

I

Steradian Definition

A **steradian** (sr) is a unit of measure of solid angle (Fig. A-1). The solid angle at the center of a sphere subtending a portion of the surface *is inversely proportional to the square of the radius of the sphere and directly proportional to the area subtended by the solid angle.* A sphere with a radius of 2 μm would have a steradian area of 4 μm².

Once the area on the spherical surface has been determined, it is possible to determine the solid (in steradians) angle by the following formula:

$$\text{solid angle} = \frac{\text{area on the surface of the sphere}}{\text{radius}^2}$$

$$= \frac{4\pi r^2}{r^2}$$

$$= 1 \text{ steradian}$$

Since both the area on the surface of the sphere and radius² have dimensions of length squared, the solid angle has no dimensions.

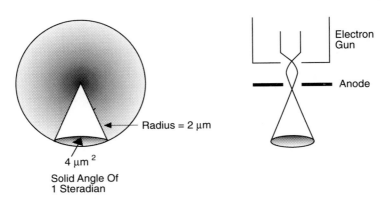

Radius = 2 μm

4 μm²

Solid Angle Of
1 Steradian

Electron
Gun

Anode

Figure A-1 Illustration of a stredian and how the terminology is applied to the electron beam.

A solid angle of one steradian is very large (there are 4π or about 12.5 sr in a sphere), much larger than the solid angle occurring in the electron beam; so the solid angle of electron beams is usually expressed in millisteradians (1/1000 steradian).

The steradian angle can also be calculated by taking the semiangle (α) of the electron beam and applying it to the formula

$$\text{solid angle (sr)} \cong \pi\alpha^2$$

A typical semiangle of the electron beam in the electron microscope is 10^{-2} rad. Calculating the solid angle;

$$\text{solid angle} \cong \pi\alpha^2$$
$$\alpha^2 = 10^{-4}$$
$$= 3.14 \times 10^{-4}\ \text{sr} \quad \text{or} \quad 0.314\ \text{msr}$$

II

Pressure Conversion Chart

To convert from: / To:	Pa	torr	atm Multiply by:	mbar	psi	kg/cm^2	μm
pascal (newtons /m^2)	1	7.5×10^{-3}	9.87×10^{-6}	10^{-2}	1.45×10^{-4}	10.2×10^{-6}	7.5
torr (mm of mercury)	133	1	1.316×10^{-3}	1.333	1.934×10^{-2}	1.359×10^{-3}	1000
atm (atmosphere)	1.013×10^5	760	1	1013	14.7	1.033227	7.6×10^5
mbar (millibar)	100	0.75	9.87×10^{-4}	1	1.45×10^{-2}	1.02×10^{-3}	750.1
psi (lb/in.^{2})	6.89×10^3	51.71	6.8×10^{-2}	68.9	1	0.070307	5.171×10^4
kg/cm^2	9.81×10^4	735.6	0.968	981	14.2	1	7.352×10^5
μm (micrometer)	0.1333	1×10^{-3}	1.316×10^{-6}	1.333×10^{-3}	1.934×10^{-5}	1.3595×10^{-6}	1

III

Wavelengths, Energies, and Critical Excitation Energies of K X-ray Lines

| Element | Z | K_α | | K_{β_1} | | |
		λ (Å)	E (keV)	λ (Å)	E (keV)	E_c (keV)
Na	11	11.91	1.041	11.62	1.067	1.071
Mg	12	9.890	1.253	9.570	1.295	1.303
Al	13	8.340	1.486	7.982	1.553	1.560
Si	14	7.126	1.739	6.778	1.829	1.840
P	15	6.158	2.013	5.804	2.136	2.143
S	16	5.372	2.307	5.032	2.464	2.470
Cl	17	4.729	2.621	4.403	2.815	2.819
Ar	18	4.193	2.957	3.886	3.190	3.202
K	19	3.742	3.312	3.454	3.589	3.607
Ca	20	3.359	3.690	3.090	4.012	4.037
Sc	21	3.032	4.088	2.780	4.460	4.488
Ti	22	2.750	4.508	2.514	4.931	4.964
V	23	2.505	4.949	2.284	5.426	5.463
Cr	24	2.291	5.411	2.085	5.946	5.988
Mn	25	2.103	5.894	1.910	6.489	6.536
Fe	26	1.937	6.398	1.757	7.057	7.110
Co	27	1.790	6.924	1.621	7.648	7.708
Ni	28	1.659	7.471	1.500	8.263	8.330
Cu	29	1.542	8.040	1.392	8.904	8.979
Zn	30	1.436	8.630	1.295	9.570	9.659
Ga	31	1.341	9.241	1.208	10.263	10.366
Ge	32	1.255	9.874	1.129	10.980	11.102
As	33	1.177	10.530	1.057	11.724	11.862
Se	34	1.106	11.207	0.992	12.494	12.562
Br	35	1.041	11.907	0.933	13.289	13.468

From S. J. B. Reed, *Electron Microprobe Analysis*, Cambridge University Press, New York, 1975.

IV

Wavelengths and Energies of L Lines and Critical Excitation Energies of L$_{III}$ Shell Electrons

Element	Z	L$_{\alpha_1}$ λ (Å)	L$_{\alpha_1}$ E (keV)	L$_{\beta_1}$ λ (Å)	L$_{\beta_1}$ E (keV)	E$_c$ (keV)
Ga	31	11.29	1.098	11.02	1.125	1.117
Ge	32	10.44	1.188	10.18	1.218	1.217
As	33	9.671	1.282	9.414	1.317	1.323
Se	34	8.990	1.379	8.736	1.419	1.434
Br	35	8.375	1.480	8.125	1.526	1.553
Kr	36	7.817	1.586	7.576	1.636	1.677
Rb	37	7.318	1.694	7.076	1.752	1.806
Sr	38	6.863	1.806	6.624	1.871	1.941
Y	39	6.449	1.922	6.212	1.995	2.079
Zr	40	6.071	2.042	5.836	2.124	2.222
Nb	41	5.724	2.166	5.492	2.257	2.370
Mo	42	5.407	2.293	5.177	2.394	2.523
Tc	43	5.115	2.424	4.887	2.536	2.677
Ru	44	4.846	2.558	4.621	2.683	2.837
Rh	45	4.597	2.696	4.374	2.834	3.002
Pd	46	4.368	2.838	4.146	2.990	3.172
Ag	47	4.154	2.984	3.935	3.150	3.350
Cd	48	3.956	3.133	3.738	3.316	3.537
In	49	3.772	3.286	3.555	3.487	3.730
Sn	50	3.600	3.443	3.385	3.662	3.928
Sb	51	3.439	3.604	3.226	3.843	4.132
Te	52	3.289	3.769	3.077	4.029	4.341
I	53	3.149	3.937	2.937	4.220	4.558
Xe	54	3.017	4.109	—	—	4.781
Cs	55	2.892	4.286	2.684	4.619	5.011
Ba	56	2.776	4.465	2.568	4.827	5.246
La	57	2.666	4.650	2.459	5.041	5.483
Ce	58	2.562	4.839	2.356	5.261	5.723
Pr	59	2.463	5.033	2.259	5.488	5.962
Nd	60	2.370	5.229	2.167	5.721	6.208
Pm	61	2.282	5.432	2.080	5.960	6.459
Sm	62	2.200	5.635	1.998	6.204	6.716
Eu	63	2.121	5.845	1.920	6.455	6.979

(*continued*)

Element	Z	L_{α_1}		L_{β_1}		
		λ (Å)	E (keV)	λ (Å)	E (keV)	E_c (keV)
Gd	64	2.047	6.056	1.847	6.712	7.242
Tb	65	1.977	6.272	1.777	6.977	7.514
Dy	66	1.909	6.494	1.711	7.246	7.788
Ho	67	1.845	6.719	1.648	7.524	8.066
Er	68	1.784	6.947	1.587	7.809	8.356
Tm	69	1.727	7.179	1.530	8.100	8.648
Yb	70	1.672	7.414	1.476	8.400	8.942
Lu	71	1.620	7.654	1.424	8.708	9.247
Hf	72	1.570	7.898	1.374	9.021	9.556
Ta	73	1.522	8.145	1.327	9.342	9.875
W	74	1.476	8.396	1.282	9.671	10.198
Re	75	1.433	8.651	1.239	10.008	10.529
Os	76	1.391	8.910	1.197	10.354	10.866
Ir	77	1.351	9.174	1.158	10.706	11.210
Pt	78	1.313	9.441	1.120	11.069	11.560
Au	79	1.276	9.712	1.084	11.440	11.919
Hg	80	1.241	9.987	1.049	11.821	12.284
Tl	81	1.207	10.267	1.015	12.211	12.658
Pb	82	1.175	10.550	0.983	12.612	13.038
Bi	83	1.144	10.837	0.952	13.021	13.424

From S. J. B. Reed, *Electron Microprobe Analysis*, Cambridge University Press, New York, 1975.

V

Wavelengths and Energies of M Lines

Element	Z	λ (Å)	E (keV)
Sm	62	11.47	1.081
Eu	63	10.96	1.131
Gd	64	10.46	1.185
Tb	65	10.00	1.240
Dy	66	9.590	1.293
Ho	67	9.200	1.347
Er	68	8.820	1.405
Tm	69	8.480	1.462
Yb	70	8.149	1.521
Lu	71	7.840	1.581
Hf	72	7.539	1.644
Ta	73	7.252	1.709
W	74	6.983	1.775
Re	75	6.729	1.842
Os	76	6.478	1.914
Ir	77	6.262	1.980
Pt	78	6.047	2.050
Au	79	5.840	2.123
Hg	80	5.648	2.195
Tl	81	5.460	2.270
Pb	82	5.286	2.345
Bi	83	5.118	2.422
U	92	3.910	3.170

From S. J. B. Reed, *Electron Microprobe Analysis*, Cambridge University Press, New York, 1975.

VI

Principle Emission and Absorption Energies of Elements

Element	Atom No.	K_{abs}	K_α	K_{β_1}	$L_{III\ abs}$	L_{α_1}	L_{β_1}	$M_{V\ abs}$	M_{α_1}
C	6	0.283	0.277						
N	7	0.399	0.392						
O	8	0.531	0.525						
F	9	0.687	0.677						
Ne	10	0.874	0.848		0.022				
Na	11	1.080	1.041	1.071	0.034				
Mg	12	1.303	1.253	1.302	0.049				
Al	13	1.559	1.486	1.557	0.072				
Si	14	1.838	1.739	1.836	0.098				
P	15	2.142	2.013	2.137	0.128				
S	16	2.470	2.307	2.465	0.163				
Cl	17	2.819	2.621	2.815	0.202				
Ar	18	3.203	2.957	3.190	0.245				
K	19	3.607	3.312	3.589	0.294				
Ca	20	4.038	3.690	4.012	0.349	0.341	0.345		
Sc	21	4.496	4.088	4.460	0.406	0.395	0.400		
Ti	22	4.964	4.508	4.931	0.454	0.452	0.458		
V	23	5.463	4.949	5.426	0.512	0.511	0.519		
Cr	24	5.988	5.411	5.946	0.574	0.573	0.583		
Mn	25	6.537	5.894	6.489	0.639	0.637	0.649		
Fe	26	7.111	6.398	7.057	0.708	0.705	0.718		
Co	27	7.709	6.924	7.648	0.779	0.776	0.791		
Ni	28	8.331	7.471	8.263	0.853	0.851	0.869		
Cu	29	8.980	8.040	8.904	0.933	0.930	0.950		
Zn	30	9.660	8.630	9.570	1.022	1.012	1.034		
Ga	31	10.368	9.241	10.262	1.117	1.098	1.125		
Ge	32	11.103	9.874	10.979	1.217	1.188	1.218		
As	33	11.863	10.530	11.722	1.323	1.282	1.317		
Se	34	12.652	11.207	12.492	1.434	1.379	1.419		
Br	35	13.475	11.907	13.286	1.552	1.480	1.526		
Kr	36	14.323	12.631	14.107	1.675	1.586	1.636		
Rb	37	15.201	13.373	14.956	1.806	1.694	1.752		
Sr	38	16.106	14.140	15.830	1.941	1.806	1.871		
Y	39	17.037	14.931	16.734	2.079	1.922	1.995		
Zr	40	17.998	15.744	17.663	2.220	2.042	2.124		
Nb	41	18.987	16.581	18.700	2.374	2.166	2.257		
Mo	42	20.002	17.441	19.599	2.523	2.293	2.394		

(*continued*)

Element	Atom No.	K_{abs}	K_α	K_{β_1}	$L_{III\ abs}$	L_{α_1}	L_{β_1}	$M_{V\ abs}$	M_{α_1}
Tc	43	21.054	18.325	20.608	2.677	2.424	2.536		
Ru	44	22.118	19.233	21.646	2.837	2.558	2.683		
Rh	45	23.224	20.165	22.712	3.002	2.696	2.834		
Pd	46	24.347	21.121	23.807	3.172	2.838	2.990		
Ag	47	25.517	22.101	24.921	3.352	2.984	3.150		
Cd	48	26.712	23.106	26.091	3.538	3.133	3.316		
In	49	27.928	24.136	27.271	3.729	3.286	3.487		
Sn	50	29.190	25.191	28.481	3.928	3.443	3.662		
Sb	51	30.486	26.271	29.721	4.132	3.604	3.843		
Te	52	31.809	27.377	30.990	4.341	3.769	4.029		
I	53	33.164	28.508	32.289	4.559	3.937	4.220		
Xe	54	34.579	29.666	33.619	4.782	4.109	4.420		
Cs	55	35.959	30.851	34.981	5.011	4.286	4.619		
Ba	56	37.410	32.062	36.372	5.247	4.465	4.827		
La	57	38.931	33.299	37.795	5.489	4.650	5.041		
Ce	58	40.449	34.566	39.251	5.729	4.839	5.261		
Pr	59	41.998	35.860	40.741	5.968	5.033	5.488		
Nd	60	43.571	37.182	42.264	6.215	5.229	5.721		
Pm	61	45.207	38.532	43.818	6.466	5.432	5.960	1.044	1.032
Sm	62	46.846	39.911	45.405	6.721	5.635	6.204	1.093	1.081
Eu	63				6.983	5.845	6.455	1.144	1.131
Gd	64				7.252	6.056	6.712	1.199	1.185
Tb	65				7.519	6.272	6.977	1.254	1.240
Dy	66				7.850	6.494	7.246	1.308	1.293
Ho	67				8.074	6.719	7.524	1.362	1.347
Er	68				8.364	6.947	7.809	1.421	1.405
Tm	69				8.652	7.179	8.100	1.479	1.462
Yb	70				8.943	7.414	8.400	1.524	1.507
Lu	71				9.241	7.654	8.708	1.586	1.572
Hf	72				9.556	7.898	9.021	1.664	1.645
Ta	73				9.876	8.145	9.342	1.725	1.702
W	74				10.198	8.396	9.671	1.803	1.775
Re	75				10.531	8.651	10.008	1.879	1.845
Os	76				10.869	8.910	10.354	1.963	1.921
Ir	77				11.211	9.174	10.706	2.040	1.988
Pt	78				11.559	9.441	11.069	2.129	2.066
Au	79				11.919	9.712	11.440	2.220	2.142
Hg	80				12.285	9.987	11.821	2.291	2.196
Tl	81				12.657	10.267	12.211	2.389	2.271
Pb	82				13.044	10.550	12.612	2.484	2.345
Bi	83				13.424	10.837	13.021	2.586	2.426
Po	84				13.817	11.129	13.445	2.681	2.502
At	85				14.215	11.425	13.874	2.780	2.582
Rn	86				14.618	11.725	14.313	2.882	2.663
Fr	87				15.028	12.029	14.768	2.986	2.746
Ra	88				15.442	12.338	14.233	3.093	2.829
Ac	89				15.865	12.650	15.710	3.202	2.913
Th	90				16.296	12.967	16.199	3.313	2.996
Pa	91				16.731	13.288	16.699	3.416	3.083
U	92				17.163	13.612	17.217	3.533	3.171

Glossary

Absorption Penetration of a substance into the body of another substance.

Absorption edge A discontinuity in an X-ray spectrum at the energy necessary to eject orbital electrons from an element in the specimen.

Absorption factor A correction factor used in quantitative X-ray microanalysis to adjust for differences in the absorption of X-rays by a specimen.

Accelerating voltage The difference in potential between the cathode and anode in an electron microscope.

Achromatic glass lens A lens that has been largely corrected for chromatic aberration.

Adsorption The surface penetration of gas molecules, atoms, or ions into a solid or liquid.

Alternating current or voltage A current or voltage that varies with time in a cyclic manner, with the current direction and voltage polarity reversing periodically.

Ammeter (ampere meter) An instrument for measuring current.

Ampere A unit of current defined as a constant current passed through a standard specified solution of silver nitrate in water that deposits silver at the rate of 0.001118 g/s.

Amplifier Apparatus that produces a magnified version of an input signal.

Anaerobic Lack of oxygen.

Analog A continuously varying electronic signal.

Analog to digital converter (ADC) An element that converts an analog voltage level to a binary quantity.

Angstrom unit (Å) A unit of length equal to 10^{-10} m or 10^{-8} cm or 10^{-4} μm. Ten Å equal 1 nm.

Anion A negatively charged ion.

Anode   An electrode that is positive relative to another electrode (the cathode).

Anticoagulant   A substance that helps prevent clotting of blood.

Aperture   A hole in a metal strip or disc that is placed in the path of an electron beam in an electron microscope to remove electrons that are not traveling directly down the column.

Apochromatic glass lens   A glass lens largely corrected for chromatic and spherical aberration.

Arc   A column of ionized gas.

Arithmetic logic unit   An element in image processor hardware that allows the combination of arithmetic and logical parameters to determine pixel brightness quantities.

Artifact   Details of a specimen that are not present in the natural condition and are induced during preparation of the specimen.

Aspect ratio   The ratio between the width and height of the image appearing on the screen of the cathode ray tube. Generally a 4 : 3 aspect ratio is used.

Astigmatism   The inability of a lens to bring light rays or electrons to a single point because of differences in the strength of the lens in the x and y direction.

Atomic number contrast   Contrast of the image in the scanning electron microscope due to the emission of different amounts of backscattered electrons and, to a lesser extent, secondary electrons from portions of the specimen of different atomic number.

Auger electron   An electron emitted from an outer orbital of an atom after the atom has absorbed an X-ray

Backing   The removal of accumulated gases from the base of a diffusion pump.

Backscattered coefficient   Number of backscattered electrons escaping from the specimen surface for each beam electron striking the specimen.

Backscattered electron   An electron that escapes from the specimen with an energy of more than 50 eV.

Backstreaming   The flow of oil or mercury vapors from a diffusion pump into the column of an electron microscope.

Bar   A unit of pressure equal to 0.98697 standard atmosphere or a pressure equal to 1 million dynes per square centimeter.

Battery   A cell or connected group of cells storing an electrical charge and capable of furnishing an electrical current.

Bias   A charge placed on a piece of equipment.

Bit   The basic digital quantity that represents either the true (1) or false (0) condition.

Black (dark) current   The current drawn off a cold cathode by the accelerating voltage.

Black level   The blackest part of the image on the cathode ray tube. The black-level knob on a scanning electron microscope is used to control the brightness of the image.

Blanking   The process by which the electron beam in the cathode ray tube is cut off during the horizontal and/or vertical retrace periods.

Bleb   A baglike extension of the protoplasm of a cell that is natural.

Blister   A baglike extension of a fixed cell that is an artifact of fixation.

Blooming   A condition that occurs when the signal producing the image is amplified so far that the electron beam in the cathode ray tube of the scanning electron microscope contains so many electrons that the phosphor of adjacent pixels is driven into the pure white.

Brightness   In reference to the electron beam, the current density per unit of solid angle (measured in amperes per square centimeter per steradian). In reference to pixels, the value of a pixel representing its gray value from black to white.

Byte   A binary unit of eight bits.

Capacitor   A circuit element that stores energy in its electric field.

Cathode   An electrode that is negative relative to a second electrode (the anode).

Cathodoluminescence   The emission of photons of ultraviolet, visible, or infrared from a specimen that is bombarded by electrons.

Cation   A positively charged ion.

Charge   An accumulation or deficiency of electrons. The unit of charge is the coulomb.

Charging   The buildup of an electrical charge on a specimen in the scanning electron microscope, resulting in alteration of the collecting field of the detection system as the electron beam travels over the specimen.

Chromatic aberration   An aberration of a lens that results in the focusing of electromagnetic radiation of different wavelengths to different focal points. In electron lenses, the shorter, more energetic electrons are focused farther from the lens than the longer wavelength, less energetic electrons.

Clipping   Amplification of an image to the point where those pixels with lighter gray levels are pushed to complete white, so there is no difference in gray level between pixels, resulting in loss of image detail. Clipping refers to *Y*-modulated line scans where the peaks are clipped at the top of the cathode ray tube.

Coagulant fixative Fixatives that produce gross precipitation of cellular macro-molecules so that the cellular components are distorted.

Cold cathode emission (field emission) The emission of electromagnetic radiation from a cold cathode.

Collimator A metal tube at the specimen side of an X-ray detector that is directed toward the specimen. The collimator is used to absorb the X-rays that are not coming from the specimen.

Composite signal The image plus the various blanking, sync, and equalizer pulses.

Condenser lens The lens in an electron microscope that controls the diameter of the electron beam.

Conduction In reference to electrical current, the passage of electrons to maintain an electrical current. In reference to heat, the transfer of heat by gas molecules.

Conduction band The state of an electron that is not bound to a particular atom. A conduction band electron is free to move about within the lattice of a substance.

Conductor A material that offers a low resistance to the passage of electrical current.

Continuous tone A term describing a photographic print and referring to the consistent and uninterrupted brightness in prints.

Continuum X-rays (white radiation, background) Noncharacteristic X-rays produced by the interaction of electrons with the nuclei of specimen atoms.

Contrast A measure of the range of brightness in an image. A high-contrast image contains mostly black and white; a medium-contrast image has a good range of gray levels from black to white; low-contrast images contain all grays. Also see *natural contrast* and *image contrast*.

Contrast enhancement Increasing or decreasing the contrast in an image to increase the definition of details not clearly visible in the original.

Contrast reversal The process of reversing the polarity of the signal to produce a negative image from a positive image, or vice versa.

Convergent (positive) lens A lens that converges electromagnet radiation toward the optic axis.

Conversion detector A detector that converts the energy of backscattered electrons into secondary electrons that are collected to produce the signal to construct the image.

Convolution Mathematical operation that implements group operations.

Convolution coefficient Number that defines the weight of a pixel within a kernel during a convolution process.

Coulomb (C) The practical unit of electrical charge. One coulomb equals one ampere per second.

Coulomb's law The force of repulsion or attraction between two electrostaticly charged bodies is proportional to the magnitude of the charges and inversely proportional to the square of the distance separating them.

Critical excitation energy Minimum energy required to overcome the binding energy of the atomic nucleus and eject an orbital electron.

Critical point That combination of temperature and pressure where the densities of the vapor and liquid phases are the same so that they intermingle and cannot be distinguished. At the critical point, surface tension forces are zero.

Cryogen A cold liquid used to freeze biological specimens.

Cryoprotectant A chemical that minimizes the size of ice crystals during freezing.

Cryostat A device that keeps an instrument cold to reduce electronic noise.

Crystal An assembly of molecules with a definite internal structure having an external form of a number of symmetrical plane faces.

Current A movement of electrical charge that is measured in amperes.

Dark (black) current The current drawn from a cold cathode by the accelerating voltage.

Deadtime The time that an X-ray detector system does not accumulate X-rays, usually because two X-rays arrive at the detector at nearly the same time.

Deconvolution The separation of two X-ray peaks whose energies are so close together that they appear as one in an X-ray spectrum.

Deflection coils An assembly of two sets of electromagnetic coils that is used to center and shift the electron beam in the column of an electron microscope.

Deflection plates An assembly of electrostatic plates used to deflect the electron beam in an electron microscope.

Dehydration The removal of water.

Depth of field (focus) The vertical distance on a specimen that is in focus at a particular magnification.

Derivative signal The first or second derivative of the amplitude of the signal from the detector. The derivative signal can be used to construct an image of the specimen.

Detector efficiency Percentage of X-rays striking the window of the X-ray detector that are converted into current pulses.

Diffusion pump A vacuum pump in which the vacuum is created by a stream of oil or mercury vapors carrying the gas molecules out of the evacuated system.

Digital image An image made up of pixels with each pixel having a number that indicates the brightness of the pixel.

Digital to analog converter (DAC) A portion of a display system in an image processor that converts a digital binary number of a pixel to an analog video voltage for display on a video monitor.

Digitization The process of sampling and quantitizing an analog video signal.

Dilator A substance that prevents vasoconstriction (narrowing) of arteries.

Diode An element with two electrodes, usually an anode and cathode.

Disc of least confusion The minimum diameter of an electron beam that has spherical or chromatic aberration, or astigmatism.

Display The device used to view the image. A typical display device is a television monitor.

Divergence angle The angle between the optic axis and the outermost portion of the electron beam striking the specimen.

Divergent (negative) lens A lens that diverges the beam of electromagnetic radiation (for example, visible light or electrons) away from the optic axis.

Dwell time The length of time that the electron beam rests on a pixel position in a scanning electron microscope.

Dynamic focusing A device that varies the focal point (crossover point) of the electron beam to correct for differences in the distance between the specimen and the objective lens.

Dynamic range The spread of gray values within an image. A high-contrast image has a wide spread of gray values and a high dynamic range. A low-contrast image has a small spread of gray levels and a low dynamic range.

Dynode A plate covered with an alkali–metal compound or metal oxide layer that emits electrons when struck by electrons.

Earth A connection to the ground. Usually considered to be zero potential.

Edge enhancement An operation that enhances the edge details in an image.

Elastic event The interaction of a beam electron with the electrical field of the nucleus of a specimen atom that results in a change in direction of the beam electron without a significant change in the energy of the beam electron.

Elastomer A plastic or synthetic rubber having some of the properties of natural rubber (for example, the inherent ability to accept or recover from extreme deformation).

Electric field An electric charge produces in the surrounding space an energy state that exerts mechanical forces on other charges.

Electrode A conductor to which or from which current flows.

Electromagnet A device consisting of an electrically conducting coil that surrounds a ferromagnet core. Passage of electrical current through the coil produces a magnetic field.

Electron A basic unit of matter that has mass and a waveform. An electron has a charge of 1.6021×10^{-19} C, a rest mass of 9.109×10^{-28} kg, and a dimension of 10^{-14} m.

Electron beam The beam of electrons produced by the electron gun that is directed toward the specimen.

Electron gun The device that produces the electron beam. With thermionic electron sources, the electron gun consists of the cathode and a Wehnelt cylinder.

Electron lens A device that produces an electrostatic or electromagnetic field, or a combination of the two, that affects an electron beam.

Electron probe microanalyzer A device that bombards the specimen with an electron beam to produce X-rays that are measured to determine the elements present in the specimen. Unlike the scanning electron microscope, the electron probe microanalyzer does not have electron image capabilities.

Electron volt (eV) The energy of an electron when accelerated by a potential difference of 1 V.

Electrostatic field A field created by oppositely charged electrodes.

Energy dispersive spectroscopy (EDS) A type of X-ray microanalysis that identifies the elements present in a specimen by measuring the energy of the X-rays emitted from the specimen.

Energy loss effect (stopping power) The likelihood of a collision between a high-energy electron and a specimen atom.

Escape peak A peak in an X-ray spectrum that is equal to the difference between a characteristic X-ray peak in the spectrum and the Kα peak of the element in the detector [for example, a silicon escape peak from a Si(Li) detector].

Escape volume The volume of the specimen from which the electrons or the electromagnetic radiation used to construct the image of the specimen escapes.

Evaporation temperature The temperature at which the vapor pressure of a substance reaches 1.33×10^{-2} mbar.

Everhart–Thornley detector The most commonly used detector in scanning electron microscopes for the detection of secondary and backscattered electrons. The detector consists of a Faraday cage, a scintillator disc, a light guide pipe, and a photomultiplier.

Faraday cage A closed box with a conducting surface that divides the universe into two independent parts, the part within the box and the part outside the box. In the scanning electron microscope the Faraday cage protects the electron beam from the high positive potential on the scintillator and attracts secondary electrons toward the detector.

Fermi level The energy level separating completely filled quantum states below from empty quantum states above in a system at 0 K. It is the highest energy level in a system.

Ferromagnetic element An element in which the number of electrons in the outer orbital spinning in one direction is not the same as the number of electrons in the outer orbital spinning in the other direction.

Field-effect transistor A device used to amplify the signal from a lithium-drifted detector in an X-ray microanalytical system.

Field emission (cold cathode emission) The emission of electromagnetic radiation from a cold cathode.

Filament A common term for a cathode made of tungsten wire in an electron gun.

Filter The number of weighted pixels used to produce a new output value for a pixel in spatial image-processing operations.

Fixation Crosslinking of cellular macromolecules into a rigid reticulated network that holds the size and shape of cells in a condition as near to natural as possible.

Fluid The liquid and gas phases of a substance.

Fluorescence The emission of electromagnetic radiation (for example, light or X-ray photons) when a substance is bombarded by electromagnetic radiation of higher energy.

Fluorescence (X-ray) yield   The probability that the creation of a vacancy in the inner shell of an atom will result in the emission of an X-ray.

Focal length   The distance from the central plane of a lens to the focal point.

Focal point   The point at which electromagnetic radiation is focused by a lens.

Focusing   In the scanning electron microscope, bringing the electron beam to crossover on the surface of the specimen by varying the current through the objective lens.

Forbidden gap   The energy levels between the valence band and the conduction band. No electrons are allowed to have energies equal to those in the forbidden gap.

Frame   A complete scan of the cathode ray tube. In television, a frame is two interlaced scans of the cathode ray tube.

Frame averaging   A multiple image function in which the gray-scale value of each pixel in an image is added to the same pixel in a subsequent image scan(s). The summed value for each pixel is divided by the number of accumulated frames to obtain the average value of each pixel.

Frame rate   The number of rasters of the cathode ray tube per second.

Frequency   The number of complete cycles of a periodic function in a unit of time.

Full-width half-maximum   The range in electron volts of X-ray energies in one-half the height of an X-ray peak.

Gain   The amplitude or strength of a signal; the voltage difference between dynodes in a photomultiplier tube.

Gas   Material in a gaseous state that is not condensable at the operating temperature; a fluid that can expand indefinitely to fill all available space. Also see *vapor*.

Gas proportional counter   An X-ray detector that consists of a gas-filled tube with a wire at the center at a positive potential. The tube produces a pulse of electrical charge that is proportional to the energy of the X-ray entering the tube.

Gaussian   Bell shaped.

Geometric manipulation   An image-processing operation that changes the spatial geometry of image. Such operations include scaling, translation, rotation, and rubber sheet transformations.

Getter   A chemically active metal that combines with residual gases so that they are fixed as compounds on the walls.

Giga   A decimal prefix indicating multiplication by 10^9.

Graphite   A form of carbon that has high conductivity of electrons.

Gray level   Brightness value of a pixel. The level can range from black, through grays, to white.

Gray scale   The number of gray levels available to a system. The gray scale represents the number of discrete gray levels in a system. Digital systems usually have an eight-bit system containing values from 0 to 255.

Ground   The same voltage potential as the ground, commonly neutral.

Group process   In spatial filtering operations, a group process is used to produce a number for a pixel based on a weighted average of the surrounding pixels.

Heat of crystallization   The amount of heat that must be removed to freeze a liquid.

Heat of fusion   The amount of heat released when a solid melts to a liquid.

Hertz   A unit of frequency equal to one cycle per second. One kilohertz (kHz) = 1000 cycles per second. One megahertz (MHz) = 1,000,000 cycles per second.

High-frequency spatial components   Areas of an image that have large changes in gray-scale levels from one pixel to the next.

High-pass filter   A type of filter used in spatial image processing that accentuates areas that have large changes in gray-scale levels from one pixel to the next.

High tension   Potential in volts (or kilovolts) between the anode and cathode in an electron microscope.

Histogram   A graph that is used in image processing to represent the number of pixels in an image that contain a particular gray-scale value.

Histogram slide   Addition or subtraction of a brightness value to all the pixels in an image. The effect is to slide gray-scale values up or down the scale.

Histogram stretch   Multiplication or division of the value of all the pixels in an image by a constant value. The effect is to stretch or shrink the gray-scale range.

Hole   A positive charge carrier in a *p*-type semiconductor.

Hollow magnification   The condition where increasing the magnification in the scanning electron microscope results in no increase in resolution because the image details are spread over more pixels in the cathode ray tube.

Horizontal blanking interval   Part of the composite video signal between the end of picture information on one line and the start of the picture information on the next line.

Horizontal sync   The part of the video signal that indicates the end of a line of video information. The video equipment uses sync pulses to maintain line synchronization with the incoming video signal.

Hygroscopic   The taking up of water by a substance.

Hypertonic   A solution of higher osmolarity than cellular protoplasm.

Hypotonic   A solution of lower osmolarity than cellular protoplasm.

Hysteresis   In electromagnetic lenses the strength of the magnetic field can have one of two values. The first value is when the electrical current through the coils is rising, the second, when the electrical current through the coils is falling. The hysteresis is due to the residual magnetism in the lens when the electrical current through the coils is falling.

Image analysis   The processing of an image to numerically tabulate an aspect of the image.

Image contrast   The contrast of the image after the image has undergone processing.

Image field   Positive field at the surface of metal that is produced when an electron escapes from the surface.

Image operation   An algorithm used to enhance, analyze, or code an image.

Incandescence   The heating of a substance till it emits reddish or whitish light.

Inelastic event   An event that occurs when a beam electron interacts with the electrical field of an electron of a specimen atom, resulting in the transfer of all or almost all of the energy of the beam electron to the electron of the specimen atom.

Infrared   Photons of electromagnetic radiation from 750-nm to 1000-μm wavelength (17 to 1.24×10^{-2} eV energy).

Insulator   Something that offers a high resistance to the passage of electrical current.

Interlace   The display of a field of odd image lines followed by the display of a field of even image lines in the production of a video image. Interlacing reduces the flickering in an image display.

Intrinsic zone   A volume where acceptor atoms are compensated by donor atoms in a Si(Li) or Ge(Li) detector.

Ion   A charged particle commonly formed by collision of an atom with a particle such as an electron.

Ion-getter pump   A vacuum pump that removes gas molecules from the system by ionizing the gas molecules and burying them in the cathode of the pump.

Ionization gauge   A gauge for measuring high vacuum. The electrons from a heated filament in a vacuum are accelerated toward a collecting electrode. The positive ions formed during the passage of the electrons are collected by a third electrode, with the current recorded varying with the pressure inside the device.

Isotonic solution   A solution of the same osmolarity as cellular protoplasm.

Kernel   In spatial image processing, the neighboring pixels of the input image that are weighted to determine the value of a pixel in the output image.

Kinetic energy   The energy of a body resulting from its motion.

Latent image   An invisible chemical site in the silver halide layer of film or photographic paper that initiates the conversion of the silver halide crystal to metallic silver. A latent image is produced when electromagnetic radiation (electrons or light) strikes the silver halide layer.

Leidenfrost phenomenon   The formation of a layer of cryogen gas around a specimen during freezing of the specimen.

Light guide pipe   A solid plastic or glass rod that has a high internal reflection of light photons.

Linear amplification   A linear amplification of the brightness of an image.

Liquid   A fluid with a definite volume at a given temperature.

Lithium-drifted crystal   A silicon or germanium crystal that has lithium drifted partially through the crystal. Used as a detector in energy dispersive spectrometers.

Live time   Real time minus the lost time when X-rays are not being recorded by an X-ray microanalytical unit.

Look-up table   A memory device containing map values for a given point process. Digital input pixel values are fed into the look-up table that produces output pixel values that are determined by the program of the look-up table.

Loss peak   A peak in an Auger electron spectrum that represents the difference between the accelerating voltage and the voltage of a generated Auger electron.

Low-frequency spatial components   Areas of an image that have little or no change in gray-scale values from one pixel to the next.

Low-pass filter   A process by which low spatial frequencies in an image are enhanced or by which high spatial frequencies are attenuated. The process brings out details in an image that are difficult to see.

Macromolecule   A very large molecule such as a protein, nucleic acid or such polysaccharides as starch, glycogen, or cellulose.

Magnet   A piece of ferromagnetic material that has the power of attracting or repelling other pieces of similar material and of exerting a mechanical force on a neighboring conductor carrying an electrical current.

Main amplifier   A device in an X-ray microanalytical system that converts the ramped, staircase output from the preamplifier (in the millivolt range) to individual pulses that are linearly amplified up to 10 V in amplitude.

Mapping function   A mathematical function used in a point-processing operation to calculate the output pixel value from the input pixel value.

Mask   The array of weighted pixels used to produce a new output value for a pixel in spatial image processing.

Mass limit   The absolute amount of an element that can be detected in X-ray microanalysis.

Mean free path   The distance a gas travels on average between collisions with other particles.

Metal gasket   A gasket made of a metal of high ductility (easily molded or drawn) and low vapor pressure that is used in areas of electron microscopes requiring high vacuums. Indium is a metal commonly used in metal gaskets.

Micrometer (μm)   A unit of length equal to 10^{-6} m. One micron (μ) equals 1 μm.

Micron (μ)   A unit of length equal to 10^{-6} m. 1 μ equals 1 μm. (Micrometer is the preferred unit of measure.)

Microphony   Transformation of sound waves into electrical currents.

Mil   A unit of length equal to one-thousandth of an inch.

Millibar (mbar)   1/1000 bar. One mbar $= 10^{-2}$ Pa $= 1.333$ torr.

Minimum detectable level   The smallest X-ray peak that can be said to be real with a particular level of certainty.

Moiré effect   Reinforcement lines on the cathode ray tube of the scanning electron microscope that result when the specimen has an inherent periodicity that approaches the periodicity of the lines that make up the scan pattern.

Molecular flow The flow of gas molecules at high vacuum where the mean free path of the gas particles is greater than the diameter of the conducting tube. See also *viscous flow*.

Monochromatic Radiation of a single wavelength.

Multichannel analyzer (MCA) A device that sorts digital values into a specified number of channels.

n-**Type semiconductor** A semiconductor that has a small amount of a donor element so that the crystal lattice contains free electrons that can act as charge carriers.

Nanometer (nm) A unit of length equal to 10^{-9} m.

Natural contrast The contrast in the image produced by the signal leaving the detector. Also see *image contrast*.

Near-real time In image-processing operations, when the images change within 3 s after the operator has changed the parameter. Also see *real time* and *nonreal time*.

Noise Unwanted electrical signals that interfere with the clarity and purity of the video image.

Noncoagulant fixative Fixative that stabilizes macromolecules in cells with very little distortion from the natural state.

Nonlinear amplification (gamma) The process of differentially amplifying the gray scale of an image.

Nonreal time In image processing, a function that takes more than 3 s to process after the operator changes the parameters. Also see *near-real time* and *real time*.

North pole The area of a magnet from which the magnetic lines of force emerge.

Objective lens (probe-forming lens) The lens in an electron microscope that controls the focusing.

Ohm (Ω) A unit of resistance. A conducting path has a resistance of 1 Ω when the passage of a current of 1 A has a potential difference of 1 V.

Optic axis The geometrical center of the column in an electron microscope.

Orthochromatic film Film sensitive to only certain wavelengths of light.

Orthogonal scanning   The electron beam scanning the specimen in alternate horizontal and vertical scans.

Outgassing   The desorption of gases from a surface.

***p*-Type semiconductor**   A semiconductor that contains acceptor elements so that the crystal lattice contains a number of holes or positive charge carriers.

Panchromatic film   Film that is sensitive to all wavelengths of visible light.

Pascal   A unit of gas pressure equal to one newton per square meter (N/m^2). One pascal equals 10^{-2} mbar.

Penning (cold cathode ionization) vacuum gauge   A vacuum gauge that measures the discharge current produced when generated electrons strike gas molecules in a vacuum.

pH   Logarithm of the reciprocal of the hydrogen ion concentration.

Phosphor   A chemical that emits light photons when bombarded by high-energy electrons.

Photodiode   A diode that becomes conductive when it absorbs electrons.

Photoelectric effect   The ejection of an electron from an atom that has absorbed a photon of electromagnetic energy.

Photoelectron   Electron ejected from an atom by the absorption of a photon.

Pico   Prefix indicating 10^{-12} of a unit.

Pirani (thermal conductivity) vacuum gauge   A gauge that measures vacuum by measuring the amount of heat transfer from a heated sensing wire.

Pixel (picture element)   The basic element of an image.

Pixel position   The coordinate used to define the horizontal (x) and vertical (y) position of a pixel in an image.

Plasma   A mixture of ionized gas and electrons.

***pn*-Junction diode**   A diode produced when *p*-type and *n*-type semiconductors are joined.

Point process   An image-processing operation in which the value of an input pixel is changed through a mapping function to the value of the output pixel.

Poisoning   The adsorption of water vapor and/or electronegative gases (such as oxygen) on a cathode surface, resulting in reduced electron emission from the cathode.

Polarization field   Field at the surface of a metal produced by atoms at the surface having forces exerted on them from underneath but not from on top.

Pole   Each terminal of a circuit between which a high voltage exists.

Pole piece   A piece of soft iron in an electromagnetic lens that is used to concentrate the magnetic field.

Polyphenyl ether fluids   Fluids consisting of chains of benzene radicals interbonded by oxygen. Polyphenyl ether fluids are commonly used in diffusion pumps.

Postfixation   Fixation subsequent to the primary fixation. Postfixation is usually done with osmium tetroxide.

Potential   The voltage difference between two points.

Potential energy   Energy that is the result of relative position instead of motion.

Preamplifier   A device that amplifies the signal from the lithium-drifted X-ray detector.

Pressure   The quotient of the perpendicular force on a surface and the area of this surface. The partial pressure of a given gas or vapor is that pressure that this gas or vapor would have if it were present alone in the container. The total pressure in a container is equal to the sum of the partial pressures of all the gases and vapors within it.

Primary fixation   The initial fixation that cells are exposed to, usually buffered glutaraldehyde.

Probe-forming lens (objective lens)   The lens in the electron microscope that controls the focus.

Projection distortion   Distortion of the image in a scanning electron microscope due to differences in the distance from the center of the objective lens to different parts of the specimen.

Pulse pileup rejector   A device that rejects an X-ray that arrives before a voltage pulse from a previous X-ray event has been processed.

Quantization   The process of converting the brightness of an analog pixel to a digital number.

Radiation   The transfer of heat in the form of waves. An intervening medium is not required.

Raster   Area of the cathode ray tube that is scanned by the electron beam.

Real time   Actual elapsed time during an analysis.

Reflection   The throwing back by a surface of light, sound, heat, and so on.

Refraction   The bending of a wave of light, heat, or sound as it passes obliquely from one medium to another of different density.

Refractive index Ratio of the velocity of a light photon or electron in the first of two media to its velocity in the second medium as it passes from one to the other.

Resistance The opposition offered by a conducting material to the passage of an electrical current. The unit is the ohm (Ω).

Resistor A device in a circuit used to primarily provide resistance.

Resolution The accuracy at which the image is divided into discrete levels.

Resolving power The ability of an optical system to form distinguishable images of objects separated by small differences.

RGB Refers to a system that handles images composed of red, green, and blue.

Roughing The change in gas pressure from 1 atm to 10^{-3} mbar (10^{-1} Pa).

Rubber sheet transformation A geometric image-processing operation that contorts an image through the use of specified control points.

Saturation point The temperature at which electron emission from the cathode of an electron gun does not increase when the temperature is increased.

Scaling A geometric image-processing operation used to enlarge or shrink an image.

Scanning electron spectrometric spectroscopy (SESM) The production of a scanned image of a specimen using a defined range of electromagnetic radiation (for example, a specific range of Auger electron voltages).

Scintillator An inorganic compound containing heavy elements that emits light photons by cathodoluminescence when bombarded by electromagnetic radiation.

Secondary electron An electron that escapes from the surface of the specimen with 50 eV or less.

Secondary electron coefficient The number of secondary electrons that escape from the surface of a specimen for each beam electron striking the specimen.

Secondary emission The emission of electromagnetic radiation from a substance being bombarded by electromagnetic radiation or atomic particles.

Semiconductor A material having electrical properties intermediate between good electrical conductors and insulators.

Shaping The processing of a signal by the main amplifier in an X-ray microanalytical unit. The time constant used determines the shape of the voltage pulse.

Signal to noise ratio A logarithmic comparison of a video signal to the noise in the system. The greater the number is, the clearer the image.

Silicones   Semiorganic fluids with a molecular backbone built on silicon–oxygen linkages. Methyl and phenyl groups are attached to the backbone.

Solenoid   A coil of insulated wire that produces a magnetic field when current is passed through the wire.

Solid angle of collection   The three-dimensional angle that the face of a detector makes with the specimen. The angle is measured in steradians.

South pole   Area of a magnet where the magnetic lines of force enter the magnet.

Space charge   The production of an electron cloud in the aperture of a Wehnelt cylinder of an electron gun.

Spatial   Referring to the two-dimensional nature of an image.

Spatial frequency   The amount of brightness change from one pixel to the next. High spatial frequencies occur when there is a large change in brightness from one pixel to the next; low spatial frequencies occur when there is a small change in brightness from one pixel to the next.

Spatial-processing operation   An image-processing operation in which the value of each pixel is considered as part of a family of pixels.

Specimen current   The current from the specimen to ground in an electron microscope.

Spherical aberration   Aberration caused when electromagnetic radiation (for example, electrons or light photons) is focused to different focal points depending on the part of the lens that transmits the radiation. Electrons close to the optic axis are focused to a focal point that is farther away than electrons that pass closer to the windings of the lens.

Sputtering   The process of knocking atoms out of a target by bombardment of the target with ions.

Stereoscopic image   A three-dimensional image produced from two micrographs taken at two positions equivalent to the distance between the eyes, or taken at two angles approximately 5° apart.

Stigmator   A device that corrects the astigmatism of an electron lens by superimposing on the magnetic field of the lens a second electrostatic or magnetic field of asymmetric and variable magnitude, which can be positioned to oppose and cancel the existing asymmetry of the lens magnetic field.

Stopping power (energy loss effect)   The likelihood of a collision between a high-energy electron and a specimen atom.

Sublimation   Evaporation from the solid to the vapor state without any intervening liquid state.

Sum peaks   Artifactual X-ray peaks in an energy dispersive spectrometer that are nearly equal to the sum of the energy of two real peaks in the X-ray spectrum.

Supercooling   The cooling of liquid below the freezing point without crystallization of the atoms or molecules.

Surface tension   The force per unit length at the free surface of a liquid by reason of intermolecular forces unsymmetrically disposed about the individual surface molecules.

Sync pulses   A series of pulses produced by the sync generator of a scanning electron microscope that coordinates the horizontal and vertical scan coils of the column with those in the cathode ray tube.

Take-off angle   The angle the detector makes with the surface of the specimen.

Thermionic emission   The emission of electrons when a substance is heated. Emission occurs from a metal or semiconductor when their electrons acquire sufficient thermal (kinetic) energy to overcome the potential energy barrier at the surface.

Thermocouple   A device that uses the Seebeck effect to measure temperature. The Seebeck effect occurs when a closed circuit is constructed of two (or more) dissimilar metallic conductors, and the junctions between the different metals are maintained at different temperatures, which causes an electrical current to flow in the circuit. The current is used as an indicator of the temperature.

Tilt correction   A device that corrects for distortions in the image of the specimen due to the tilt of the specimen.

Tilt distortion   Distortion of an image in the scanning electron microscope due to differences in the distance from the center of the objective lens to the different parts of a tilted specimen. Those parts of the specimen closer to the lens will be magnified more than those parts of the specimen farther from the objective lens.

Topographic contrast   Contrast in the image produced by a scanning electron microscope due to different areas of the specimen producing different amounts of secondary and backscattered electrons depending on the topography of the specimen surface. The rougher the topography, the greater is the emission of electrons.

Torr   A unit of gas pressure able to maintain a mercury column 1 mm high at 0°C. One torr = 0.75 mbar.

Transitional fluid   The fluid that is used to pass the specimen through the critical point in critical point drying.

Translation   A geometric image operation used to move an image from one location to another; the movement of the specimen in the scanning electron microscope so that the operator can view a different area of the specimen.

Triode   An electron tube with an anode, cathode, and control grid.

Triple point   Temperature and pressure at which there is an equilibrium between the solid, liquid, and gas phases of a substance.

Tunneling   Pulling of electrons through the potential energy barrier at the surface of a substance.

Turbomolecular pump   High-vacuum pump that consists of a high-vacuum gas turbine with a rotor and stator with blades.

Ultraviolet   Photons of electromagnetic radiation from 10- to 400-nm wavelength (1240- to 30-eV energy).

Vacuum   A space from which gas has been removed.

Valence band   The ground state of the outer electrons of an atom.

Vapor   Material in the gaseous state that is condensable at operating temperature. Also see *gas*.

Vertical sync   The part of the video signal that indicates the end of a raster of a field. The vertical sync pulse is used to orient the electron beam of the cathode ray tube at the start of a raster.

Viscous flow   The flow of gas molecules in a poor vacuum where the mutual interactions of the gas particles determine the character of the flow. Also see *molecular flow*.

Visible light   Photons of electromagnetic radiation from 400- to 750-nm wavelength (30 to 17 eV in energy).

Viton   A elastomer that is a copolymer of perfluorpropylene with vinylidene fluoride. Viton has a very low outgasing rate and is a common elastomer used in vacuum O-rings.

Volt (V)   The potential difference that is produced when a current of 1 A is applied to a conductor with a resistance of 1 Ω.

Voltage   Difference in electrical potential, expressed in volts.

Wavelength dispersive spectroscopy (WDS)   A type of X-ray microanalysis that identifies the elements that are present in a specimen by measuring the wavelengths of the X-rays emitted from the specimen.

Wehnelt cylinder   The cap over the cathode in the electron gun. The Wehnelt cylinder is usually biased negatively to the cathode, so the electrons emitted from the cathode are concentrated at the aperture in the Wehnelt cylinder to produce a concentrated cloud of electrons.

Work function   The potential in volts that must be overcome for an electron to escape the surface of a metal.

Working distance   Distance from the bottom of the objective lens in a scanning electron microscope to the top of the specimen.

X-ray   Electromagnetic radiation with a wavelength between 10^{-3} and 10 nm.

Y-modulated scan   A scan of the cathode ray tube of a scanning electron microscope with the brightness modulated in the vertical direction so that a line or lines with peaks and valleys occur on the cathode ray tube.

Z-modulated scan   A scan of the cathode ray tube of a scanning electron microscope with the brightness of the image modulated toward the operator. The normal mode of operation of the scanning electron microscope.

Bibliography

BARTLETT, A. A., and BURSTYN, H. P. 1975. A review of the physics of critical point drying. *Scanning Electron Microscopy/1975* (Johari, O. and Corvin, I., eds.), IIT Research Institute, Chicago. pp. 306–16.

BAXES, G. A. 1984. *Digital Image Processing*. Cascade Press, Denver, Colo. p. 186.

CARPENTER, L. 1970. *Vacuum Technology: An Introduction*. American Elsevier, New York. p. 130.

CHAPMAN, S. K. 1986. *Working with a Scanning Electron Microscope*. Lodgemark Press, Chislehurst, Kent. p. 93.

CHESCOE, D., and GOODHEW, P. J. 1990. *The Operation of Transmission and Scanning Electron Microscopes*. Oxford University Press, New York, p. 86.

COHEN, A. L. 1974. Critical point drying. In *Principles and Techniques of Scanning Electron Microscopy* (Hayat, M. A., ed.), Van Nostrand Reinhold Co., New York. Vol. 1, pp. 44–112.

CREWE, A. V. 1971. A high-resolution scanning electron microscope. *Sci. Amer.* 224:26–35.

———. 1983. High-resolution scanning transmission electron microscopy. *Science* 221:325–30.

———, ISAACSON, M., and JOHNSON, D. 1970. Secondary electron detection in a field emission scanning microscope. *Rev. Sci. Instr.* 41:20–28.

———, and others. 1968. Electron gun using a field emission source. *Rev. Sci. Instr.* 39:576–84.

DOVE, D. B., and SAHNI, O. 1988. Thermal printing. In *Output Hardcopy Devices* (Durbeck, R. C., and Sherr, S., eds.), Academic Press, New York, p. 277.

EVERHART, T. E., and HAYES, T. L. 1972. The scanning electron microscope. *Sci. Amer.* 226 (Jan.): 55–69.

GABRIEL, B. L. 1982. *Biological Scanning Electron Microscopy*. Van Nostrand Reinhold Co., New York. p. 186.

GEDCKE, D. A. 1972. The Si(Li) X-ray energy analysis system: Operating principles and performance. *X-Ray Spectrometry* 1:129–41.

GILKEY, J. C., and STAEHELIN, L. A. 1986. Advances in ultrarapid freezing for the preservation of cellular ultrastructure. *J. Electron Microscopy Technique* 3:177–210.

GILMORE, C. P. 1972. *The Scanning Electron Microscope*. New York Graphics Society, Greenwich, Conn. p. 159.

GOLDSTEIN, J. I., and others. 1981. *Scanning Electron Microscopy and X-ray Microanalysis*. Plenum Press, New York. p. 673.

GOMER, R. 1961. *Field Emission and Field Ionization*. Harvard University Press, Cambridge, Mass. p. 453.

GONZALEZ, R. C., and WINTZ, P. 1987. *Digital Image Processing*. Addison-Wesley Publishing Co., Reading, Mass. p. 503.

HART, R. K. 1975. Scanning electron spectrometric microscopy. In *Principles and Techniques of Scanning Electron Microscopy* (Hayat, M. A., ed.), Van Nostrand Reinhold Co., New York. Vol. 4, pp. 174–223.

HAWKES, P. W. 1972. *Electron Optics and Electron Microscopy*. Taylor & Francis Ltd., London. p. 244.

————. 1985. *The Beginnings of Electron Microscopy*. Academic Press, New York. p. 633.

HAYAT, M. A. 1970. *Principles and Techniques of Electron Microscopy*. Vol. 1. Biological Applications. Van Nostrand Reinhold Co., New York. p. 412.

————. 1978. *Introduction of Biological Scanning Electron Microscopy*. University Park Press, Baltimore, Md. p 323.

HEINRICH, K. F. J. 1981. *Electron Beam Microanalysis*. Van Nostrand Reinhold Co., New York. p. 578.

HOWARD, V. 1990. Stereological techniques in biological electron microscopy. In *Biophysical Electron Microscopy* (Hawkes, P. W., and Valdre, U., eds.), Academic Press, New York. pp. 479–508.

JOY, D. C., ROMIG, D., and GOLDSTEIN , J. I. 1986. *Principles of Analytical Electron Microscopy*. Plenum Press, New York. p. 405.

KLEMPERER, O., and BARNETT, M. E. 1971. *Electron Optics*. Cambridge University Press, New York. p. 506.

LAPELLE, P. R. 1972. *Practical Vacuum Systems*. McGraw-Hill Book Co., New York. p. 238.

LIM, J. S. 1990. *Two Dimensional Signal and Image Processing*. Prentice Hall, Englewood Cliffs, N. J. p. 694.

LYMAN, C. E., and others. 1990. *Scanning Electron Microscopy, X-Ray Microanalysis, and Analytical Electron Microscopy. A Laboratory Workbook*. Plenum Press, New York. p. 407.

LYNE, M. B. 1988. Paper requirements for impact and non-impact printers. In *Output Hardcopy Devices* (Durbeck, R. C., and Sherr, S., eds.), Academic Press, New York. p. 383.

MOODY, M. F. 1990. Image analysis of electron micrographs. In *Biophysical Electron Microscopy* (Hawkes, P. W., and Valdre, U., eds.) Academic Press, New York. pp. 145–287.

MORGAN, A. J. 1985. *X-Ray Microanalysis in Electron Microscopy for Biologists*. Oxford University Press, New York. p. 79.

NEWBURY, D. E., and others. 1986. *Advanced Scanning Electron Microscopy and X-Ray Microanalysis*. Plenum Press, New York. p. 454.

NIBLACK, W. 1986. *An Introduction to Digital Image Processing*. Prentice Hall, Englewood Cliffs, N.J. p. 215.

PASZKOWSKI, B. 1968. *Electron Optics*. Iliffe Ltd., London. p. 305.

POSTEK, M. T., and others. 1980. *Scanning Electron Microscopy*. Ladd Research Industries, Burlington, Vermont. p. 305.

REED, S. J. B. 1975. *Electron Microprobe Analysis*. Cambridge University Press, New York. p. 400.

ROBARDS, A. W., and SLEYTER, U. B. 1985. Low temperature methods in biological electron microscopy. In *Practical Methods in Electron Microscopy*, (Glauert, A. M., ed.), Vol. 10. Elsevier Publishing Co., New York. p. 551.

ROBINSON, D. G., and others. 1987. *Methods of Preparation for Electron Microscopy. An Introduction for the Biomedical Sciences*. Springer Verlag, New York. p. 190.

SAPARIN, G. V. 1980. Cathodoluminescence: New methods in scanning electron microscopy. In *Biophysical Electron Microscopy* (Hawkes, P. W., and Valdre, U., eds.), Academic Press, New York. pp. 451–78.

SHELTON, E., and MOWCZKO, W. E. 1977. Membrane blebs: A fixation artifact. *J. Cell Biol*. 72:206a.

———, and ———. 1978. Membrane blisters: A fixation artifact. A study in fixation for scanning electron microscopy. *Scanning* 1:166–73.

SJÖSTRAND, F. S. 1967. *Electron Microscopy of Cells and Tissues*. Academic Press, New York. p. 462.

STATHAM, P. J. 1981. X-ray microanalysis with Si(Li) detectors. *J. Microscopy* 123:1–23.

STEINBRECHT, R. A., and MULLER, M. 1987. Freeze substitution and freeze-drying. In *Cryotechniques in Biological Electron Microscopy* (Steinbrecht, R. A., and Zierold, K., eds.), Springer Verlag, New York. pp. 149–72.

TAYLOR, N. J. 1972. The technique of Auger electron spectroscopy in surface analysis. In *Techniques of Metals Research* (Bunshah, R. F., ed.), Wiley-Interscience, New York. Vol. 7, pp. 117–59.

THORNTON, P. R. 1968. *Scanning Electron Microscopy*. Chapman and Hall, London. p. 368.

TREBBIA, P. 1990. Electron–specimen interaction. In *Biophysical Electron Microscopy* (Hawkes, P. W., and Valdre, U., eds.), Academic Press, New York. pp. 35–61.

VAN ATTA, C. M. 1965. *Vacuum Science and Engineering*. McGraw-Hill Book Co., New York. 1965.

WELLS, O. C. 1974. *Scanning Electron Microscopy*. McGraw-Hill Book Co., New York. p. 421.

ZWORYKIN, V. R., and others. 1945. *Electron Optics and the Electron Microscope*. John Wiley & Sons, Inc., New York. p. 766.

Index